MAMMALIAN
COMMUNICATION
A BEHAVIORAL ANALYSIS OF MEANING

MAMMALIAN COMMUNICATION
A BEHAVIORAL ANALYSIS OF MEANING

ROGER PETERS
FORT LEWIS COLLEGE

BROOKS/COLE PUBLISHING COMPANY
MONTEREY, CALIFORNIA

To my wife and parents,
for all the times I was too busy

Consulting Editor: *Edward L. Walker*

Brooks/Cole Publishing Company
A Division of Wadsworth, Inc.

Printed in the United States of America

10 9 8 7 6 5 4 3 2 1

Library of Congress Cataloging in Publication Data

Peters, Roger.
    Mammalian communication.

    Bibliography: p.
    Includes indexes.
    1. Mammals—Behavior.   2. Animal communication.
I. Title.
QL739.3.P47   599.05'9   80-15229
ISBN 0-8185-0388-2

Acquisition Editor: *C. Deborah Laughton*
Manuscript Editor: *Grace Holloway*
Production Editor: *Cece Munson*
Interior Design: *John Edeen*
Cover Design: *Katherine Minerva*
Typesetting: *Graphic Typesetting Service, Los Angeles*

PREFACE

This book presents a comparative analysis of mammalian communication. There is a large body of research on the communication of a great variety of mammals. Investigators have naturally adopted descriptive categories appropriate to the particular species, behavior, or function they are most interested in. The result is a hodgepodge of terminology that complicates the task of the comparative theorist. This book presents a comparative analysis based on research on mammals in seven different orders. The categories of comparison are functional and are based on the meaning of the messages encoded in various communicative behaviors.

The book is organized by phylogeny rather than by message for two reasons. First, each mammal's communications form a system. Only when communicative behaviors are presented together can the complex interactions and overlappings among them be explicated. Second, communication is part of the social and ecological systems peculiar to each mammal. Explanation of the adaptive significance of a particular form of communication often requires reference to these systems. Such references would be repetitious were they introduced anew for each message category for each mammal.

The mammals discussed were chosen on the basis of two criteria. The first was diversity. I wished to include representatives of as many orders as possible in a short book. At the same time, I wished to include more than one representative of some orders to illustrate the range of variabilities and similarities within orders. The second criterion was availability. It was necessary to select mammals for which there were fairly complete ethograms, preferably not already published in English in one generally available work, and based on naturalistic observation.

These mammals use very different sets of signals, but beneath this diversity in behavior there are commonalities in meaning. I hope the definitions and descriptions that follow will elucidate rather than obscure this fundamental unity.

A note of thanks to all those who made this book possible, especially Edward Walker, who started it all with his coffee machine and his fascination with hedgehogs; Bonnie Clements, whose editorial and theoretical comments were as sharp as her pencil; James Everett, a one-man research and reprint service; my typists, Betty Perry and B. Thunder-Cometti, for their work and encouragement; and, finally, my students at Fort Lewis College, with whom I engaged in endless dialectic over countless campfires and quizzes. Foremost among these students were Mike Huffman, who researched the chapter on vervets; Alan Morton, whose cordon bleu camp cooking cheered many a cold evening in the field; and Heidi Reynolds and Jeremiah St. Ours, whose drawings appear in Chapters 2, 6, and 8. I also want to thank David Barash of the University of Washington, Roger Fouts of the University of Oklahoma, and Gary Mitchell of the University of California at Davis for their many substantive criticisms, but of course any errors that remain are mine alone.

Roger Peters

CONTENTS

1 INTRODUCTION 1

MAMMALIAN ADAPTATIONS 3

MAMMALIAN COMMUNICATION 6

DEFINITIONS OF COMMUNICATION 6
MESSAGE TYPES 9

HISTORY OF THE STUDY OF MAMMALIAN
COMMUNICATION 13

CRO-MAGNON ART 13
ARISTOTLE 14
PLINY THE ELDER 16
CHARLES DARWIN 17
20TH-CENTURY WRITERS 20

2 HEDGEHOGS, SHREWS, AND TREE
SHREWS 23

HEDGEHOGS, GENUS *ERINACEUS* 25

NEONATAL COMMUNICATION 26
INTEGRATIVE MESSAGES 27
AGONISTIC MESSAGES 28
SEXUAL MESSAGES 29
SUMMARY 30

SHREWS 31

NEONATAL COMMUNICATION 32
INTEGRATIVE MESSAGES 34
AGONISTIC MESSAGES 36
SEXUAL MESSAGES 38
SUMMARY 40

TREE SHREWS, FAMILY TUPAIIDAE   41

NEONATAL COMMUNICATION   42
INTEGRATIVE MESSAGES   43
AGONISTIC MESSAGES   45
SEXUAL MESSAGES   47
SUMMARY   48

3   RATS AND BEAVERS   51

RATS, GENUS RATTUS   52

NEONATAL COMMUNICATION   54
INTEGRATIVE MESSAGES   55
AGONISTIC MESSAGES   57
SEXUAL MESSAGES   63
SUMMARY   65

BEAVERS, SPECIES CASTOR CANADENSIS   66

NEONATAL COMMUNICATION   68
INTEGRATIVE MESSAGES   69
AGONISTIC MESSAGES   72
SEXUAL MESSAGES   76
SUMMARY   77

4   RABBITS AND PIKAS   79

THE EUROPEAN RABBIT, ORYCTOLAGUS
CUNICULUS   80

NEONATAL COMMUNICATION   83
INTEGRATIVE MESSAGES   85
AGONISTIC MESSAGES   87
SEXUAL MESSAGES   91
SUMMARY   93

PIKAS, GENUS OCHOTONA   94

NEONATAL COMMUNICATION   96
INTEGRATIVE MESSAGES   97
AGONISTIC MESSAGES   99
SEXUAL MESSAGES   100
SUMMARY   101

5 WAPITI AND DEER 105

WAPITI, CERVUS ELAPHUS 106

NEONATAL COMMUNICATION 112
INTEGRATIVE MESSAGES 114
AGONISTIC MESSAGES 119
SEXUAL MESSAGES 126
SUMMARY 134

DEER, GENUS ODOCOILEUS 135

NEONATAL COMMUNICATION 138
INTEGRATIVE MESSAGES 141
AGONISTIC MESSAGES 149
SEXUAL MESSAGES 156
SUMMARY 159

6 VERVET MONKEYS 161

NEONATAL COMMUNICATION 165
INTEGRATIVE MESSAGES 168
AGONISTIC MESSAGES 174
SEXUAL MESSAGES 180
SUMMARY 183

7 BOTTLENOSED DOLPHINS 185

NEONATAL COMMUNICATION 197
INTEGRATIVE MESSAGES 200
AGONISTIC MESSAGES 206
SEXUAL MESSAGES 208
SUMMARY 210

8 WOLVES AND CATS 213

WOLVES, CANIS LUPUS 214

NEONATAL COMMUNICATION 218
INTEGRATIVE MESSAGES 220
AGONISTIC MESSAGES 232
SEXUAL MESSAGES 244
SUMMARY 248

CONTENTS

X

THE DOMESTIC CAT, *FELIS CATUS*   249

NEONATAL COMMUNICATION   252
INTEGRATIVE MESSAGES   255
AGONISTIC MESSAGES   259
SEXUAL MESSAGES   263
SUMMARY   265

9 NONVERBAL COMMUNICATION AND LANGUAGE   267

NONVERBAL COMMUNICATION IN HUMANS   268

NEONATAL COMMUNICATION   269
INTEGRATIVE MESSAGES   269
AGONISTIC MESSAGES   271
SEXUAL MESSAGES   272

THE EVOLUTION OF LANGUAGE   273

FEATURES OF LANGUAGE   273
THEORIES ON THE EVOLUTION OF LANGUAGE   276

10 SUMMARY AND CONCLUSIONS   281

REVIEW OF THE MESSAGE TYPES   283

NEONATAL COMMUNICATION   283
INTEGRATIVE MESSAGES   287
AGONISTIC MESSAGES   292
SEXUAL MESSAGES   295
RECAPITULATION   297

SOME PROPERTIES OF MAMMALIAN
COMMUNICATION   298

RELATIONSHIPS AMONG MESSAGE CATEGORIES   299

GLOSSARY   305

REFERENCES   309

NAME INDEX   331

SUBJECT INDEX   335

1
INTRODUCTION

What's in a name?

WILLIAM SHAKESPEARE

We humans have probably always been interested in the messages exchanged by our fellow mammals. The idea that we could understand these messages if we could only find the appropriate code is at least as old as the legend of Solomon, whose magic ring supposedly allowed him to speak with animals. Konrad Lorenz's (1952) charming *King Solomon's Ring* and a host of other books have popularized the notion that it is possible and worthwhile to translate animal (including our own) nonverbal communication into words and sentences. Unfortunately, attempts to translate animal messages into human language almost invariably obscure what they mean to the creatures who send and receive them. E. O. Wilson (1975) goes so far as to describe understanding what an animal "is really trying to communicate" as the "grail of zoosemiotics." *Zoosemiotics* is the analysis of animal communication in terms of meaning (Sebeok, 1963). Wilson believes that this grail is in principle unattainable. This book is not a search for Solomon's ring, but it is a crusade for the zoosemiotic grail.

The book is an attempt to understand the communications of some of our fellow mammals, not in terms of verbal translations, but in ways that make sense from psychological and evolutionary points of view. I hope to show that there is a rather small number of meanings underlying the incredible diversity of signals used by several mammals and that these meanings provide categories that can be used to compare the communications of mammals of different orders.

My working hypothesis is that the problems faced by mammalian communication systems are similar enough to allow comparison. There is a striking contrast between the diversity of mammalian form, *phylogeny*, and ecology, on the one hand, and the similarities of the postures, sounds, movements, and even smells with which mammals communicate, on the other. Asses sometimes rest their chins on each other's backs during social interactions, just as dogs do. Some species of deer groom each other with the same sort of nibbles used by the Bahamian hutia, a rare herbivorous rat. The short-chain fatty acids that have been claimed to induce male sexual behavior in rhesus monkeys occur in the anal sacs of wolves and are frequently sniffed

2

by other pack members. These occasional convergences suggest a common underlying structure in mammalian communication.

The rest of this chapter will present some basic information about the problems and adaptations of the "typical" (furry, warm-blooded, *viviparous*) mammal, the types of messages that have evolved, and the history of attempts to understand them. Succeeding chapters will deal with the communication systems of several kinds of mammals.

There are more than 4000 species of living mammals, organized into about 20 large groups, or orders. This book discusses only a few (see Table 1-1). Chapter 2 deals with three species of *insectivores*, small, solitary creatures considered by some to be representatives of some of the earliest mammals. Chapter 3 deals with two *rodents*, the rat and the beaver, and Chapter 4 with two *lagomorphs*, the rabbit and the pika. Chapter 5 examines two even-toed ungulates, or *artiodactyls*: the wapiti and the deer. Chapter 6 is about a *primate*, the vervet monkey, and Chapter 7 is about dolphins. Chapter 8 deals with two *carnivores*: one, the wolf, highly social, and the other, the cat, somewhat solitary. Chapter 9 is an introduction to nonverbal communication and the evolution of language in humans. These mammals are usually called by their common names. In some cases common names are imprecise, but I will always use them to refer to the *taxon* used at the beginning of each mammal's section. In the last chapter I will evaluate the success of this crusade.

MAMMALIAN ADAPTATIONS

To understand mammalian messages, it is necessary to know something about the adaptive contexts within which they function. There are two fundamental mammalian adaptations. The first is a high level of parental investment in a small number of viviparous young. The second is a high level of activity. Although there are traces of both these adaptations early in the Age of Reptiles, they did not flower until the Mesozoic era, about 200 million years ago.

Compared to reptiles, mammalian parents invest heavily in their young. A "typical" reptile lays scores of eggs, leaving the embryos to complete most of their development on their own. In contrast, the "typical" mammal harbors only a few young within her body, bearing them at a relatively advanced stage of development. She then cares for her young by nursing and, in some cases, by sheltering and teaching them.

TABLE 1-1
PHYLOGENY OF MAMMALS INCLUDED IN THIS BOOK

Kingdom Animalia
  Phylum Chordata
    Subphylum Vertebrata
     Class Mammalia
      Order Insectivora
       Family Erinacidae (hedgehogs)
        Diet, animal matter; habitat, widespread; predators, small carnivores; social organization, solitary.
       Family Soricidae (shrews)
        Diet, insects; habitat, widespread; predators, small carnivores; social organization, solitary.
       Family Macroscelididae (jumping, or elephant, shrews)
        Same as *Soricidae.*
       Family Tupaiidae (tree shrews)
        Diet, invertebrates.
      Order Rodentia
       Family Muridae (Old World rats and mice)
        Rattus rattus
        Diet, omnivorous; habitat, widespread near humans; predators, small and medium carnivores; social organization, colonial.
        Rattus norvegicus
        Same as *rattus.*
       Family Castoridae (beavers)
        Castor canadensis
        Diet, bark and branches; habitat, forest; predators, medium and large carnivores; social organization, colonial.
      Order Lagomorpha
       Family Leporidae (rabbits and hares)
        Oryctolagus cuniculus
        Diet, grass; habitat, widespread; predators, medium carnivores; social organization, colonial.
       Family Ochotonidae (pikas)
        Ochotona spp.
        Diet, grass; habitat, alpine; predators, raptors and smaller carnivores; social organization, solitary.

The second basic mammalian adaptation, high activity level, is shared with birds and was perhaps a characteristic of some dinosaurs. It allowed radiation to new niches and successful competition in old ones. This adaptation was made possible by two developments. The first was *homoiothermy,* or internal temperature regulation, which allows mammals to maintain a constant body temperature independent

Order Artiodactyla
  Family Cervidae (wapiti and deer)
    Cervus elaphus
        Diet, grass and browse; habitat, meadow; predators, large carnivores; social organization, herd.
    Odocoileus hemionus
        Diet, grass and browse; habitat, widespread; predators, large carnivores; social organization, herd.
    Odocoileus virginianus
        Diet, grass and browse; habitat, forest; predators, large carnivores; social organization, herd.
Order Carnivora
  Family Canidae (wolves, coyotes, and dogs)
    Canis lupus
        Diet, meat and some vegetable matter; habitat, widespread but reduced; social organization, family groups.
  Family Felidae (cats)
    Felis catus
        Diet, meat; habitat, widespread; social organization, solitary.
Order Cetacea
  Family Delphinidae (dolphins and porpoises)
    Tursiops spp.
        Diet, fish; habitat, pelagic; predators, sharks and killer whales; social organization, herd.
Order Primata
  Family Cercopithecidae (Old World monkeys)
    Cercopithecus aethiops
        Diet, omnivorous; habitat, widespread in Africa; predators, medium and large carnivores, birds, and snakes; social organization, troops.
  Family Hominidae (humans)
    Homo sapiens
        Diet, omnivorous; habitat, widespread; social organization, family groups.

of their environment. This development made it possible to operate in the cold and, perhaps more important, to remain active without overheating. The other development that made a high activity level possible was increased efficiency in nutrition through specialized teeth, which could grab, slice, and grind high-energy foods or low-energy foods in great quantity. In turn, for some mammals, a high activity level made it possible to pursue prey. Viviparity and homoiothermy were not the only mammalian adaptations, but they were the fundamental ones.

MAMMALIAN COMMUNICATION

DEFINITIONS OF COMMUNICATION

Some of the difficulties underlying discussions of nonhuman communication are that the term *communication* means different things to different people (see Burghardt, 1970); that, as humans, we are liable to see communication only when it resembles our own in modality and complexity; and that we tend to interpret messages verbally, even when there is no appropriate verbal translation. Therefore, it is worthwhile to specify exactly what we mean by communication.

Frings and Frings (1964) defined communication as "the giving off by one individual of some chemical or physical signal, that, on being received by another, influences its behavior" (p. 3). They further required that the sender use a specialized organ or structure and that the sender and receiver be of the same species. Sebeok (1972) subsumes animal communication under semiotics, the study of signs and their meanings. He restricts *communication* to situations in which "a small amount of energy or matter in the source . . . brings about a large redistribution of energy or matter in the destination" (pp. 286–294). W. N. Tavolga (1970) follows the usage of Frings and Frings and distinguishes six levels of interaction among animals: vegetative, tonic, phasic, signal, symbolic, and language. Vegetative interactions are purely physical. For example, one beaver may interact with another by bumping it into a pond. The effect of the first beaver's behavior on the second can be understood in terms of purely physical constructs, such as force and mass. This level of interaction is clearly excluded by Sebeok's definition because the effect on the second beaver is exactly proportional to the energy expended by the first. It is excluded from the Fringses' definition because no specialized structure was used.

Tavolga's second level of interaction, the tonic, includes situations in which continuous metabolic processes in one organism influence the behavior of another. If the day is warm, the body heat of one beaver may induce the other to dive into the pond to cool off, and a tonic interaction has occurred. Note that this level of interaction is included in Sebeok's definition but excluded from the Fringses'.

Tavolga himself excluded these first two levels of interaction from true communication and reserves the term for the four highest levels: phasic, signal, symbolic, and language. Communication at the phasic level occurs when discontinuous, more-or-less regular stages or events in the development of one animal, acting across a broad spectrum of energy or through several different channels, influence the behavior of another. If one of the beavers under consideration is an adult male, his response to the presence of the other will be determined by the other's size, demeanor, and odor. If these are characteristic of a young (less than 2 years old) family member or estrous female, it will be tolerated. Otherwise, it may be chased away.

Communication by signal involves specialized structure, a single effective channel, and, usually, a restricted portion of the spectrum of that channel. For example, when one of our beavers is in the water, the slapping sound of its flat tail will generally cause the other to dive in. Although the tail slap is also visible, only the auditory channel seems to be effective.

We must leave the shores of the beaver pond for examples of Tavolga's fifth level, the symbolic. Unfortunately, he does not define this level, but he does state that pointing by the chimpanzee is symbolic. If the criterion is the use of a symbol, definition of this level will depend on definition of *symbol*, which usually means a conventional (rather than universal) sign. Therefore, if some groups of chimps use gestures other than pointing to direct attention, pointing by chimps is a good example. Pointing by setters is not, because all setters point in the same way, regardless of their experience, so pointing in this case is universal, not conventional. A clearer example of symbolic communication would occur in a group of people who have agreed to go wherever their leader points. If the leader points to the beaver pond and they dive in, symbolic communication has occurred.

Finally, language. Definitions of language were once the bugaboo of comparative psychologists who wished to convey that many animals express complex information about the environment, but without using words, and the bane of anthropologists who wished to discriminate between human speech and fluent use of American Sign Language, plastic tokens, and computer terminals by chimpanzees. Today there is near-universal agreement that the term *language* should be reserved for human speech, writing, and other artificial codes, perhaps including those used by some specially trained chimpanzees. Tavolga's criterion for language is the communication of abstract ideas, and there is little doubt (Fouts, 1973) that chimps can be trained to communicate ideas.

Our ability to use language did not replace older systems but incorporated them. These older systems provided the context within which language evolved, and they may provide some insights into how this process took place. Moreover, we still share with other mammals most or all of the types of communication that will be described in the following chapters. We have much to learn about our own non-verbal communication from the messages of animals, which, though similar to us in fundamental ways, are sufficiently different that we can observe them with some degree of detachment.

Another writer, Bateson (1966), has brought the number of interaction levels to seven. He proposed the term *metacommunication* for messages that modify the meaning of other messages. For example, a playful dog may wag its tail as it nips at another dog's heels. The tail wag, according to Bateson, involves a new level of meaning because it is a comment about the message implied by the nips.

There remains one last point of definition: how seriously are we to take the requirement that it must be *another* animal whose behavior is influenced? If we take this requirement seriously, we must exclude the effects of an animal's scent marks or some other sign on its own later behavior. If we ignore the requirement, we must include echolocation by bats and cetaceans as communication. My approach has been to include some kinds of communication by an animal with itself. I exclude *echolocation,* in which an animal emits high-frequency sound and perceives objects by their echoes, but include reflexive effects of scent marking. There are three justifications for this decision. The first is substantive: olfactory and visual signs left by mammals often influence members of their own species, or *conspecifics.* The second and third are theoretical. Echolocation seems to be mainly perception, not communication, and is thus beyond the scope of this book. Finally, since one major theory of the evolution of language (Jerison, 1975) states that the original function of language was *mne-monic* rather than communicative, it seems worthwhile to include messages that, in some mammals, sometimes serve as mnemonic devices (Henry, 1976).

The definition of communication used here is purposely broad. It includes all forms of information transmission in the senses used by the Fringses and Sebeok. It excludes some forms of energy emission, like echolocation, where an animal communicates with itself (on the grounds that perception is a different function from communication) but includes tonic and some vegetative forms (for example, hip slams in canids) that are excluded from Tavolga's definition. The present state of understanding what is and what is not communication encourages a definition that is as broad as possible without including fundamentally different kinds of processes.

MESSAGE TYPES

In the following chapters the unit of analysis is the message type, a set of behaviors with a similar meaning. Message types are defined operationally. They are grouped into four message systems: neonatal, integrative, agonistic, and sexual.

My classification is what Sebeok would call zoosemantic because it is based on the kind of information transmitted as inferred from the form of the message, the ecological and behavioral context, and the typical response of the receiver. I combined under one message type all behaviors that had been assigned similar meanings. In general, I split rather than lumped, separating behaviors when the type of information transmitted, the context, the sender, or the receiver was different.

Table 1-2 lists 22 message types, not counting *neonatal* versions, in which, during the first few days after the birth of offspring, the parents, especially the mother, and the offspring influence each other through a complex of tonic, phasic, and signal interactions. These include messages that appear later in life and that are described in the rest of this section. They sometimes have a special meaning or form in the neonatal context, where they form the basis for later communication. Neonatal messages increase fitness mainly by allowing parents and offspring to influence each other's behavior in ways that increase the offspring's chances for survival.

Later on in life many mammals must coordinate their behavior with the behavior of others. Considerable advantage often accrues to individuals that can rest, eat, travel, and flee with conspecifics. Integrative messages facilitate coordination of the behavior of individuals, however tenuous their association. Play, an integrative message type that develops early in life, is a form of communication because in social play each animal influences the behavior of its playmates. Moreover, play often involves special signals notifying associates that what transpires is not in earnest. Contact signals are low-intensity, short-range sounds, smells, or gestures that, by announcing location, promote cohesion in group-living mammals.

Since some mammals live in groups, it is not surprising to find a variety of behaviors involved in the formation, maintenance, and expression of attractions. In black-tailed deer, for example, maternal odors promote an affiliative bond attracting each fawn to its own mother (Müller-Schwarze, 1971). Sexual behaviors emitted when neither of the mates is physiologically capable of reproduction are also classified as affiliative because they express nonsexual attractions. *Allogrooming,* care of the hair and skin of a conspecific, is also an im-

TABLE 1-2
DEFINITIONS OF MESSAGE TYPES

MESSAGE SYSTEM	MESSAGE TYPE	TRANSMITTER	FORM	RECEIVER	RESPONSE
INTEGRATIVE	Play	often young	exaggerated, interrupted	any	increased activity
	Contact	any	frequent	any	orientation approach
	Affiliation	any	gentle contact	associate	approach, satisfaction
	Assembly	isolated	loud	group members	approach
	Identity	any	individually distinctive	any	investigation
	Familiarization	any	mark	same as transmitter	increased confidence
	Solicitation	in high drive state	variable	any	related to drive
	Alarm	any	salient	any	fleeing, hiding, aggregating
	Distress	injured	loud, strong, repeated	any	variable
	Satisfaction	drive reduction	variable	?	reinforcement?

AGONISTIC				
Territory advertisement	"owner"	mark, display, loud call	intruder	withdrawal
Submission	lower rank than receiver	renders unimposing	higher rank	decreased aggression
Defensive threat	defensive	preparation for fight	aggressor	decreased aggression
Offensive threat	aggressive	preparation for fight	any	withdrawal
Dominance	high ranking	renders imposing	lower rank	withdrawal
Fighting	any	liable to injure	any	reciprocation, approach
SEXUAL				
Male advertisement	sexually ready male	mark, display, loud call	female	approach
Female advertisement	sexually ready female	mark, call, posture	male	approach
Courtship	prospective mates	play, aggression, affiliation	prospective mate	increased sexual motivation
Synchronization	female?	olfactory	female	synchronized ovulation
Suppression	female?	alarm or agonism	female	decreased fertility
Copulatory signal	copulating pair	cry, tie	?	investigation

portant expression of affiliation. Allogrooming is a mammalian adaptation to social living and probably arose early in mammalian evolution. It has been regarded as the basis of the social structure of some primates and is a central construct in some recent theories of human verbal interaction (Berne, 1964). Zajonc (1971) has shown that mere repeated exposure to stimuli can increase their attractiveness. On this basis many behaviors that expose mammals to one another's odors can be classified as affiliative.

The other types of messages in the integrative system can be described more briefly. Assembly signals are those to which group members respond by aggregating. Identity signals allow mammals to identify the species, sex, age, or rank of the transmitter. Familiarization is a form of *autocommunication* in which a mammal applies a mark to a conspecific or to the environment, thus making it easier to recognize at later encounters. Familiarization has both cognitive and motivational aspects. Solicitation includes behaviors that, like panting with faint vocalization in dogs and wolves, express a desire for a response from the receiver. Alarm signals are emitted at the appearance of potential danger and cause conspecifics to take defensive action, which may include hiding, fleeing, or aggregating. Mammals give distress signals when lost, hungry, cold, or in pain. Conspecifics often respond to these signals with approach, sometimes with succor. Satisfaction, the opposite of distress, signifies drive reduction.

Paradoxically, the agonistic system, containing messages associated with conflict, acts mainly to ameliorate its effects. Territory advertisement allows mammals to avoid one another altogether. Submission often deters aggression on the part of the receiver. Threats act as substitutes for fighting and are given by both antagonists. Some forms of threat are given only when the transmitter is under attack. These warnings are called defensive threats. Dominance is distinguished from threat because, in most social mammals, some displays are performed primarily by the adults that "win" in most agonistic encounters but do not themselves involve serious conflict. High-ranking bighorn sheep, for example, exhibit two horn displays ("low-stretch" and "twist") that do not seem to involve serious conflict, since the subordinates only rarely withdraw (Geist, 1971). Even fighting, which is sometimes fatal, is often highly stylized in ways that reduce the potential for injury.

The sexual system includes advertisements that allow potential mates to find one another. Courtship then ensues, its messages coordinating the mates' physiology and behavior or simply keeping them together without fighting until conception is possible. In addition, many social mammals transmit messages that tend to bring females into estrus at about the same time or to suppress fertility in either sex.

Such messages are termed *synchronization* and *suppression*, respectively.

The same behavior may appear in several different message types, depending on what kind of information is being transmitted. Mounting, for example, can transmit information about physiological readiness for copulation, but it commonly occurs in other contexts in many different mammals, including rats, wolves, baboons, and dogs. In canids, for example, mounting may involve sex in one context (during breeding) and dominance in another (when two males are involved). During agonistic encounters, females of many species often mount a companion and display rapid pelvic thrusting difficult to distinguish in form from that displayed by males during normal intercourse. This behavior is not pathological; depending on the context, it may express sexual arousal, social dominance, or simple assertiveness. Tinbergen (1952) refers to this sort of separation between the original meaning and later meanings of an expression as emancipation. Bateson (1966) points out that these later meanings typically express the relationship (that is, dominance or dependency) between mammals.

At times the system of classification used in this book may seem unnecessarily refined, the distinctions confusingly specious. They have often seemed so to me. At such times it is a good idea to remember the goal—an understanding of the meanings of particular communicative acts performed under particular circumstances. The goal is not to know what a display means in general; displays have no general meanings, but specific acts do have meanings. These are the meanings I have tried to explain.

Classifying behaviors in terms of meaning does not imply that mammalian messages always "stand for" some meaning that exists independently of the message, as is often the case in speech. Lorenz (1966) puts it quite succinctly when he says that friendly behavior is not the expression of a bond—it *is* the bond. This McLuhanesque aphorism can be generalized to most message types—the expression *is* the message.

HISTORY OF THE STUDY OF MAMMALIAN COMMUNICATION

CRO-MAGNON ART

The permanent record of our interest in mammalian messages begins with the Cro-Magnon cave paintings of central and southern France. From about 35,000 to 12,000 B.C., the inhabitants of this region left collections of carvings, paintings, and engravings: finely detailed, accurate representations of deer, horses, bison, and cattle, often with archers and spearers in pursuit. Since the artists depended on hunting with short-range weapons for much of their food, it is not surprising that they, like contemporary hunters/gatherers, were experts on the expressions of their prey.

In any of the collections of Paleolithic art (for example, Lerio-Gourhan, 1968), one can find reproductions and descriptions of the great assemblies of paintings at Lascaux and Les Eyzies. Many of these representations depict "exact proportions of head, body, limbs, hooves, horns, etc.; others show posture, locomotion, fighting, herding and migration" (Locy, 1925, p. 15). There are only a few recognizable patterns of expression, but some of these display a familiarity with the subject that has been equaled only by the technical illustrators of the past few decades, most of whom work with the aid of photographs. Figure 1-1, a restoration by E. Ray Lankester (1875) of an engraved antler found in a cave near Lortet, shows three red deer. The hind legs of the deer on the left suggest *spronking* (an alarm display with a characteristic bouncing run). The middle deer is in a low-stretch posture, often associated with threatening. The deer on the right displays a *flehmen*, a retraction of the lips often associated with olfactory investigation. The engraver of this piece knew the subject and was able to perform a remarkable cylindrical projection of the scene onto the antler. Although a reliable date for this piece is unavailable, the style, classically Magdalenian, places it in the period from 17,000 to 12,000 B.C. Representations like this one suggest that our distant European ancestors achieved a level of unity with art and science that has only in modern times, if ever, been equaled.

ARISTOTLE

A gap of more than 10,000 years separates the art of the Cro-Magnons from the science of Aristotle, who produced the oldest available written record of mammalian communication. Most of his writings on animal communication can be found in *Historia Animalium*, Books 4–9. Fortunately, there is a translation of this work by the great biologist D'Arcy Thompson (1910) that presents an authoritative interpretation of passages that were rendered obscure by early translators more interested in systematizing Aristotle than in understanding his biology.

In Books 4 and 5 of the *Historia*, Aristotle accurately describes the vocalizations of elephants and, surprisingly, porpoises. He men-

FIGURE 1-1   A Magdalenian carved antler found
near Lortet, France, in 1873. The shaded area was
reconstructed by Lankester (1875). From Locy
(1925).

tions the mating calls of goats, swine, and sheep and vocal changes
associated with reproductive maturity in deer, horses, and cows. He
notes that, in general, at maturity the voice of the female is higher
and "sharper toned" than that of the male. More-or-less accurate
descriptions of copulation postures in hedgehogs, deer, elephants,
camels, wolves, dogs, seals, bears, and cats accompany more-or-less
fanciful descriptions of associated courtship rituals and sexual competition.

   In Book 6 Aristotle presents a description of a harem-like
breeding system in horses that includes olfactory recognition by the
male of the mares he herds. Aristotle mentions an increase in frequency of urination accompanying estrus in cows, horses, and domestic dogs. He was a more careful observer than many moderns when
it comes to the last of these species, for he notes that estrous bitches
occasionally lift their legs when they urinate. He notes the friendliness
of lions toward their pride mates and describes the altruistic behavior
of porpoises, who sometimes support weak or newborn conspecifics at
the surface, enabling them to breathe.

   Although Aristotle lived more than 2000 years before the
ethologists of the late 19th and early 20th centuries, his descriptions
of mammalian messages are thoroughly modern in their detail, emphasis on reproduction, recognition of the importance of olfaction,
and, above all, emphasis on survival value as an explanatory principle.
Contrary to scholastic exegesis and modern misconception, "Aristotle's

conception of natural ends and natural teleology is very remote from the conception of final causes familiar in the religious tradition" (Randall, 1960, p. 226) and is quite close to modern evolutionary functionalism.

PLINY THE ELDER

Pliny the Elder's scholarly avocation was ultimately more dangerous as well as more illustrious than his military career. An admiral in command of a fleet, he died not at sea, but while observing at close quarters the eruption of Vesuvius that destroyed Pompeii in A.D. 79. His major work, *Historia Naturalis*, was a kind of "Believe It or Not" of the day. He was not totally uncritical, remarking at one point, "A wonder it is to see to what pass these Greeks are come in their credulity" (Pliny, 1962, p. 95). He was unjustly accused (by Thomas Browne, 1658) of promulgating the notion that cornered beavers habitually bite off their testicles (to what end is never made clear). In fact, Pliny dismisses this theory as nonsense and proceeds to describe in excellent detail the probable origin of this myth—the beaver's use of an odorous substance called castoreum, secreted from a scrotum-like pouch, as a territorial advertisement. We shall encounter this substance later, but Pliny's description is worth presenting here: "a certain liquor resembling waxy honey, standing much upon wax . . . a strong and rank smell; a bitter, hot and fierce taste . . . (p. 342). His is also the first detailed written description of urine marking with a raised leg in dogs, and he notes the association of this behavior pattern with sexual maturity. However, unlike Aristotle, he mistakenly reports that females never urinate in this posture.

Many of Pliny's descriptions are taken from other authors, including Aristotle, and so many of these are apocryphal that his work remains primarily of historical interest. Such descriptions show the antiquity of many popular beliefs about mammalian communication and the habit of inference from small or anecdotal evidence. Among Pliny's misrepresentations are the claim that unfaithful lionesses wash off the odors of their paramours in order to avoid the wrath of their mates and the theory that vaginal licking by male mice is responsible for conception. Although each of these stories contains a germ of truth (lionesses do enter water, and orogenital contact is an essential part of mouse courtship), they should serve primarily to heighten our skepticism about conventional wisdom when applied to animal behavior.

CHARLES DARWIN

Charles Darwin (1859, 1872) was the founder of modern biology and comparative psychology. All attempts to understand types of mammalian communication and relationships among them must take his theory of natural selection into account; indeed, most descriptions and analyses of mammalian communication use natural selection as the paradigm for explanation.

The term *natural selection* refers to the fact that individuals vary in reproductive success because of hereditary differences among them. This principle can be explained in Darwin's (1859) own words in the form of a syllogism whose premises are observable facts and whose conclusion is inescapable. Major premise: "If variations useful to any organic being ever do occur, assuredly individuals thus characterized will have the best chance of being preserved in the struggle for life . . ." (p. 128). The *struggle for life* refers not to fights between individuals but to competition for resources: food, habitat, and mates. That such a struggle exists is clear because not all organisms survive long enough to reproduce.

Darwin illustrates this premise by a calculation showing that a single pair of elephants, "the slowest breeder of all known animals," would, if each pair of descendants produced only six young, in 750 years produce 19 million living descendants. Alcock (1975) has calculated that, were all descendants of a pair of house flies to survive, 1.9×10^{20} descendants would be produced each summer, and the earth would be completely blanketed by flies.

Minor premise: "Organic beings present individual differences in almost every part of their structure . . . and from the strong principle of inheritance [like breeds like], these will tend to produce offspring similarly characterized" (p. 128). Conclusion: "This situation, continuing differential reproduction of variations suited to a given environment, leads to the improvement of each creature species in relation to its organic and inorganic conditions. . . . Nevertheless low and simple forms will long endure if well fitted for their simple conditions of life" (p. 128).

Thus, any heritable structural characteristic or behavioral tendency that improves an individual's reproductive success will be selected, spread throughout a population, and perhaps be exaggerated until its benefits are offset by other demands of existence. For example, according to one theory of the evolution of light rump patches in some mammals, the early ancestors of modern cervids (deerlike animals) probably varied in rump coloration. Those with lighter rumps were more easily followed by their young. The advantages that accrued to them and their offspring through ease of following or grouping resulted in the spread of this characteristic throughout the breeding group. Selection for increasingly lighter rumps would, however, be countered by the opposing pressure of easy detection by predators. What is benefited is not necessarily the species, or even the individual, but the genes responsible for the characteristic or behavior in question. In this paradigm it is necessary to explain a genetically based pattern of communication by showing that it benefits the genes of the sender—that is, maximizes their representation in following generations. This may occur through kin selection, which selects the gene via its expression in a relative. Thus, a monkey that utters an alarm cry may expose itself to predation (endangering its genes) but may save the replica of those genes in its kin. Such a pattern could be selected for and would be expected to spread as long as the benefits to relatives are greater than the costs to the emitter of the signal.

Functional explanation—that is, explanation in terms of survival, reproductive success, and natural selection—does not consist of simply imagining some way in which a pattern of behavior might benefit its emitter. It requires demonstrations that the behavior has a genetic basis and that it is adaptive—that is, promotes the reproductive success of the individual or its close relatives.

The principle of natural selection was proposed in *On the Origin of Species,* published in 1859. In 1872 Darwin published *The Expression of the Emotions in Man and Animals,* the first modern work in the field of animal communication. Much of the book deals with forms of communication in various kinds of mammals, and these sections will be reviewed later, in the appropriate chapters. Darwin's first three chapters deal with three general principles of emotional expression, which form his second major theoretical contribution to the study of animal communication.

The first of the three principles is the principle of serviceable associated habits, which Darwin describes as follows: "Certain complex actions are of direct or indirect service . . . and whenever the same state of mind is inducted, however feebly, there is a tendency . . . for the same movements to be performed. . . . In certain cases the checking of one habitual movement requires other slight movements and these are likewise expressive" (p. 28). Scratching the head, rubbing the eyes, and coughing are presented as examples of this first principle, the person with a mild intellectual irritation acting in a manner appropriate to the relief of a mild physical one. This principle also explains why animals that fight with their teeth put back their ears when expressing a tendency to attack. Originally associated with the physiological state of rage, this serviceable habit, which protects the ears, appears with the physiological state even when attack is not imminent.

The second principle is antithesis, which states that when a state of mind leads to a habitual expression according to the first principle, the opposite state of mind produces "a strong and involuntary tendency to the performance of movements of a directly opposite nature, though these are of no use . . ." (p. 50). For example, a dog in a hostile mood displays various postures serviceable in attack: he walks upright with tail erect, hair bristling (all of which make him appear larger, hence more formidable and which may thus scare away his opponent), ears forward and eyes staring (to attend to his opponent). (See Figures 1-2 and 1-3.)

> Let us now suppose that the dog suddenly discovers that the man he is approaching is not a stranger but his master . . . the body sinks downwards or even crouches; his tail . . . is lowered and wagged from side to side; his hair instantly becomes smooth; his ears are depressed and drawn backward . . . and the eyes no longer appear round and staring [p. 51].

In fact, the eyes are often markedly directed away from the master, with whites showing and emphasizing the act of looking away, the opposite of staring. This display is sometimes useful in inhibiting the aggression of an attacker, but this benefit may not have been clear to Darwin.

The third principle of emotional expression is the direct action of the nervous system. This principle includes forms of expression that cannot be described in terms of association with serviceability or antithesis but that stem from an excess of interruption in "nerve-force." Darwin was aware that in the formulation of this principle he was on shaky ground, but he felt that "it is always advisable to perceive clearly our ignorance." Included under the third principle are trembling in fear, writhing in pain, blushing in anger, frisking in joy, and rocking in depression. Given the state of neurophysiology in Darwin's day, he can be forgiven his hydrostatic conception of nerve-force and should be commended for including in his description of emotional expressions "many points . . . [that] . . . remain inexplicable."

All three of the principles assume that behavior, like structure, can be transmitted genetically. Since there are individual differences in behaviors rendering them more or less appropriate to environmental conditions, behavior is subject to natural selection. A modern Darwinian would thus seek to explain the origins of various behaviors in

FIGURE 1-2   A dog with hostile intentions. From Darwin (1872).

terms of natural selection. Darwin was not, however, as "Darwinian" as his successors and supposed that some expressions arose in a Lamarckian manner—that is "through long continued and inherited habit" (p. 67). Therefore, although the principles are valid generalizations about the form of many emotional expressions, as theories about their origins they are victims of a discredited genetic hypothesis in which acquired habits are genetically transmitted.

20TH-CENTURY WRITERS

By 1859 the foundation for the functional investigation and explanation of mammalian communication had been laid. By 1872 Darwin's *Expression of the Emotions* had already begun the task of organizing and analyzing the communication of humans, monkeys, and several domestic animals. The next major systematic attempts to analyze communication among groups of mammals did not occur for 60 years, when Zuckerman's (1932) *The Social Life of Monkeys and Apes* was followed in 1934 by Carpenter's monograph on howler monkeys. Throughout the 1930s and 1940s (with a hiatus during the war), such studies were published at about the rate of one major monograph every two years (Darling, 1937; Allee, 1938; Burt, 1940; Allee, 1942; Schenkel, 1947; Hediger, 1949).

FIGURE 1-3   The same dog in a humble and
affectionate frame of mind. From Darwin (1872).

In 1950 the rate of publication accelerated sharply with books
and papers by Allee, Eibl-Eibesfeldt, Leyhausen, and Lorenz. The
recent period of research had begun, and its characteristics were as
exciting to some scholars as they were confusing to others who knew
what biology was, knew what psychology was, and knew that they
were different.

Psychologists, whose discipline is usually defined as the study
of behavior, had not been studying behavior at all but rather pieces
of behavior isolated from their natural contexts (Barker, 1963), and
they found, to their dismay, that their flank had been overrun by
hordes of biologists with binoculars and stopwatches who were willing
to get out into the field and actually watch animals behave. Meanwhile, biologists were discovering that the territory they were con-
quering without struggle had been visited by a psychological scout or
two (T. C. Schneirla, Frank Beach, and Daniel Lehrman foremost
among them) whose thinking on the development of behavior had
avoided the nature/nurture skirmishes of preceding decades and whose
knowledge of the conceptual terrain was sophisticated and profound.

The result of this confusing period in the history of the two
disciplines is an emerging research tradition—eclectic, functional,
ecological—based on field observation as well as laboratory experiment. Recent works by Ewer (1968, 1973), Fox (1971), Geist (1971),

Kruuk (1972), Mech (1970), Mykytowycz (1968), Schaller (1963, 1967, 1972), and Theissen (1973), combining ecological and behavioral studies with analyses of communication in laboratory and field, are landmarks in this tradition. So are the theoretical syntheses by Sebeok (1972) and E. O. Wilson (1975). Psychologists are overcoming an extreme environmentalism and are learning to appreciate the utility of an ecological/functional approach. They bring with them methodological sophistication, an arsenal of concepts for the objective discussion of behavior and its causes, and, in particular, a profound respect for the influences of experience on development. Biologists, on the other hand, approach the topic armed with an understanding of evolutionary and ecological mechanisms, appreciation of the importance of studies conducted under natural conditions, and expertise in natural history and the mechanics of field work. There are now many investigators involved in a painful but productive identity crisis, unsure what to call themselves, but quite sure of what to study and how to go about it.

A climax in research in animal communication occurred in 1975, with the publication of E. O. Wilson's *Sociobiology*. As I mentioned at the beginning of the chapter, Wilson's prognosis for understanding mammalian communication by using functional categories that cross orders or families is pessimistic. He throws cold water on the zoosemioticist's hope of demonstrating that "message categories are not endlessly proliferated in evolution, and that animals are only able to say a few things to each other" (p. 216). Wilson's argument is based on the indisputable fact that categorizing behaviors of animals of different species, genera, families, or orders "is increasingly a matter of judging analogy rather than homology" (p. 217). By this he means that such classifications must be based on common function rather than on common descent. He goes on to say that such judgments are necessarily subjective and hence liable to "collapse in a bewildering debris of contradictory definitions and arcane terminology" (p. 217). Here the perspectives of the zoologist and the psychologist differ. To a zoologist, any attempt to objectify the subjective may be a futile crusade for a holy grail, but to a psychologist, such attempts are as common as coffee cups. This does not mean, of course, that such attempts are assured of success. The following chapters are an attempt to apply objective and consistent definitions to the communications of mammals of seven different orders. Judgments of the success or failure of the enterprise are left to the reader.

2
HEDGEHOGS, SHREWS, AND TREE SHREWS

The fox has many tricks.
The hedgehog has but one.
But that is the best of all.

ERASMUS

The hedgehog's trick (rolling into a ball when alarmed) is the best not because it is clever but because it works. The same can be said for the entire behavioral repertory of the insectivore order, whose simple tricks have permitted evolutionary success without increases in size of body or brain. Insectivores are of special interest to the student of mammalian communication because they are considered by some to be living representatives of the ancestral mammals from which all modern forms evolved (Romer, 1954). Insectivores are small and subsist primarily on invertebrates, which they must consume voraciously to avoid starvation. Most have elongated, pointed snouts, a small brain, and a nasty disposition. Although there are many exceptions, insectivores are generally found together only as breeding pairs or as mother with young. Their message systems have a relatively simple theme whose statement precedes, by millions of years, the complex variations found in other orders. Even though several genera of insectivores have undergone some specialization (for example, for subterranean food finding in the mole), contemporary species are presumed to bear a much closer relationship to their ancestral forms than do most other groups of living mammals. It is therefore reasonable to use the behavior of living insectivores as clues to the evolutionary origins of mammalian communication.

There are nine major families, with almost 400 species, of insectivora. Hedgehogs, of which there are 19 species, form the family Erinacidae. Shrews (265 species) are the most numerous family. They are extremely small, secretive, and active, consuming their body weight in insects daily. Another family of shrewlike creatures, the *tupaiids*, with 18 species, are squirrel-sized. Because of their arboreal habits and the bony orbit surrounding their large eyes, these mammals are sometimes grouped with primates, for the first primates were evidently very similar to modern tupaiids. Nevertheless, the diet, shape, and communication typical of this family are much more similar to insectivores than to primates, and some fairly recent research (Campbell, 1966; McKenna, 1966) place them in this order. Tree shrews avoid most enemies by fleeing into the trees.

24

The remaining six families of insectivores contain moles, tenrecs, and 41 other species of rare, secretive, and obscure creatures about whose communication very little is known. Their small size, phylogenetic position, and threatened ecological niches recommend them as subjects for study in captivity. Several of these species, like the Cuban solenodon, are geographically confined, slow to reproduce, and threatened by carnivores introduced to their habitat by humans.

In spite of their key position in mammalian phylogeny, relatively little is known about the communication of insectivores other than hedgehogs, moles, and shrews. One reason is that they lack the complex behavioral specializations that fascinate the ethologist. As Morris (1965) puts it:

> In almost all insectivores the eyes are poorly developed, the brain is small, and the teeth are unspecialized. The fingers and toes are always clawed, but the first digit is never opposable. In other words—insectivores are incapable of grasping anything awkward, either physically or mentally, and represent a living reminder of the sort of animals that were the ancestors of our more advanced groups [p. 67].

This attitude does not do the insectivores justice. Although their communication system is *relatively* simple, there are complexities and mystery sufficient to keep several generations of ethologists in a state of mild surmise. For example, Poduschka (1977) has found evidence that hedgehogs can echolocate.

HEDGEHOGS, GENUS *ERINACEUS*

There are 19 species of hedgehogs found only in Europe, Asia, and Africa. All are between 15 and 20 centimeters long. They have short, pointy faces that peer from beneath a spiny neck and back (see Figure 2-1). These spines ordinarily lie pointing backward, but at the slightest disturbance they are erected so that those on the neck and back of the head point forward.

Hedgehogs' spiny ball is so effective a defense that they can move around quite boldly. Although predominantly nocturnal, in broad daylight they often rustle through dry leaves, bustle through brush, or snore loudly in hedgerows.

The hedgehog's defense not only deters predation but also conserves energy. When rolled up, they can remain immobile in their

FIGURE 2-1   A hedgehog of the genus *Erinaceus*.
Photograph by Roger Peters.

winter burrows for days at a time, so minimizing their heat loss.
Hedgehogs are thus exempt from the constant search for food that
dominates the daily routine of many insectivores.

When hedgehogs do eat, however, they are not choosy. They
eat snails, insects, worms, feces, and any other animal matter that
they can find within a few hundred meters of their burrows. In some
areas they are blamed for predation on pheasant eggs and chicks and
are thus considered pests. They sometimes kill and eat snakes and are
at least partially immune to their venom.

Like Greta Garbo, hedgehogs want to be left alone. Except
they want to be left alone most of the time. When not breeding or
rearing young, they ignore one another at rare chance encounters,
each hedgehog proceeding on its way without the risk of social involvement. The hedgehog's life is a paradigm of successful solitude.

A brief, almost literally whirlwind courtship culminates in
fertilization, followed by a two-week honeymoon. During this period
captive hedgehogs are almost sociable, enduring if not cultivating each
other's company. The young are born five or six weeks after conception
and remain with the mother 40 to 60 days.

NEONATAL COMMUNICATION

Hedgehog birth is noisy. In the summer, according to Herter
(1957), litters of from two to ten blind, naked young are born in a
small nest. Almost immediately they fight their way toward the nipples,

peeping shrilly and trying to push one another away. Since the peeps abate when each infant is at a nipple, they probably inform the mother whether anyone is being left out. The peeps can be considered distress signals. The young remain on the nipple continuously for the first 24 hours of life. For the next few days the mother leaves the nest only briefly, after covering the litter with materials collected in advance. When left alone, they huddle in a pyramidal pile. When they touch the mother's belly hairs, she arches her back, allowing them to nurse. She moves the litter from a disturbed nest and retrieves strays by mouth. Mothers lick their young, inducing elimination and cleaning them simultaneously (Herter, 1957; Poduschka, 1977). Poduschka mentions a "quiet squeaking" with ultrasonic components, possibly the same as peeps; like purring and lip smacking, these squeaks act as contact or hunger signals.

After about three weeks, the infants' teeth erupt, their eyes open, and they begin to forage near the nest. For the next six weeks or so, the litter maintains a loose association, foraging as a group with the mother a short distance behind them. According to Herter, this association is maintained by vision, olfaction, and audition. When a young hedgehog loses contact with the group, it emits a twittering "lost whistle." At this distress call, heard only during the period of loose association, the mother immediately approaches the youngster, sniffs it, and leads it back to the group.

INTEGRATIVE MESSAGES

Poduschka mentions a piercing cry given by hedgehogs *in extremis*. This call is an extreme form of a graded vocalization that at lower levels of intensity is a snort.

During the period of loose association, the sexually immature captive hedgehogs engage in precocious sexual play, males and females both showing many of the elements and much of the frenzy of adult copulation. They poke their noses into each other's back spines, claw each other's backs with their forelegs, and mount in a variety of orientations, with pelvic thrusting and (in males) penile erection.

Since they are nocturnal foragers, hedgehogs probably have a keen sense of smell. Thus, the characteristic odor of hedgehog feces, which Herter compares to oiled leather, probably acts as a contact signal.

Hedgehogs can detect one another visually—Herter's tame hedgehog, Eri, often approached another behind a pane of glass, and in 14 out of 20 trials, he attacked his own image in a mirror. He also jumped at a plastic hedgehog, which suggests that it was the shape, not the motion, of the stimulus that mattered.

Quay (1965) notes that some species rub the corners of their mouths on the ground even outside of breeding season. This behavior may be related to the puzzling behavior called self-salivation displayed by hedgehogs of both sexes (Herter, 1957), in which novel odors are applied to the spines and sometimes transferred to other hedgehogs. Because the stimulus is always novel and because the results are contact with and sharing of the novelty, this behavior probably promotes familiarity with and recognition of odors and the individuals to which they are applied. The importance of novelty in stimulating self-salivation is shown by the fact that young hedgehogs are stimulated to self-salivate by foods (fat, smoked fish, mealworms) that do not so stimulate adults who eat them regularly.

In self-salivation a hedgehog licks or chews a novel odorous substance, salivates copiously, and applies the odor-bearing saliva to a small patch of its spines. The procedure is repeated two to 50 times, and usually the saliva is applied to a different spot each time. Herter reports that self-salivation is a consistent response to glue, tobacco, human sweat, flowers, soap, perfume, tincture of valerian, varnish, tanned leather, fur, wool, tar, rotten eggs and crabs, feces, dog urine, toad skin, and other (living) hedgehogs. Sometimes a hedgehog licks another, salivates on itself, and is licked by a third. Group salivation occurs with equal frequency among same-sexed and different-sexed animals.

Herter dismisses three possible functions of self-salivation, pointing out that the saliva is not poisonous, does not clean the spines, and did not prevent olfactory detection by a dog in 20 controlled trials. There are analogues to self-salivation in the behavior of other mammals, particularly among canids and felids. Poduschka and Firbas (1968) suggest that self-salivation is analogous to flehmen in ungulates, introducing substances to the vomeronasal organ, an olfactory receptor that may be involved in the detection of sexual pheromones. The significance of this unique behavior for the hedgehog, however, remains a mystery.

According to Poduschka (1977), European hedgehogs (E. europaeus L.) leave deposits of feces at particular places. These deposits could act as territorial advertisements or simply as landmarks. Juveniles paste their feces on vertical surfaces (Poduschka, 1977).

AGONISTIC MESSAGES

Hedgehogs do not always simply pass in the night. Sometimes they greet each other by butting each other on the flanks. Sometimes they bite and sometimes they "box," alternately striking and hugging each other with their forelegs. Boxing is accompanied by hissing and spitting threats, the latter often performed by a retreating hedgehog, which suggests that it may be a defensive threat (Herter, 1957).

In captive hedgehogs these fights often lead to serious injury because usually one hedgehog attacks while the other does little or nothing to resist. In nature the less aggressive hedgehog probably flees, an option denied it in captivity.

When kept in cages, groups of hedgehogs eventually form boxing hierarchies in which they box only animals of lower rank. The relevance of this observation to free-ranging animals, which rarely, if ever, form groups, is uncertain. Interestingly, however, boxing dominance does not seem to depend on size or sex (Herter, 1957). In any case, there are no known dominance signals differentiated from fighting. Hedgehogs do, however, threaten by quickly moving spines on the head forward and giving sharp snorts. These signals often cause the other hedgehog to withdraw or to gape, displaying an intention to bite (Poduschka, 1977).

SEXUAL MESSAGES

The intensity of hedgehog courtship was known to Aristotle. The breeding season for hedgehogs occurs in the summer. It begins with a long, high-pitched cry by the male, which Herter interprets as a signal of sexual readiness. In nature this call probably functions to allow the female to find a mate, establishing a pattern of female initiative and male pursuit.

Poduschka (1977) describes *scent-marking* in sexual contexts. A male hedgehog scent-marks by arching his spine, placing his hind legs together, extending his penis, and rocking from side to side. This action is usually performed behind the female and leaves a white secretion whose source is either urine or accessory sexual glands around the penis. Perhaps substances in these scent marks are the functional stimuli for self-salivation, but I have found no descriptions of those behaviors during sexual encounters.

Once the couple gets together, usually in a small clearing, the male attempts to sniff the female's rear, but she either jumps away or turns, erects her spines, and attacks the male, boxing and biting. When the male persists, the result is the hedgehog "roundabout." The female controls the progress of this dual whirlwind, moving it back and forth across the clearing, sometimes changing the rules by butting the male away until he retreats, hissing. According to Poduschka, the roundabout brings the female into repeated contact with the male's scent marks, which act as a releaser for sexual cooperation. After a quarter of an hour or so, she leads her suitor into the privacy of the surrounding vegetation, where the rustling leaves, snapping twigs, and spitting threats suggest either intensified ardor or renewed aggression. Privacy is insufficient for consummation, however, for after about five minutes of commotion, the couple whirls out into the clearing, and the merry-go-round continues.

Suddenly the female flies off on a tangent, runs about a meter, drops flat on her stomach, and stretches her hind legs out behind her, exposing her vulva in an unambiguous posture of sexual readiness. The male mounts quickly but carefully, avoiding her spines by placing his forefeet lightly on her back. He rattles his spines together loudly, and both snort noisily. The male inserts his penis quickly and thrusts strongly about a dozen times as the female, transformed, lies quietly. The male then steps away and sits quietly, as does the female. Sometimes several mounts are necessary to achieve intromission. In *Hemiechinus auritus*, the eared hedgehog, male and female snuggle together and brush their cheeks on each other's bodies before mounting (Herter, 1957; Poduschka, 1977).

Aristotle recorded the notion that hedgehogs copulate belly to belly. His opinion is reasonable, given the length of the female's spines and the frequency and duration of boxing episodes, but it is not correct. Aristotle failed to reckon with the hedgehog penis, probably the longest (in proportion to body size) and certainly the most anterior in any mammal.

Ewer (1968) points out that females of spiny mammal species take the initiative in courtship and switch suddenly from rejection to receptivity. She suggests that this behavior is necessitated by the defensive effectiveness of the spines—the male cannot take the initiative, and were the female's transition to receptivity gradual, the male might be impaled.

The hedgehog's ardor can be ascribed at least partly to the necessity of inducing ovulation in the female. In other induced ovulators, courtship is quite violent, and states of high arousal seem to be necessary for the release of the ovum.

SUMMARY

To summarize, it seems that hedgehogs have a fairly simple repertory of 12 different types of messages, including neonatal distress peeps; olfactory, visual, and auditory contact among young; a distress whistle by lost young; and maternal recognition of young by sniffing. Integrative messages include sexual play, adult contact odors, visual identification of conspecifics, and familiarization by self-salivation and perhaps by defecation. Agonistic messages include boxing and a hissing defensive threat. Sexual messages include male sexual advertisement by a mating cry; prolonged courtship, with chasing, scent marking, and wrestling; and male spine rattling during copulation.

SHREWS

Shrews of family Soricidae include the smallest mammal, the pygmy shrew (genus *Suncus*). Most of the approximately 265 other species of shrews are also small. The typical shrew's small size places it at the mercy of thermodynamics, and its life is a race against cold and starvation. A rough index of the tempo of life for a shrew is its heartbeat—in the common shrew *(Sorex araneus)*, rates of more than 1200 beats per minute have been recorded (Burt & Grossenheider, 1964). Shrews consume their weight in insects and worms every 24 hours, and for some shrews this period has been estimated at three hours (Bertin, 1967). Crowcroft (1957) reports that shrews, like rabbits, ingest their feces. This practice, called refection, presumably allows the shrew to extract additional nutrition on the second passage. Shrews alternate three-hour periods of frantic hunting and exhausted rest, day and night, winter and summer.

The teeth of shrews are specialized for grabbing, puncturing, and crushing the chitinous bodies of insects, but those of the water shrew *(Neomys fodiens)* work well on fish 60 times its own weight (Lorenz, 1962). The specialized insectivore teeth are housed in a long, pointed face that gave the family its German name, *Spitzmaus*, or "pointy mouse." The pointed snouts bear whiskers, or vibrissae, that are as sensitive to touch as the nose itself is to odors.

Shrews have successfully adapted to a wide range of habitats and climates without major changes in morphology. In North America, for example, a single species, the short-tailed shrew (genus *Blarina*), is found from southern Florida and Texas through the United States east of the Mississippi to Lake Winnipeg and Hudson Bay. Other species are successful in the Arctic, in deserts, and at high altitudes. Shrews are, however, subject to predation from local small carnivores. Their only defenses are fleeing and hiding.

Because shrews are so small and secretive, observations of free-ranging ones are few, and we must depend on studies of captive shrews and infer the extent to which their behaviors resemble those of shrews in the wild. Captivity observations and trapping studies of the common European shrew, *Sorex vagrans*, and other species suggest that they defend small territories. Captive shrews defend "their" parts of the cage (Crowcroft, 1957), and the intensity of this defense seems to vary with duration of occupation (Bunn, 1966). Rankin (1965) describes the territory of the elephant shrew *Nasilio* as consisting of a burrow, several feeding areas, and the trails linking them. According to Shillito (1963), a male *Sorex* briefly enters the female's territory for breeding. Several species, including some members of genera *Crocidura* and *Suncus*, are semisocial, tolerating one another's presence even when not breeding or rearing.

NEONATAL COMMUNICATION

Although shrews have only six teats, the litters of some species are sometimes as large as ten. Newborn shrews are so vociferous that Herter (1957) remarks "There is nothing as noisy as young shrews" (p. 37). Their repertory includes two distress vocalizations. One is peeping, which in *Crocidura* allows either parent to locate and retrieve stray young from more than a meter away (Gould, 1969). The other, a rattle, probably signifies distress of higher intensity, for the rattle typically occurs when young shrews are picked up by a human. Other vocalizations are a high, quiet peeping, made by small nestlings, which Herter refers to as a contact and contentment call, presumably because it is uttered after nursing; a greeting twitter or trill upon awakening and on meeting parents or siblings; and a variety of noises whose significance is not known but probably help maintain contact between mother and young. These include "zitt," "click," and "tuckern."

In *Crocidura* shrews there is a unique form of vegetative interaction between mother and young—the "caravan" (see Figure 2-2). It is seen in all species of the *Crocidura* genus but in no other mammal. Meester (1960) describes the caravan as "a minute furry railroad train with the mother as the engine and its destination the safety of the nest" (p. 37). Herter quotes a description of a caravan: "a single being with one head and will and obedient legs" (p. 37). Caravaning is stimulated by unfamiliar surroundings, sudden noise, strange odors, cold, or dampness. The caravan breaks up when it nears the nest, even before the mother enters.

Cohesion in the caravan is maintained by each shrew's biting onto the fur of the hindquarters of the shrew in front of it. This procedure allows drill-team precision as the members of the caravan accelerate, turn, and stop together. However, this ideal of coordination is not always attained; sometimes the caravan goes double file or degenerates into a scrambling bunch. Such disarray is evidently intolerable to the maternal martinet, for, after traveling a short distance, she reforms her platoon into single file and proceeds. In some species the mother gives a command in the form of a twitter, upon which the young immediately form up behind her (Meester, 1960). Sometimes this does not suffice, and she grabs one by the snout, pulls it toward her, and presents her hindquarters.

FIGURE 2-2 A caravan of shrews of the genus *Crocidura*. Drawing by Heidi Reynolds.

Three stages of development precede orderly caravaning. In the first stage, seen also in elephant shrews and other insectivores, young are dragged along while attached to the mother's nipples. The nipples of the rare Caribbean insectivore solenodon, which are located on the buttocks (Morris, 1965), may be specially adapted to this practice.

The second stage is usually seen five or six days after birth, immediately after the young have been carried away from the nest in the mother's mouth. As soon as she puts one down, it tries to grab her as she moves away, briefly biting the fur of her haunches, but to no avail.

In the third stage, during the second week of life, young *Crocidura* begin to follow their mother, and if a mother and young are placed in unfamiliar surroundings, caravan formation takes place. Caravaning without human intervention soon follows, and it lasts until about the 18th day after birth.

Crowcroft (1957) observed no caravaning in shrews of the genus *Sorex*, but he did describe a behavior from which caravaning may have evolved. When a mother *Sorex* leads her young to the nest, she pauses frequently to let stragglers catch up. It is easy to imagine the tendency to nurse generalizing to a tendency to hang onto the most accessible part, and from hanging onto the mother to hanging onto whoever is closest.

INTEGRATIVE MESSAGES

Lorenz (1962) provides us with a brief description of play in young water shrews (genus *Neomys*). In the water, as on land, their play consists of excited chases with loud twittering. This twittering evidently differs from that heard during fighting and may signal non-serious intent.

Crowcroft (1957) heard a short squeak uttered at accidental encounters between adults. The circumstances suggest that this is the alarm call mentioned but not described by Herter (1957). Shrews may also have an olfactory alarm signal. Dryden and Conaway (1967) found that house shrews (*Suncus murinus*) release a musky scent from glands on their necks and behind their ears when chased or otherwise disturbed. Lorenz (1952) describes a violent, deadly fight amongst a family of water shrews (*Neomys*). The small box that held them was filled with a "strong, musk-like odor." This odor could be a signal of distress or alarm or may simply have been released by the frantic activity of the fighting shrews. When irritated, *Blarina* shrews produce sharp, high-pitched squeaks (Poduschka, 1977).

Paradoxically, contact signals are used not to stay close to but to avoid other shrews. Mutual withdrawal was the most common response when Eisenberg (1964) placed *Sorex* shrews together in novel surroundings. Shrews stay separate with the aid of keen noses and strong smells. The rear third of the flanks of some species of shrews is covered with special glands that evidently trap the substances produced by the glands around the head (Dryden & Conaway, 1967). Odors applied to these glands are probably perceived directly, through the nose-to-nose and nose-to-body contacts that occurred in encounters arranged by Eisenberg. Presumably, odors from these glands are also applied to areas over which the shrew crawls.

The significance of the indirect transmission of odor by application to the environment is neatly illustrated by a comparison of two groups of elephant shrews. Those inhabiting arid, open regions (genus *Elephantulus*) travel by hopping and have a scent gland on the bottom of their tails, so that a spot of odor is deposited on each landing (Morris, 1965). Those living in moister areas crawl through small tunnels in the grass and have the same sort of gland on top of their tails, so that the odor is rubbed onto the roof of the tunnel (H. Lang, 1923). Evidently, effective application of odor to the environment is sufficiently important to these shrews that there has been selection for position of the tail gland. These odors presumably inform individuals that a route has been used recently.

Some species of shrews supplement the indirect action of their flank glands with the odor of feces. In the tempo of their other activities, active *Sorex minutus* and *Neomys* defecate about once a minute, usually at special locations some distance from the nest. This practice has been termed localized elimination. Shrews of the species *Crocidura russula* place bales of their feces on objects, usually 5 to 7 centimeters above the ground (Crowcroft, 1955). This practice may allow the odors to spread more readily by getting them out of the boundary layer of still air along the ground and into air currents.

Odors from glands and feces probably figure in a tendency observed by Crowcroft for shrews to avoid trails recently used by another. Some species emit a soft, high-pitched twitter whenever foraging. Shrews are reputed to have keen ears. Should the odor-avoidance system fail during the shrew's frantic foraging, the twitter could serve as a short-range backup system, transmitting the information that another shrew is in the vicinity.

In a small percentage of the encounters staged by Eisenberg (1964), shrews changed course to approach each other but made no contact. In a few cases one shrew followed another for a short distance. This response may be a vestige of neonatal following. In any case, even shrews show traces of affiliation. Eisenberg concluded that shrews can learn to recognize one another because encounters between strangers resulted in higher levels of aggression than encounters between animals who lived in the same tub. Nasonasal, nasoanal, and nose-to-body contacts may be involved in such recognition.

AGONISTIC MESSAGES

Usually, however, shrews are not at all friendly, as is shown by the following duel described by Crowcroft (1957). Each of two captive male *Sorex* "owned" part of an enclosure that was small enough so that encounters were frequent. The first male to notice the other raised his nose and squeaked, at which the second male generally withdrew. The head-up posture with squeak thus probably functions as an offensive threat. Crowcroft suggested that ultrasonic signals may accompany threats. Sales (1972a), however, could detect no such signals. When the second male did not withdraw, he responded with a threat squeak similar to that given by the first and approached him face to face. Then both emitted a scream that is probably a higher-intensity offensive threat. At this point, either may withdraw and may be followed but will not be attacked.

Eisenberg (1964) observed two types of agonistic interaction without fighting. In the first, encounters between subadults led to nasoanal contact and following. These behaviors often preceded a fight and may therefore function as threats. The second type of interaction was a chase. Poduschka (1977) states that gapes act as threats by signaling an intention to bite. He also notes that *Crocidura* shrews threaten with a single, sharp, metallic squeak and that members of the genus *Blarina* withdraw at the chatter of a conspecific.

In fighting, shrews rear up on their hind legs, squeal, strike with their heads and necks, attempt to bite, and clutch and box at each other with their forelegs. Eisenberg reports fighting in a "clinching scuffle," following a rush or pounce. The miniature combat rages furiously until one shrew throws itself on its back, screaming and striking out with its legs. This tantrum often results in withdrawal by the other shrew. According to Bunn (1966), the shrew that throws itself on its back first is the victor. This observation suggests that this behavior, often considered a sign of submission in other mammals, may be a sign of dominance in shrews. Sometimes, however, the response to the tantrum is not retreat but another tantrum by the other shrew, and both shrews lie on their backs in a screaming duel. This duel is terminated by the withdrawal of one of the shrews, a

These two shrews threaten each other with the head-up posture. Photograph by Rick Search, courtesy of the Brookfield Zoo.

return to the boxing phase, or a whirling bout of mutual tailbiting (Crowcroft, 1957).

When two of Herter's (1957) captive *Sorex araneus* met, they usually took off in different directions. Sometimes one or both would squeak, one would jump into the air, and the other would run away. When the encounter took place in the territory of one of the shrews, the resident often froze and made a staccato squeak, suggestive of a small bark, and the intruder usually fled. When the intruder remained, a tug of war ensued, each shrew biting the other on the tail. The tug of war usually ended with the flight of the intruder but sometimes escalated to tantrums and tailbiting as described by Crowcroft. Eisenberg (1964) noted that when a shrew approached an occupied nest box, its occupant often defended its nest with squeaks and short rushes, which usually drove off the intruder.

Fights in other species of shrews are very similar, but each species has its own dialect of threat. *S. minutus*, for example, often grinds its teeth, and *Crocidura* shrews bare their teeth by raising their lips in doglike snarls (Herter, 1957).

Eadie (1938) and Pearson (1946) have suggested that the sebaceous ventral gland of *Blarina* may be used in territorial scent marking. It would be interesting to observe responses to ventral scent marks in natural or seminatural conditions to test this hypothesis.

SEXUAL MESSAGES

The exception to the general tendency for adult shrews to avoid one another occurs during the summer breeding season, when a male's customary aversion to fresh shrew scent becomes an attraction if the scent was left by a mature female. *Sorex* males' squeaks are staccato, females' higher, more continuous, and piercing (Crowcroft, 1957). These differences may allow identification of the sex of the transmitter before courtship begins.

Shrew courtship begins in a typically shrewlike manner, with a noisy fight. Differences in the quality of the shrieks are usually the observer's first clue that what he or she is watching is love, not war—to humans, males and females appear identical. The male's fighting cry, a staccato, barklike squeak, is, however, distinguishable from that of the female, a "continuous, bright, penetrating tone" (Herter, 1957, p. 30). Male *Blarina* produce noisy clicks as they pursue females; if a female is unreceptive, she produces shrill squeaks and chatters. The difference between male and female vocalizations is probably no coincidence, and almost certainly provides the combatants with information about each other's sexual identity.

Eadie (1938) suggested that the flank glands of male *Blarina* shrews produce an olfactory sexual attractant because these glands are largest during the breeding season. When introduced to a cage containing a female, a male of the genus *Suncus* uses one odor in two different ways. First, the glands around the head emit a musky odor similar to the alarm scent. This odor spreads and dissipates quickly. Then he scent-marks by rubbing his throat and belly on the floor and his flanks on the walls. This odor, presumably from substances emitted by the head glands and trapped in sebaceous complexes on the ventral and lateral surfaces, lasts longer than the odor emitted directly into the air (Dryden & Conaway, 1967).

In *Crocidura*, courtship begins with the female's mating call, a high, continuous peeping, which Herter (1957) represents as "tji-tji-tji-tji . . . (long 'i')." A fight ensues but quickly abates and is replaced by mutual genital sniffing accompanied by a friendly-greeting call, a "delicate, quick trill or twitter that is best depicted 'zititititititi' (long 'i') " (p. 30). This call is given by both partners.

As courtship progresses, the female begins to move with stiffly extended hind legs, raising her genitals in a posture known as lordosis. The male alternately sniffs her genitals and attempts to mount. Mounting consists of pressing his hindquarters against those of the female while biting her hard on the nape of the neck and twittering in the greeting call. Pelvic thrusting has not been observed. The penis seems to become locked in the vagina, as in dogs. The couple remains motionless for about a minute and a half, while the female emits a copulatory call, a high, tinkling chatter, represented as "trrrr trrrrt trrrr. . . ."

After copulation the shrews separate and groom themselves. The female initiates subsequent couplings by sniffing and licking the male and uttering the mating call. Though unresponsive at first, he soon begins to sniff her vulva, emits the friendly-greeting call, and mounts.

Gould (1969) provides an excellent description of courtship in two other genera:

> During initial phases of courtship in both *Suncus* and *Blarina*, the male appears to play a passive role—approaching the female, rubbing the substrate, tolerating bites from the female without biting her—while rubbing and exuding odor in new areas and gradually increasing the receptivity of the female. The female repels the male with high intensity chirps and buzzes; the male is easily repelled by the female's loud vocalizations and bites. Some males emit frequent "put" while courting. Orientation of the female's head and body toward the male was particularly prominent when the male emitted frequent and loud "put." The male responds to bites and loud chirps by closing his eyes and ears and exposing his gland-covered neck. The male rubs his venter over the substrate and simultaneously over his body. His glandular odor, immediately detectable by the observer, is emitted after 2 or 3 minutes (Dryden & Conaway, 1967). The female's body is pervaded with the male's odor through the following means: rubbing of the substrate by the male followed by the female walking over the rubbed areas and toileting herself; occasional fights; the

female rubbing her tail against the male's neck as he positions himself behind her. Female *Suncus* reduce biting after the male fur is covered by glandular secretions. (Distortions of lips and tongue-smacking after biting indicate that glandular secretions have a noxious taste to the female.) The male bites the female on the flanks, rump and tail and as she becomes more receptive he follows closely behind her, oriented in a manner similar to the caravan formation prevalent during infancy. Continual advances and increases in click rates by the male (*Suncus* only) are followed by a receptive chirp or twitter by the female *Suncus* and a series of clicks by the female *Blarina.* Copulation follows [p. 271].

After breeding, interactions between mates are few and far between, but Herter (1957) cites an experiment in which a male was reintroduced to his mate in three weeks (roughly the period of gestation) and the mates "seemed to recognize each other" (p. 31). The male and female approached each other, gave greeting twitters, and copulated.

SUMMARY

Shrews are considered solitary, but the demands of reproduction and rearing have selected for a message system with 19 discernible message types. Neonatal messages include infantile distress calls, infantile contact peeps and noises, affiliative greeting twitters, and a caravaning-assembly command. Integrative messages include playful chasing, with a play twitter; distress and alarm squeaks and odors; contact odors and calls; quasi-affiliative following; familiarization by localized defecation; and identity by recognition sniffing. Agonistic messages include offensive and defensive threats by rushing and squeaking, fighting by boxing and tug of war, a dominance posture, and territory advertisement with glands. Sexual messages include male advertisement by odor, directly and through scent marks, and by voice; courtship; and a female copulatory vocalization.

Some of these message types are difficult to distinguish in the natural behavior of the shrew, as the forms are similar and the contexts are ambiguous. However, since shrews have not been thoroughly investigated and since there are a variety of vocalizations whose significance is unknown, these 18 types must be regarded as a minimal lexicon for the shrew.

TREE SHREWS, FAMILY TUPAIIDAE

Tree shrews look so much like squirrels that Malaysian natives use the same word, *tupai*, for both. The tree shrew's face, however, is an insectivore's, not a rodent's; the nose is long, conical, and pointed (see Figure 2-3). Tree shrews' eyes are larger and their ears rounder than squirrels', but tupaiid is their family name. There are 18 living species of tree shrews in rain forests from Southeast Asia to the Philippines.

Tupaiids represent a pivotal stage in mammalian evolution when, more than 60 million years ago, an opportunistic group of

FIGURE 2-3   Tree shrew. Photograph by Roger Peters.

insectivores took to the trees. In this arboreal Eden they were safe from most mammalian predators, but two dangers remained: snakes and falling. Both of these encouraged activity during daylight hours, when tree shrews could see. Daylight also provided complex information about the identity of food and conspecifics, fostering two adaptations based on keen vision: binocularity effective enough to judge distance and a brain enlarged to handle the knowledge gained. Daytime activity in the trees also provided the basis for the evolution of other adaptations that now characterize the primate order: grasping hands and small litters. For these reasons some taxonomists classify tree shrews with primates. The dominant opinion today, however, is that they are insectivores (Hill, 1972). In particular, Martin (1966) has argued against inclusion of tupaiids in the primate order.

Tree-shrew societies vary in complexity and coherence much more than do the societies of shrews and hedgehogs. Nevertheless, all tupaiid social organization is an accommodation to tupaiid personality, which has been described by Eimerl and DeVore (1965) as "wildly irate, gluttonous, and libidinous" (p. 21). During their frequent periods of excitement, the adrenalin in their blood reaches concentrations 100 times greater than those found in humans (Schwaier, 1975). This characteristic makes them ideal subjects for studies of socially communicated stress.

Some species (*T. minor, T. chinensis,* and *T. palawanensis*) are as solitary as shrews and hedgehogs, but one species (*T. montana*) lives in groups of about 12, separated from other groups by about 1 kilometer (Sorenson, 1970). In captivity some species (*T. glis, T. chinensis*) show *linear-dominance hierarchies* in which high-ranking animals harass and are groomed by lower-ranking ones.

NEONATAL COMMUNICATION

Like shrews and hedgehogs, tupaiids are born blind and naked. However, unlike other insectivores, their litters are small, containing one to four. The young seem to form strong emotional attachments to whatever they contact within two hours after birth. This phenomenon is called *imprinting.* The usual object upon which young become imprinted is their mother, but they seem to imprint on a human who handles them almost as easily. They express their attachment by approaching the imprinting object whenever they can and by giving chirps and clicks of contentment when in contact with it. These same contentment vocalizations are given after they have been fed.

Tree shrew. Photograph by Rick Search, courtesy of Brookfield Zoo.

Shortly after giving birth, the mother eats the placenta and in some species scent-marks the young with her sternal (chest) gland. In other species (such as *T. belangeri*), scent marking is performed by the father. This practice probably facilitates imprinting by providing the young with a sample of parental odor. It also allows the mother to recognize her young by smell, because when, in periods of social stress, she fails to scent-mark her young, she is liable to eat them, as will any other adult that enters the nest. According to Martin (1966), the mother nurses her young only once every 48 hours.

INTEGRATIVE MESSAGES

Play among young tree shrews is frantic, but silent and formalized. One game is a chase, a silent scramble without biting. In follow the leader, a chain of scurrying youngsters pursues an elaborate course over, under, and around various obstacles, which may include their mother. Ambushing involves a sudden rush from behind and may precipitate chases or follow the leader. Young tupaiids also play peekaboo and keep-away, using pieces of nest materials.

Tree shrews are highly vocal, so the various vocalizations described in this section ordinarily suffice to inform them of one another's presence. According to Williams, Sorenson, and Thompson (1969), individuals of *T. palawanensis* engage in antiphonal contact calling, or duetting. When silent, many species stay in touch visually by flicks of their bushy tails, performed every few seconds whenever they are active.

Affiliative messages express friendly intent. Tree shrews devote considerable time and energy to one of the most widespread and intimate expressions of mammalian affiliation, mutual grooming. In fact, they are one of the few mammals to evolve a body structure with social grooming (allogrooming) as a major function. Tupaiids' lower incisors stick out (this condition is called procumbent) and are used like little combs. One tupaiid presents itself for grooming by backing or sidling up to another (Hill, 1972), which alternately licks and combs an area on the groomee's body until it is clean, then moves on to a new area. According to Sorenson (1970), being groomed appears pleasureable. Like presentation, grooming occurs in every sexual combination.

For animals as sensitive to temperature as tree shrews, grooming is no mere luxury—clean, unmatted fur is essential for insulation. Some parts of a tree shrew's body are far more accessible to another than they are to the owner. Mutual grooming is thus an efficient strategy for care of the integument, the organ composed of skin and fur.

Allogrooming is not, however, just a matter of hygiene. It allows tree shrews to maintain close contact without aggression. Since dominant animals are more likely to be groomed than to groom, grooming can be considered a signal of submission. Grooming is also common among the supposedly rankless *belangeri* tree shrews, so, at least in this species, it expresses affiliation without submission.

Because they are small and adapted to the relatively constant high temperatures of the tropics, captive tree shrews huddle in groups of two to five when resting if the ambient temperature drops below about 24° C. This vegetative interaction saves body heat and makes it advantageous to inhibit aggression, at least temporarily. Even when the temperature is about 24° C, tupaiids often sleep in small groups in their nest boxes. Sometimes huddling tree shrews bury their noses in another's fur. Consorting among females is common, among males, rare. Group sleeping may be merely an artifact induced by restricted or luxurious accommodations, but this practice reveals a capacity for peaceful proximity unknown in other insectivore adults and thus can be regarded as at least inchoate affiliation.

Behavior while feeding also reveals a potential for peaceful coexistence. Members of the eight species studied by Sorenson (1970) often ate side by side. Even when killing and devouring a mouse, they did not display aggression toward one another. On several occasions Sorenson saw one tree shrew take food from another without ensuing violence. Another form of food sharing, suggestive of the rituals of social canids, is the immediate ingestion of vomited food by all tupaiids in the immediate vicinity.

At any disturbance the tail of the tree shrew can bristle, nearly doubling in size. Von Holst (1974) has investigated this phenomenon, which he calls *Schwanzstrauben*, or SST, and finds that although it is somewhat variable in intensity, it is basically a binary signal: the tail either bristles or it doesn't. According to Von Holst, in SST the hair-raising muscles are activated by sympathetic fibers. Sympathetic fibers are the outputs of the part of the autonomic nervous system that prepares organs for fight or flight.

By comparing behavior and reproductive physiology at different SST values, defined as the percentage of the time an animal was displaying SST, Von Holst found that high frequencies of SST communicate social stress and limit reproduction. SST values vary directly with population density and inversely with rank. When the SST value for a pregnant female was below 20%, reproductive behavior was normal. When it rose above 20%, birth was normal, but mothers no longer sternal-marked their young, which consequently were often eaten, sometimes by their own mothers. At higher SST values, mothers lost weight and no longer nursed, and their mammaries shrank.

The effects of high SST values on males were equally disastrous. At 50% SST, males' testes withdrew, and at 80% they shrank. High SST values also increased the rate at which males sternal-marked their environment. Sternal marking by one male often leads to sternal marking and SST by another. Thus, SST and sternal marking created a stressful sort of positive feedback. Sternal marks of castrates had no such effect and thus may have lacked some essential ingredient.

In captivity most tupaiids are "localized eliminators," with two or three definite latrine areas. In nature this behavior would facilitate familiarity with the home range by providing olfactory landmarks, but, as with shrews, there is no direct evidence for this hypothesis.

AGONISTIC MESSAGES

Tree shrews are most eloquent when angry, and they are often angry. Their most common expression is a rigid crouch, with neck stretched forward, head up, and mouth open. In some species this display is augmented by hissing, whistling, barking, screaming, or chattering. Nuances in these expressions are difficult to distinguish, but in one of the more vocal species (*T. longipes*), continuous screams with an open mouth are combined with staring and may denote rage; barking and hissing are combined with lunges and can be considered offensive threats; and rigid crouching with head up, neck stretched, and staccato barks is associated with retreat and can be considered a defensive threat. This defensive threat is often offered by a female when approached by a male.

At the other extreme of tree-shrew vocalization, the most taciturn tupaiid, the tiny, omnivorous, solitary *T. minor*, has only two calls, both of them threats. One is a rapid, high-pitched chatter with open mouth and crouched legs. The other is a muffled bark given when the animal is cornered. These can be provisionally characterized as offensive and defensive threats, respectively (Sorenson, 1970).

Social ranking limits aggression by "fixing" fights—the victor is determined in advance and can win the token displays of strength and viciousness. As in Saturday night wrestling, threats and displays of rank look dangerous but rarely are.

According to Sorenson, most species of tree shrews show linear-dominance hierarchies based on aggressive interactions. In a linear-dominance hierarchy each animal has a different rank. For example, in *T. glis* there is a male despot, the alpha, which harasses all the others in the group without fear of retaliation; a beta, which harasses all but the despot; and so on, down to a bottom-ranking omega, which is pushed around by all and which bothers no one. In another species, *T. chinensis*, the females have their own hierarchy, separate from that of the males.

Among the privileges attendant upon high social rank are priority in access to food, rest areas, and grooming by females. Tree-shrew despots have the ability to disrupt, by their mere presence, all sexual behavior by others.

A dominant tree shrew asserts its rank by offensive threats and by supplantation, in which it walks toward a subordinate, which moves off, leaving the dominant one at least momentarily in its place. Dominants often supplant subordinates at nests or food, but the displacement is sometimes purely ritualistic, with no particular attraction or resource involved. Mounting is another common expression of dominance. These mounts are not sexual, for they are unaccompanied by genital sniffing, intromission, or other sexual behavior. They occur in all sexual combinations, and dominant females often mount low-ranking males.

Dominant male *longipes* tree shrews leave an enduring olfactory record of their victories by rubbing their sternal glands on prominent objects in the vicinity. As previously mentioned, these sternal marks seem to alarm and stress others that detect them. High-ranking *T. palawanensis*, among the least social of tree shrews, chirp to express their rank. Schwaier (1975) studied 25 groups of *T. belangeri* for more than two years but found no evidence of rank order, so linear-dominance hierarchies are not universal among tupaiids.

Tree shrews are not always irritable. They have several ways of conveying friendliness or at least nonaggression. Many of these messages complement and appear as responses to dominance displays and are termed submission. One form of submission, grooming, has already been mentioned. In some species anal presentation complements dominance mounting by minimizing effort by the mounter. According to Sorenson (1970), both male and female *T. tana* and *longipes* present themselves for mounting when approached by an animal of higher rank. By doing so, they spare themselves the stress of a dominance contest. The mount is often perfunctory, and it is probably more humiliating than painful.

Like most other mammals, tupaiids threaten impressively but kill rarely. Observations of their behavior in the wild are few, but Schwaier and her colleagues never observed severe fighting with lunges and bites among wild-caught animals, even after their presumably stressful journey from Bangkok to Frankfurt. Sorenson found no aggressive fatalities in his colony in two years.

The situation is quite different for tupaiids born in captivity. Schwaier's captive-bred animals were extremely aggressive, and even brothers and sisters eventually killed one another when kept in the same cage. Schwaier attributes this aggression to a specific deprivation due to captive rearing, combined with an innate aggressive response to the odor of a stranger. Parental scent marking does not seem to be involved in the inhibition of this response because intralitter aggression is high even when parental marking occurs.

In wild-born tree shrews, once a bout of screaming, lunging, and chasing escalates into a real fight, ritualized combat seems to moderate aggression. For example, in *T. montana*, the most social species, fighting takes the form of boxing as seen in hedgehogs and shrews. Boxing, in which combatants spar face-to-face, limits the target of the most effective weapon, the teeth, to the mouth of the opponent. Only the relatively harmless paws come into contact with less well-protected parts of the body.

SEXUAL MESSAGES

In most species of tree shrews, courtship is associated with uterine bleeding and increased frequency of scent marking. Most of our information about tupaiid courtship comes from Sorenson's (1970) captivity observations of eight species. In *T. longipes*, males and females rub and slide against each other. As in other species, imprinting affects sexual responsiveness, for *longipes* imprinted on humans will display sexual rubbing against a human hand. As courtship progresses, the normally imprinted male *longipes* begins to groom the female from a mounting posture. I have observed captive *T. glis* engaging in peaceful snuffling, resting with one's snout in the neck fur of the other.

In most other species of tupaiids, the male grooms the female's flanks, attempts to mount, biting her on the neck and shoulder, and is repulsed. The female then mounts the male, with pelvic thrusts, and is herself repulsed. Sexually aroused male *T. lyongale* express their excitement with a series of quacks culminating in a shrill chirp at the moment of ejaculation.

According to Schwaier (1975), *belangeri* tree shrews form pair bonds, exclusive associations between mates that last for several seasons. These are maintained by frequent copulations, even during pregnancy, and reach their highest expression in a rare form of male parental care. In this species, males build a nest of leaves well in advance of the birth of young, which makes them among the most liberated of solitary mammals.

SUMMARY

Tree shrews have the most elaborate social systems found in insectivores, but with the exception of *T. montana*, they are basically solitary. It is not surprising, then, that they have 21 different types of messages, approximately the same number of types as shrews, albeit with a greater variety of forms in each type. Neonatal messages include sounds of satisfaction and parental scent marking of neonates, which facilitates imprinting and parental recognition of young. Integrative messages include play, mainly aggressive, but with vocalization suppressed; visual contact by tail switching; affiliation, expressed by allogrooming, food sharing and huddling, solicitation of grooming, and olfactory identification; and alarm expressed by tail bristling. Agonistic messages include two types of threats, offensive and defensive, each with distinctive postures and vocalizations; dominance, signaled by supplantation, mounting, and scent marking; submission, expressed by anal presentation; and fighting. Sexual messages include female sexual advertisement by scent; male advertisement by vocalization; courtship; a variety of copulatory vocalizations; and suppression of reproduction by SST.

This chapter has investigated three families of predominantly solitary insectivores, ranging from hedgehogs, which get together only to reproduce, through shrews, which at least seem to recognize each other several weeks after their separation, to tree shrews, which include one more-or-less group-living species. Hedgehogs use about 12 different types of messages. The larger number of messages in shrews (19) and tupaiids (21) may be in part attributable to the larger number of species discussed in these groups.

3
RATS AND
BEAVERS

"Is there any chance for me
to get onto the language of the rats?"

CARL SANDBURG

Rats and beavers are members of the order Rodentia, which is composed of about 3000 species of gnawing mammals. The tremendous diversity within this group makes it impossible to speak of a "typical" rodent, but there are some characteristics that are common to almost all members of the order. All rodents have two long, continuously growing incisors at the front of each jaw. These teeth are enameled only on the front, so that wear creates a sharp, beveled edge. The incisors are an adaptation to gnawing large amounts of hard substances such as wood or the husks of seeds and nuts. Rodents have no canine teeth. Instead, they have a gap, or *diastema,* through which they can fold the inner skin of their cheeks. This allows them to avoid swallowing the materials they gnaw. Both the rat and the beaver illustrate two behavioral characteristics, building and hoarding, that are common to many rodent families. Both characteristics are more fully developed in the beaver than in the rat, which has taken advantage of the fact that humans also build and hoard.

RATS, GENUS *RATTUS*

There is no mammal more closely associated with or thoroughly despised by humans than the rat. This contempt is bred not by familiarity but by ignorance, for in spite of their propinquity, rats tend to remain hidden within the walls that gave them their family name, *murid.* In 1963 Barnett pointed out that a standard psychological handbook on the rat relegated social interactions to the chapter on abnormal behavior. The handbook further stated that communication, by our definition, does not exist, for "rats are not especially influenced by each other's actions" (p. 55). Both these notions are wrong. Rats are normally highly social and do communicate in a variety of ways.

These ways have been neglected in part because they are very different from ours. Rats are macrosmic (big on smell), and their vision is keen. Their communication is dominated by touch, pain, and smell. They also use ultrasonic vocalizations, which we can detect only with special equipment.

The behavior of captive members of only two of the 137 species of *Rattus* is well-known. These species are R. *rattus*, the house rat, and R. *norvegicus*, the Norway rat, albino strains of which are to the psychologist as fruit flies are to the geneticist.

Rats live in colonies, permanent groups containing varying numbers of males, females, and young, all using common nests, feeding sites, and routes linking them. These locations and routes lie in a highly restricted area. Calhoun (1962) studied home ranges of wild R. *rattus* in Baltimore by capturing, marking, and recapturing them. He found that 80% of his recaptures were within 20 meters of the original capture site. When he dyed rat bait, he found all colored rat feces within 30 meters of each bait station. Therefore, rats rarely travel more than 20 to 30 meters from their feeding and resting areas.

There is considerable evidence that rats not only have restricted home ranges but that they are territorial as well, attacking unfamiliar male (but not female) rats that intrude into their ranges. Two male strangers placed in a cage simultaneously will not fight, but if one is allowed to familiarize himself with the cage for as little as 10 to 15 minutes before the second is added, the resident will attack the newcomer. Females defend the area around the nest, and males defend a somewhat larger area, but whether the larger area of male defense coincides with the home range is not known. An intruder will not attack a resident even if the latter is much smaller.

In R. *norvegicus*, violence toward intruders contrasts with harmony among residents. Familiars not only share the same space but stay close to one another, especially when sleeping. Conflicts between males are regulated by a system of ranks, not a linear-dominance hierarchy, as in tree shrews, but a system of castes. Rat Brahmins are called alphas. They are the most confident and belligerent rats in the colony, attributes that suit them well, for they are also the largest. Betas are second-class citizens, dominated by the alphas but resigned to their station—in captivity they gain weight and seem healthy. The untouchable omegas are seen only in laboratory colonies. They are distinguished by sluggish movements, an unkempt appearance, and a retiring manner. In captivity they lose weight and in nature probably emigrate or die. Females may be exempt from the rank system, for they are not attacked by males, and their aggressive tendencies vary dramatically with their reproductive condition (Ewer, 1968), which would make a rank system unstable. According to Barnett (1963), to whom we owe this description of male social structure, rank is absent from captive populations of R. *rattus*.

NEONATAL COMMUNICATION

The first stimulation received by a newborn rat is, as in most mammals, a maternal lick. This vegetative and phasic interaction not only cleans the infant and the nest but, as the mother ingests the placenta, stimulates maternal behavior (Rosenblatt, 1970). Reyniers and Ervin (1946) have shown that maternal licking in the infants' anogenital region is essential for the development of elimination, which must be stimulated at first. Since the young do not eliminate without stimulation and the mother ingests their excreta, the nest is kept fairly clean. The nest and the young are far from odor-free, however, because, according to Beach and Jaynes (1956), mother rats recognize their young with the aid of "nest odor," presumably a product of glandular secretions and excreta. According to Schapiro and Sales (1970), infant rats respond as though they recognize their mother's odor. Infants not only recognize but prefer maternal odors. This preference is strong and persistent enough to be considered a form of olfactory imprinting. Marr and Gardner (1965) showed that infants raised in nests treated with artificial odors later displayed sexual preference for partners with similar odors.

Young rats reciprocate by eating maternal feces, thereby innoculating their digestive tracts with flora necessary for the assimilation of solid food and, possibly, establishing food preferences (Harder, 1949). Behse (1977) showed that during weaning, materials in maternal feces ingested by infants suppress the infants' tendency to approach maternal odors in unfamiliar environments. Behse concluded that this process plays an important role in the eventual emancipation of pups from their mother.

Nursing, like reciprocal *coprophagy*, involves mutual vegetative and phasic interaction, as the mother adopts a posture that facilitates access to her nipples and the young perform a "milk tread," stimulating the flow of milk.

The first true signals sent by young rats are quiet squeaks produced in chorus when they are cold, hungry, or lost. Their mother responds by approaching them and, if they have strayed from the nest, by retrieving them. These quiet squeaks of distress are made only by young rats.

Infant albino rats also produce ultrasonic chirps, each about .1 second long, with most acoustic energy around 32,000 hertz. These signals are termed distress calls. The maternal response to a pup's chirps is approach to and retrieval of the pup, which usually quiet it (Noirot, 1968). Allin and Banks (1972) showed that lactating *R.*

norvegicus orient their heads and bodies toward a source of recorded pup distress calls, approach the source, and act as though searching for a pup.

INTEGRATIVE MESSAGES

The rat's domination of the literature on animal behavior (Beach, 1950) does not extend into the area of play. Barnett's (1963) classic work on the rat, for example, does not even have the word *play* in the index, and Hinde's (1970) basic text on animal behavior does not mention rats in the section on play. A description by Ewer (1968), however, suggests that rats' capacity for play may be underestimated. She describes a game that began with mock attacks by young rats on a feather duster. These attacks became a tug of war, with the rats on one side, Ewer on the other, and the duster in the middle. Another tug of war began when the rats attempted to steal a handkerchief from Ewer's pocket. After playing this game a few times, her rats learned to bring the handkerchief to her in an invitation to play. Young rats in our laboratory often play by chasing one another's tails when released in a large room. This behavior alternates with playful boxing, in which rats sit up and spar with their forepaws.

Adults produce an alarm signal distinguishable in volume, circumstances, and consequences from the distress squeaks of infants. It is a chattering squeal most often produced as a response to handling or attack, and it often causes other rats to flee or hide.

Valenta and Rigby (1968) were able to teach rats to discriminate the odors of shocked and unshocked conspecifics. Whether this ability was involved in the crowding studies of Calhoun (1948) and Christian (1959) is not known. It does, however, raise the possibility of an olfactory distress signal.

Several behaviors involving close contact familiarize rats with one another and thus form (Zajonc, 1971) as well as express affiliations among members of a colony. Huddling exposes rats to olfactory and tactile stimulation from one another. Members of a colony and even rats that do not know one another will in unfamiliar surroundings, after a little mutual sniffing, settle down and form a pile. The pile is never completely still, for rats are continually joining the pile, leaving it, or readjusting themselves. Huddling may be a mechanism for conserving heat, but rats huddle even when it is warm, and huddling is just as common in the subtropical *rattus* as in the more temperate *norvegicus*. Benedict (1938) showed that huddling is not an efficient method of energy conservation, so all this suggests is that huddling is a social at least as much as a thermodynamic phenomenon.

A second way in which rats are exposed to one another's odors is indirect and involves autocommunication, in which an animal communicates with itself. Rats leave trails of genital secretions and urine (Reiff, 1952) whenever they move about. These trails are visible to humans as a faint, dark stain whenever rats travel on a light surface, but their significance to rats lies in their odor, for they are investigated by sniffing and following even when invisible to humans, whose eyesight is more acute than that of rats. Although Gawienowski (1977) has found exceptions, rats tend to follow their own trails and those left by others. Rats in their own range are therefore usually in contact with odor from their associates or from themselves. Barnett (1963) suggested a mnemonic, or familiarization, function for urine trails and proposed that a rat uses its own odor trail to remember its way about. This, too, would facilitate cohesion.

A third form of odor exposure is called walking over (see Figure 3-1), a behavior observed in our laboratory most frequently in conflict situations. It is not an incidental effect of locomotion, for a rat often detours, walks over another, and then returns to its original course (Barnett, 1963). This behavior probably transfers odors from the rat that walks over to the one walked on. Mutual tactile stimulation is certainly involved, and rats display no behavior suggesting that being walked over is aversive. According to Müller-Schwarze (1974), the rat that crawls over another often places a drop of urine on the other. He interprets this practice as osmollaxis, the sharing of odors.

Rats can discriminate the familiar odors of their fellows re-

FIGURE 3-1 The rat on the left is walking over the rat in the middle. Photograph by Roger Peters.

sulting from mutual contact from those of strangers. A new rat introduced into a group is not only sniffed but also initiates recognition
sniffing (Barnett, 1963) among all of the rest of the rats, as though
they are reassuring themselves about who is who. In our laboratory
R. norvegicus groups sniff one another's anogenital regions upon introduction of a stranger (see Figure 3-2). Recognition sniffing directed
toward fellow colony members is often combined with nosing (Barnett,
1963), in which a rat approaches another and gently nuzzles it, usually
near the neck. Nosing is often combined with gentle pawing of the
fur next to the area being nosed (see Figure 3-3). Krames (1970)
showed that, under appropriate circumstances, estrous female rats can
discriminate odors from different males. The relationship of nosing,
pawing, and recognition sniffing to this ability has not been explored.
Males can evidently identify the sex of strangers by their odors. Alberts
and Galef (1973) found that an anaesthetized male in a plastic bag
was attacked only if the bag had holes. After sniffing an intruder's
anogenital region, a resident male will attack if the intruder is male,
but not if it is female (E. C. Grant & Chance, 1958).

AGONISTIC MESSAGES

When he attacks an intruder, the resident male almost invariably wins, a classic case of the motivational effects of territory
defense. Reiff (1952) found that captive rats tend to follow trails left

FIGURE 3-2 The rat in the middle is sniffing the
anogenital region of a newly introduced stranger, on
the left. Photograph by Roger Peters.

FIGURE 3-3   The rat on the left is nosing the other rat. Photograph by Roger Peters.

by strangers. Other studies show that males spend more time on surfaces used by strange males (Stevens & Koster, 1972) or ones that have been treated with urine from strange males (Richards & Stevens, 1974) than on clean surfaces. These tendencies would seem to increase the probability of detection and repulsion of intruders. Gawienowski (1977), on the other hand, found that males avoid urine voided by other males, an observation suggesting that urine can sometimes act as a territorial advertisement. Price (1977) found that rats can discriminate urine marks of different ages, and Krames, Carr, and Bergmann (1969) found differences in tendencies to approach areas used by rats of different rank. These studies may account for the discrepancy between Gawienowski's results and those of Stevens and his colleagues.

The high degree of cohesion promoted by affiliative messages makes conflict among colony members inevitable. Occasional contacts with members of other colonies may also produce conflict. The major source of conflict within the colony is competition for food, mates, or space, and when a stranger is involved, territorial aggression also comes into play. It is clearly to a rat's advantage to have some means of reducing the aggressive consequences of these conflicts. Submissive messages accomplish this by removing stimuli for aggression, by providing an alternative to fighting, or both.

The major stimuli for attack are motion and unfamiliar male odor and form in familiar surroundings. In one common form of appeasement often performed by intruders, a rat about to be attacked by a resident throws himself on his back, legs extended, and remains motionless (Barfield, Busch, & Wallen, 1972). The resident typically sniffs the intruder but does not attack. This display removes the motion stimulus, produces a striking change in appearance from a convex dark form to a concave light one, and presents anogenital and sternal odors for investigation. Thus, this display not only replaces some of the stimuli for aggression with their antitheses but provides sniffing as an alternative to attack (see Figure 3-4).

Another form of appeasement, descriptively termed crawling under, is seen most often in conflict situations (Barnett, 1963). One rat lowers his head and pushes it under the other's belly. The other rat sometimes cooperates by raising himself slightly, and the first inserts himself further, sometimes passing completely underneath. Crawling under not only provides an alternative to fighting but also probably transfers odor from the upper rat to the lower, making his odor less

FIGURE 3-4   The rat on the left is grooming the other rat, which is in a submissive posture. Photograph by Roger Peters.

unfamiliar and possibly thereby further reducing the probability that he will be attacked (see Figure 3-5).

Another alternative to fighting is allogrooming, in which one rat delicately but thoroughly nibbles a small area of another's fur. It is seen mainly in situations in which there is a potential for conflict and may reduce aggression by promoting familiarity and pleasurable tactile stimulation.

Neither *rattus* nor *norvegicus* males will attack females of their own species, but males of either species will attack females of the other (Barnett, 1963; E. C. Grant & Chance, 1958). It is thus likely that there is a species-specific female odor that inhibits aggression. That male *norvegicus* sometimes mount rather than attack male *rattus* suggests that the male *rattus* odor resembles the female *norvegicus* odor (Barnett, 1963). *R. norvegicus* sometimes adopt a submissive, crouching posture in response to approach (E. C. Grant & Mackintosh, 1963), but in our laboratory, at least, this does not generally deter attack.

Seward (1945) observed long exhalations, lasting several seconds, in submissive rats. Sales (1972b) showed that these exhalations were associated with ultrasonic pulses averaging about 26,000 hertz in frequency and about 3.4 seconds in length. These pulses were initially emitted by rats as they crouched submissively but often con-

FIGURE 3-5   The rat on the right is crawling under the rat on the left. Photograph by Roger Peters.

tinued for 15 or 20 minutes as the rats fed or groomed themselves. The pulses seemed to deter attack. Aggressive behavior was much less common in arranged encounters in which these long pulses were detected than in encounters in which they were not.

Submission is not always effective in deterring aggression, but rats have a backup mechanism that may prevent fighting even after appeasement has failed. Adults often peep or whistle upon approach by another rat. Females are especially likely to peep when approached by a strange male, and both sexes are likely to peep when they are approached while in or near their nests. The usual response to this mild defensive threat is withdrawal, but sometimes it is ineffective. In such cases the peeping increases in volume and begins to resemble the alarm squeal. At the highest levels of intensity of this defensive threat, it is virtually indistinguishable from the alarm signal. Further approach by the aggressor will lead to attack by the defensive rat.

Rats rarely attack without warning in the form of a distinctive threat posture in which the aggressor presents his flank, arches his back by bringing stiffly extended front and rear legs together, and minces about his opponent with short steps (Barnett, 1963). As in most intimidation displays, the result is an increase in the rat's apparent size, sometimes enhanced by bristling the fur. This display is often associated with other signs of sympathetic arousal: urination, defecation, tail lashing, and tooth chattering.

If the receiver of this threat display is of lower rank or is an intruder, he will usually withdraw. If both rats are on home ground or of equal rank, the receiver is likely to respond in kind, sometimes adding a tactile channel to the message by pressing his flank against the other rat as they mince around in a tight circle. After such a mutual display the opponents usually separate without further conflict (Barnett, 1963).

Threatening rats produce ultrasonic pulses ranging from 3 to 65 milliseconds in duration. These long pulses are sequences of very short emissions. In one study their frequencies varied considerably even within a pulse, but usually the frequencies lay between 40,000 and 70,000 hertz. Band widths of up to 50,000 hertz were recorded. Short pulses typically occurred in association with other threat displays, such as those just described. These pulses were common when male rats were introduced to lactating females, which are highly aggressive. In such cases the pulses were synchronous with the head movements of the female (Sales, 1972b).

Fighting Norway rats are a blur of writhing fur, but high-speed motion pictures allowed Barnett (1958) to resolve this whirling confusion into several components. A fight usually begins with an assault by one of the opponents, in which the attacker hurls himself through the air, repeatedly punching with his forelegs and simultaneously biting at ears, limbs, or tail. After each bite the jaws open rapidly, and the attacker jumps back. If, however, he is extremely enraged, he will continue his assault, hanging onto his opponent with his jaws or limbs. Each assault is followed by frantic scrambling and tumbling as the attacker presses on and the other tries to writhe away.

After a few seconds of tumbling over each other, the combatants appear to become disoriented, for they pause, right themselves, look around, and resume a semisitting crouch, with mouths open, heads extended toward each other, and forelegs raised off the ground. From this stance they can separate, begin a new bout of leaping, biting, and writhing, or stand erect on their hind legs with heads raised and hurl themselves at each other in a manner very similar to Herter's (1957) description of hedgehog boxing. Rats box palm to palm on their hind legs. In silhouette, they form a squealing Rorschach of flailing paws and flashing teeth (Barnett, 1958).

Almost all fighting observed in captivity is among males, but females become quite belligerent when pregnant (Ewer, 1968) or lactating. R. rattus fighting is similar to that by norvegicus, but the excitement spreads to other colony members, which frantically run and jump around the combatants (Barnett, 1963).

According to Barnett (1963), dominance position, or rank, is determined by fighting, at least among captive rats. Rats raised together from infancy fight only rarely, so once rank is determined, dominance and submission messages prevent further strife.

There are two displays that seem to express dominance in rats because they are typically performed by the winner of a fight. In intervals between assaults, one rat, usually the one that will win, often mounts the other as he would mount an estrous female (Barnett, 1958). He may also mount the loser in later encounters, reestablishing his dominance without attack. A variation of this posture has been reported by E. C. Grant and Mackintosh (1963). They observed dominant rats to stand at right angles to the subordinate and to place their paws on the inferior's back.

According to Ewer (1968), victorious males sometimes celebrate their triumph with enuration, or marking their defeated opponent with urine. The dominant rat's urine contains aversive substances not found in the urine of subordinates (Krames et al., 1969) or castrates (Gawienowski, 1977). The loser thus bears for some time tangible evidence of his defeat, an insult that may spare him further injury.

SEXUAL MESSAGES

The sexual behavior of rats has been thoroughly researched, probably more thoroughly than that of any other mammal. Calhoun (1962) provides a comprehensive description of behavior in the advertisement phase. His description is particularly valuable because, unlike most accounts of rat behavior, it is based on observations in a natural environment. Just before coming into estrus, which occurs every 4 days and lasts about 10 hours, females begin to wander off their home ranges and to rub their flanks and genitals on prominent objects, particularly those most likely to be encountered by other rats, so the objects were probably chosen for their proximity to scent trails and their junctions. Females also drag their genitals along the ground near such objects, lowering themselves and dragging themselves forward with their forelegs, often wriggling from side to side as they do so. These behaviors transmit the estrous females' odors, which adult male rats can distinguish from those of nonestrous females (Le Magnen, 1951), to the marked objects and their surroundings.

A male rat often arrives at one of these scent marks within a few minutes and, after sniffing it, marks like the female. His marking deposits odors that could inform other males that the female's mark has already been discovered. The male then follows the female's scent trail until he discovers another of her marks, where he repeats his performance. If he encounters a fresh female scent mark at the entrance of her burrow, he is likely to rub and roll in her odor deposit. This male response to female scent is one of the most commonly observed male sexual behaviors.

Eventually the male's following of the female's trail brings him into contact with her, and the courtship phase begins. The two touch each other's noses and sniff each other's genitals. Experiments by Orsulak and Gawienowski (1972) show that males and females are attracted to odors from each other's preputial glands. If she is not fully receptive, the female then moves away, and the male follows briefly. Should he attempt to mount, she expertly kicks him off with a swift blow from a hind leg and retreats.

If the female is fully receptive, ensuing events depend on the number of males present (Steiniger, 1950). When several males have located the female, they all pursue her, and an extended period of chasing ensues. When the female finally stands for one of them, he clasps her about the flanks with his forepaws and shivers. She eventually responds with lordosis, that is, raising her rear, and copulation begins. The process may then be repeated as many as 400 times with as many as eight males. When females at the height of receptivity are sexually stimulated, their ears quiver (Wells, 1977).

If the pair is alone, courtship is more elaborate, with mutual crawling under, walking over, nosing, and genital sniffing but without following. The female then stands, the male mounts, and copulation begins.

Ultrasonic pulses synchronous with courtship movements of the male have been reported by Sales (1972a). She detected these ultrasounds when the male entered the female's cage, sniffed her genitals, and chased and mounted her. Short pulses (3–50 milliseconds) at the beginning of a sexual interaction were largely supplanted by longer ones with frequency drifts during later sniffing and mounting. Many species of rodents are sensitive to ultrasounds in the frequency range of 40,000–116,000 hertz reported by Sales(A. M. Brown, 1970), but the communicative significance of these emissions for rats is open to further investigation.

Copulation consists of sequences of intromissions and pauses between them. Each intromission lasts about 3 seconds and consists of several shallow penile thrusts followed by a deep one. After each deep thrust the male dismounts, sits, and licks his penis for 45–60 seconds—longer if ejaculation has occurred. Bermant (1961) noted that females were ready for copulation sooner than males after an ejaculation had occurred.

Intromission is repeated, sometimes scores of times, over a period of an hour of more. The number of intromissions that precede ejaculation varies, but the average number per ejaculation is five (Barnett, 1963). According to J. R. Wilson, Adler, and Le Boeuf (1965), nonejaculatory intromissions may facilitate sperm transport in the vagina and secretion of progesterone in the uterus, both of which are essential for conception. Sales (1972a) reports ultrasonic pulses emitted by the male during intromission and ejaculation.

Barfield and Geyer (1972) suggest that the longer rest typically following ejaculation can be attributed to another ultrasonic cry emitted by the male during his postcopulatory refractory period (a short interval within which ejaculation is impossible). This vocalization presumably inhibits the female's initiation of postejaculatory coitus.

A number of experiments with rat sexual behavior suggest parallels with that of some humans. A. E. Fisher (1962) showed that introduction of a new female rekindled ardor in a sexually exhausted male, but the effect was small for older males. On the other hand, he also showed that a male's interest in the same female could be piqued if a discotheque-like atmosphere with flashing lights and intermittent tones was provided. This effect was small, however, compared with that obtained by introducing a new female.

SUMMARY

Rats have a complex message system, suited to the demands of colonial life. They display 22 different message types, including neonatal tonic and phasic interactions between parents and young, such as anogenital licking, coprophagy, nursing, and a weaning *pheromone*; neonatal sonic and ultrasonic distress vocalizations; neonatal and maternal recognition by odor; and olfactory imprinting. Integrative messages are play, in the form of tug of war and chasing; alarm vocalization, a loud, chattering squeak, and an alarm odor; affiliation signals involving odor sharing, including nosing, huddling, and production of scent trails; a distress vocalization; individual recognition by odor; and contact and familiarization by urine trails. Agonistic messages include defense and advertisement of territory; submissive displays, including lying on the back, crawling under, walking over, allogrooming, distinctive female odors, and ultrasonic pulses; defensive threat in the form of a whistling peep, graded in intensity; offensive threats with arched back, mincing walk, tail lashing, tooth chattering, and ultrasonic pulses; fighting, with striking, wrestling, and boxing; and dominance, shown by mounting and enuration. Sexual messages include sexual advertisement, including scent marking and genital and facial odors; courtship; copulation, with lordosis, ear quivering, and ultrasonic pulses; and male suppression of female solicitation.

BEAVERS, SPECIES CASTOR CANADENSIS

The beaver is the second-largest living rodent. Only the capybara is bigger (Bartlett & Bartlett, 1974). Adult beavers usually weigh about 20 kilograms. The beaver's face is shaped like a rat's but is distinguished by bright-orange upper incisors, which work against lower ones, maintaining an edge that allows a beaver to gnaw through saplings up to 10 centimeters in diameter in less than 20 minutes (Rue, 1964) and to fell trees up to 1.7 meters in diameter (Hatt, 1944).

Beavers' morphology, like their social and message systems, are adaptations to a complex ecological niche based on life in the water. The beaver walks with a rolling waddle but swims with undulating grace, using its flat, scaly tail. Folds of skin behind the incisors seal the mouth, allowing a beaver to gnaw underwater, while other flaps of skin seal the nostrils and ears. Transparent membranes cover the eyes, allowing underwater vision (Rue, 1964). The hind feet are webbed and have split nails on the inside toes. These nails are used to comb the fur, wiping away water and spreading waterproof lipids. Castor glands (described in Chapter 1) empty into the *cloaca*, an opening just under the tail. In both males and females, the cloaca is used for elimination, reproduction, and scent marking.

Beavers are found on forest lakes and streams throughout North America, except for Florida and coastal and southern California (Burt & Grossenheider, 1964). Their wide distribution is largely attributable to their engineering accomplishments. They often create their own habitat by gnawing trees until they fall and partially block a stream. The trunks are then laced together with branches and the interstices sealed with mud. The pond that forms behind the dam allows beavers to elude most predators by diving or swimming. The pond also enables them to float large trees for the dam or for food. The shore of the pond is sometimes penetrated by canals dug by beavers. These canals facilitate movement ashore, especially when the beaver is burdened with a load of branches.

Whether living on a lake or on a pond, beavers construct a lodge of sticks and mud. Some lodges are built away from shore, but others are built against a bank, with tunnels extending inland for more than 100 meters. Wilsson (1968) noted that such lodges usually have several rooms, some apparently used for eating and others for sleeping.

Bark and twigs are the staple of a beaver's diet. They prefer aspen, willow, maple, and poplar and its relatives. Beavers store branches and pieces of trunk underwater near their lodge (Burt & Grossenheider, 1964). Beavers are active mainly at dusk, night, and dawn. One would therefore expect that their messages would be mainly auditory and olfactory, rather than visual. The beaver's major predators are coyotes, bears, wolves, lynxes, bobcats, otters, and dogs (Rue, 1964).

The beaver's social unit is a family, or colony, composed of a pair of adults, one or two yearlings, and two to four kits. The adults are monogamous and remain together until one dies. If the survivor is a male, he may pair with a daughter (Wilsson, 1968). Young beavers leave or are driven away as they approach sexual maturity at about 2 years of age (Burt & Grossenheider, 1964). Apart from the intolerance that may precede dispersal and an occasional conflict over food, family members are closely bound by ties of mutual affection (Tevis, 1950).

Tevis noted deferential avoidance of the adult male by year-lings. Tevis concluded, however, that in the only important agonistic situation, competition for food, there was no evidence of a dominance hierarchy. All age class took and gave up food with approximately equal frequencies. Typically, beavers surrendered food to whichever beaver approached. Tevis attributed this habit to the relative vulner-ability of the feeding animal. Wilsson noted, however, that adult females dominated their mates and were more aggressive toward strangers than males were.

The home range of the beavers studied by Tevis was generally restricted to the area immediately around their pond, but he reported excursions of more than 100 meters from the pond. My own obser-vations of cuttings and tracks in Colorado, Michigan, and Minnesota support this finding. Beavers about 2 years old, which leave their natal pond, often travel considerable distances before settling down. Beer (1955) reported that a tagged beaver traveled more than 200 kilo-meters, but Burt and Grossenheider (1964) reported the average dis-persal distance as under 10 kilometers.

Beer suggested that the female is the focus of a family's attachment to its home range. When an adult female was drowned in a trap, the pond was abandoned by the rest of the family. Conversely, when a similar accident befell the adult male, a new mate took his place, and the group remained.

Bradt (1938) and Burt and Grossenheider (1964) state that beaver groups defend territories. Aleksiuk (1968) reported no instances of either trespassing or defense and concluded that beavers "voluntarily avoid occupied territories" (p. 760). Many beavers studied by Beer (1955) and by Wilsson (1968) wandered between occupied ponds, but most such visits were friendly. Thus, there may be some variation in the territorial motivation of beaver groups, possibly depending on their degree of relatedness.

NEONATAL COMMUNICATION

Birth can occur only once a year, between April and July. The usual litter size is two to four. Kits are born in the large sleeping chamber of the lodge with open eyes, erupted teeth, and thick fur. According to Wilsson (1968), both parents eat the placentas. Wilsson (1968) noted that the mother, vocalizing softly, often sits up to nurse her young. She grooms them frequently, rolling them in her paws as she combs their fur with her teeth.

After a few weeks the kits are weaned, and the mother brings leafy branches to the lodge for them to eat (Tevis, 1950). She is met with cooing and high, ultrasonic piping by the kits (Wilsson, 1968). Tevis describes this greeting as "sharp, petulant whining" (p. 55). It is probably this vocalization that Leighton (1933) calls a cry and describes as similar to the short cries of a human infant. Leighton heard the cry in several different situations, including disputes over food.

Kits apparently express distress at separation or lack of attention with the "whine," which is "thin, scarcely audible," and "ventriloquial" (Leighton, 1933, p. 28). Wilsson described the whine as whimpering. Leighton noted that whines often led into the cries just described. He heard whines when a kit swam around a yearling, perhaps seeking attention. Wilsson's tame kit whined when Wilsson began to walk away. This kit also whined or whimpered when soliciting contact by cuddling and when being dragged by its mother.

Dragging by the tail is a common means of transporting kits on land. Sometimes a mother moves her kits by carrying them one by one in her mouth, occasionally walking on her hind legs. A week or so after birth, a kit is able to swim (Bartlett & Bartlett, 1974), and the mother swims behind it with her chin on its tail. On one occasion such guidance was interrupted when the kit turned to rub its chin against its mother's head (Tevis, 1950). Mills (1922, cited in Tevis, 1950) observed a kit riding on its mother's back as she swam. Tevis also saw a kit attempt, unsuccessfully, to ride on its father's back. Later on in the kits' first summer, the mother leads her litter by swimming in front of them. They form a line behind her, each in the wake of its predecessor.

Just as carrying gives way to leading, so leading gives way to imitation. As the end of their first summer draws near, kits imitate adults as they float idly, select particular foods, and deposit castoreum on the shore (Tevis, 1950; Wilsson, 1968). By this time the mother and kits vocalize to one another in typically adult fashion, and grooming is reciprocated. A beaver raised and groomed by Wilsson never formed social bonds with other beavers (Wilsson, 1968).

INTEGRATIVE MESSAGES

Beavers of all ages are extremely playful. There are four basic types of play, but mixtures of these types are common. Warren (1922, cited in Tevis, 1950) described mutual pushing by kits and yearlings. They were in the water and used their shoulders and heads. Leighton (1933) observed similar behavior involving a kit and a yearling on land. They stood on their hind legs, put their forepaws on each other's shoulders, and pushed with their chests. According to Wilsson (1968), this form of play often looks like human boxing.

A second form of play is purely aquatic. Beavers swim side by side, chase one another, dive, and engage in vigorous tussles (Tevis, 1950). Riding on the back of a swimming beaver is another form of aquatic play performed by kits, yearlings, and adults. Usually, smaller beavers ride on larger ones, but there is no clear destination, as there is when a mother transports her kit.

The fourth form of play is particularly interesting because it has often been confused with copulation. In this form of play, two animals roll in the water while embracing each other, or they embrace while lying or standing. Tevis observed this behavior outside the breeding season, when copulation is impossible. He concluded that this behavior is a stereotyped form of play, not copulation, and that the ventral-copulation hypothesis needs confirmation.

Beavers solicit play, grooming, sexual interactions, and even fights by what Wilsson (1968) calls dancing. In dancing, one beaver approaches another, flings its head from side to side, and then pivots or lunges sideways. If play is desired, the dancer may frolic in the water, then dance. The dancer solicits grooming by grooming the receiver at the end of the dance; the receiver usually reciprocates. Courtship often begins with dancing followed by wrestling. A beaver sometimes solicits a fight by rearing on its hind legs at the end of the dance.

Beavers' food-gathering activities often take them ashore or into the lodge where they cannot easily be seen or smelled by other members of the colony. It is to be expected, therefore, that these highly gregarious creatures would use sound to inform one another of their continued presence. Since beavers rarely leave the immediate vicinity of the pond, such contact signals need not be loud or provide precise information about location, which could be used by predators.

Beavers do in fact exchange calls that Leighton (1933) called cries and described as ventriloquial and similar to the short cries of a human infant. In Colorado and Minnesota I found them difficult to localize. They seemed to come first from the lodge, then from the dam, then from the surrounding woods. They are faint, intermittent, and rise and fall in pitch and volume. According to Leighton, cries cease immediately at any disturbance.

Cries are heard in a great variety of contexts, "when the beaver is [both] pleased and displeased" (Leighton, 1933, p. 28). This suggests that cries do not serve to communicate any particular mood. Leighton heard cries while beavers swam, ate, squabbled, and met. Often the cries were emitted with no apparent stimulus. These observations suggest that the cry is primarily a contact signal. Leighton believed that the category "cry" probably includes several different kinds of calls that he could not distinguish. This hypothesis should be tested by correlating sonagrams with behavior to determine whether all cries have a particular meaning or whether there remains a residual "contact" category.

The strong social bonds within a beaver family are mediated by a variety of affiliative messages. One such form is allogrooming. Tevis (1950) observed allogrooming by adults during the summer, when sexual drive is presumably at a minimum. The male apparently solicited grooming by floating in shallow water in front of the female with only his hindquarters above the surface. The female dropped the stick she was gnawing and, grasping his rump with her front paws, carefully groomed his back with lateral movements of her head "while opening and closing her lips" (p. 62). Then the male reciprocated by briefly grooming her with his mouth. When beavers are ashore, allogrooming is sometimes solicited by the dance described as a solicitation (Wilsson, 1968). Wilsson also observed mutual allogrooming in which beavers simultaneously groomed corresponding parts of each other's fur.

Nasonasal sniffing or contact is another expression of social attraction. Leighton (1933) observed such behavior at the meeting of two kits that seemed especially close and that showed several other signs of affection. One of these signs was the "soft churr," a vocalization that occurred only when a beaver greeted or looked for a sibling or parent.

A fourth expression of affiliation is "talking," mutual vocalization that Wilsson (1968) heard mainly between mother and kits and between mates. Talking is soft in volume, varied in pitch, and extended in duration. It resembles human conversation heard at a distance or through a wall. My students have noted that talking is the most common vocalization at a Colorado beaver pond. It is often difficult to localize but often clearly comes from the lodge.

The beaver's most characteristic message form is the tail slap. It is often the first evidence that a pond is inhabited by beavers. There is no other forest sound quite like it. It is an explosion, sometimes two in quick succession, with a geyser of water thrown 3 or 4 meters into the air. It is almost always followed by a dive. The tail slap is an alarm signal that alerts other beavers and sends them scurrying or swimming to the safety of deep water. The slap may also intimidate predators, for the effect belies the size of the creature that causes it.

Tevis (1950) noted that beavers have three types of dives. When not alarmed, a beaver simply submerges, with barely a ripple. Startled beavers, on the other hand, perform the fright dive, which is abrupt but not noisy. Thus, the tail slap of the third type, the warning dive, is not essential to rapid submerging.

Tevis observed tail slaps on 36 occasions. Of these 36 slaps, 30 were preceded by a detectable disturbance. In the four cases in which a receiver was in deep water, none responded to the slap. In the 14 cases in which a receiver was in shallow water, ten responded by seeking "refuge," by which Tevis probably meant the lodge or deep water. In each of three cases in which a beaver was on land when a tail slap occurred, the beaver ran into the water. Tevis concluded that beavers' responses to tail slaps depend on their vulnerability.

Leighton (1933) kept an adult female beaver in a large pen that extended into a lake. A kit and a yearling, both males and probably her offspring, were kept in a similar, adjacent pen. On many occasions the female, and once the yearling, gave a call Leighton refers to as the "summons, . . . a plaintive, rather high pitched straining call" (p. 29). It was sometimes repeated as many as eight times over a period of a minute or more. Whether the female or the yearling gave the call from land or water, the response of the other beavers was the same. They approached quickly and immediately. Leighton noted that all of the summonses issued by the female were produced while her food supplies were restricted. This was not the case, however, for the yearling, which also produced the call.

Leighton's male yearling often inhaled so as to produce a nasal sound similar to that made by dogs when they are pleased. The yearling only, but not always, made the sound when swimming to the shore to feed. Leighton concluded that the sound expressed "good feeling and perhaps joyful anticipation" (p. 29).

When Leighton handled the yearling shortly after capturing him, the beaver produced a "spluttering, . . . guttural sound" (p. 29). This sound is not reported, however, in other contexts such as fights, in which the loser might be expected to be distressed.

AGONISTIC MESSAGES

The predominantly peaceful quality of life in the pond is occasionally shattered by squabbles over food. Even in such cases, however, fighting is generally averted by threats. Tevis (1950) sometimes heard yearlings and adults exchange low, quiet whines. On one such occasion two beavers were engaged in a tug of war over a cherry branch. Tevis noted that one of them emitted the low-pitched whine at 6-second intervals until the conflict was over. He concluded that this whine expressed a low-intensity threat. A threat of higher intensity is described by Wilsson (1968), who noted that when a beaver approached a second beaver that was gnawing on a tree, the second snarled and the first withdrew.

Hissing through constricted nostrils is evidently an even more serious threat, for it sometimes precedes biting (Leighton, 1933). Leighton heard hissing when he handled, confined, or frustrated his beavers. He noted that hissing by one beaver caused another to withdraw.

The highest level of threat is expressed by a combination of visual and auditory signals. Wilsson (1968) describes an encounter between strangers in which one, an adult female, slapped her tail sharply against the ground, grated her teeth, and constricted her facial muscles.

Beavers' predilection for technological solutions to ecological problems is exemplified by the manner in which they announce their possession of a territory. Ernest Thompson Seton described the scent mounds of beavers in 1929 and called them mudpie telegrams. Scent mounds are rounded, conical piles of mud impregnated with castoreum. They vary in height from a few centimeters to more than 1.5 meters (Wynne-Edwards, 1962). They are always located on the shore and are usually quite easy to locate. (See Figure 3-6.)

Several characteristics of these mudpie telegrams suggest that one of their main messages is territorial advertisement. First, they are strategically placed. In the Mackenzie River delta Aleksiuk (1968) noted that most scent mounds were built at the edge of a colony's territory, where they would be encountered before a trespass had occurred. When a lodge was newly established and its ownership perhaps liable to question, it was often marked with a scent mound.

By investigating the odors emanating from scent mounds and tracks around them, Aleksiuk determined that mud and castoreum were applied almost every day. Each of a family's two to seven mounds was visited by all members, but least frequently by the adult female. In Sweden, however, Wilsson (1968) observed the opposite. The adult female in a semi-tame family deposited castoreum more often than the males, which often merely went through the motions of scent marking, each dragging his everted *cloaca* over the mud. There are at least two possible explanations for the difference in these observations.

FIGURE 3-6   A beaver scent mound. Photograph by Roger Peters.

Since Aleksiuk's observations, unlike Wilsson's, were confined to summer months, the females watched by Aleksiuk may have been too busy gathering food to participate fully in scent-mound maintenance. Alternatively, there may be a difference in the social ranks of females in Canadian and Swedish families. Any inferences about sexual differences in behavior must, of course, be interpreted in the light of the great difficulty of determining the sex of beavers at a distance.

A second form of evidence for a territorial function of scent mounds comes from a fortuitous observation by Aleksiuk. He had just applied a preparation containing castoreum to a scent mound and moved about 15 meters away when an adult beaver swam directly to the mound. It sniffed the mound, hissed in threat, and then deposited castoreum on the mound. The sound made by the depositing of the castoreum was clearly audible. The beaver then swam, hissing continually, to each of four others of Aleksiuk's scent deposits and repeated its ritual. Aleksiuk was convinced that this behavior is typical of responses to deposits made by transient beavers. The hissing indicates that the mood associated with this form of scent marking is threatening and thus lends credence to the notion that scent mounds act as warnings.

Wilsson (1968) does not mention the hissing described by Aleksiuk, but he often observed castoreum deposition in agonistic contexts. For example, his female beavers customarily deposited castoreum between "rounds" in fights with strangers. They began by scratching the ground with their forelegs, making a small mound. They then humped their backs and "strained," repeatedly kicking with a hind leg, "audibly forcing a stream of castoreum" (p. 61) onto the mound.

Aleksiuk live-trapped and examined 150 beavers in the course of his study. He found no evidence of trespassing and only one wound. These data, combined with the aggressive response to odor from unfamiliar conspecifics, convinced him that transients voluntarily avoid ponds marked with fresh castoreum. Paradoxically, however, the mounds themselves seem attractive. Seton (1909) describes use of scent mounds by neighbors as follows:

> When two beaver lodges are in the vicinity of each other the animals proceed from one of them at night to a certain spot, deposit their castoreum, and then return to their lodge. The beavers in the other lodge, scenting this, repair to the same spot, cover it over with earth, and then make a similar deposit

on top. This operation is repeated by each party alternately, until quite a mound is raised, sometimes to a height of four or five feet [p. 442].*

Wilsson's (1968) description, based on observations made 60 years later and an ocean away, is similar. He noted that castoreum deposits are not repulsive. Rather, they attract visitors, which deposit their own castoreum on them. Wilsson's hypothesis about the function of castoreum deposition is that it operates only indirectly as a territorial advertisement. Its primary function is to familiarize beavers with their territory, increasing their confidence and thereby stimulating defense. The low frequency of serious wounds reported by Aleksiuk (1968) suggests that this defense is more commonly done through threat than fighting. Indeed, Wilsson describes as scolding the vocalizations regularly performed by adults at the boundary of their territory and directed at their neighbors.

Castoreum itself is a mixture of at least 45 different substances, including "alcohols, phenols, ketones, organic acids, and esters, as well as salicylaldehyde and castoramine ($C_{15}H_{23}O_2N$)" (Ledered, 1950, cited in Wilson, 1975, p. 188). Like the ancient Greeks, Native Americans considered castoreum a panacea (Bartlett & Bartlett, 1974). Salicylaldehyde, which is similar in structure to aspirin, may have been responsible for some of its alleged curative properties.

Tevis (1950) found no evidence of a dominance hierarchy in the family he studied. Wilsson (1968), on the other hand, concluded that females customarily dominate their mates. Among the beavers he studied, castoreum deposition seemed to be a dominance display, performed in full form only by females. In one paradigmatic set of interactions, a 1½-year-old female was placed in a cage in which a male, of the same age but considerably larger, had resided for some time and regarded as home. The male so frightened the female that he was removed for 2 days. In the interim the female marked the enclosure with castoreum. When the male was reintroduced, dominance relations were reversed. In intervals between ensuing wrestling matches, the female deposited castoreum. Wilsson concluded that the castoreum stimulated her to dominate the male. In another experiment

*From *Life Histories of Northern Mammals*, by E. T. Seton. Copyright 1909 by Charles Scribner's Sons. Republished in 1974 under the title *Lives of Game Animals* by Doubleday & Company, Inc. This and all other quotations from this source are reprinted by permission of Doubleday & Company, Inc.

in which a female was placed in an enclosure with a male that was larger than she, the female established dominance immediately.

Apart from the occasional deferential avoidance of the adult male by yearlings reported by Tevis (1950), there seems to be only one expression of submission by beavers. When strangers were introduced in cages to the enclosure "owned" by one particularly aggressive female, she would threaten them with tail slapping and tooth gnashing. They would not threaten back but lie flat on their bellies (Wilsson, 1968). These intruders were always protected by their cages, so it was impossible to determine the effectiveness of this posture in deterring attack.

Aleksiuk's (1968) live-trapping data show that fighting among beavers is rarely serious enough to result in injury. Wilsson has, however, observed rough-and-tumble actions too serious to be classified as play. Fights between beavers begin with violent wrestling, which usually culminates in the withdrawal of one of the combatants. Only when the fight is prolonged, as when confinement prevents escape, does combat escalate to biting (Wilsson, 1968). Wilsson describes several fatal fights among confined animals. In one such fight a pair of beavers broke into the cage of another pair; all four were injured, and two died. In another case a female killed a beaver newly introduced into her enclosure by biting through the spine just above the base of the tail. The teeth of the beaver can clearly be formidable weapons, and their use is clearly inhibited in most agonistic situations.

SEXUAL MESSAGES

Members of pairs need not advertise to find each other because they work, feed, and sleep together and are never more than a few hundred meters apart. The female may announce her readiness to copulate by odor or behavior, but little or no reliable information is available on how this process occurs in nature.

The attractive properties of castoreum may assist unpaired beavers to find each other. Wilsson (1968) presumes that when a dispersing female finds a suitable homesite, she deposits castoreum to attract males. This hypothesis should be tested by observations in areas in which beavers are being reintroduced.

The courtship phase of reproduction usually occurs when pairs are formed, generally several months before mating. On the basis of seven experimental introductions in captivity, Wilsson concluded that courtship typically begins with a fight, which either establishes the female's dominance or results in "complications," which sometimes include her death. As soon as the male submits, amicable interactions in the form of dancing, talking, and allogrooming form and maintain a lifelong bond.

According to Reed (1946), there have been several more-or-less reliable accounts of ventroventral copulation by beavers. Beach (1947), on the other hand, is skeptical of such accounts of nonprimates. His skepticism is justified in the case of beavers, whose sex is nearly impossible to determine from a distance. Bourlière (1964) states that beavers copulate ventrodorsally.

SUMMARY

In keeping with their highly gregarious nature, beavers use 18 different message types, including tonic and phasic interactions: neonatal consumption of placentas by both parents, nursing, grooming, provisioning, bipedal carrying, riding, leading, and imitation; and neonatal signals for greeting and distress. Integrative messages include play by pushing, boxing, wrestling, and riding; solicitation by dancing; contact by ventriloquial cries; affiliation by allogrooming, nasonasal sniffing, the "soft churr," and talking; alarm by tail slapping; assembly by a high-pitched call; contentment by a nasal sound; and distress by spluttering. Agonistic messages include threat by hissing, snarling, tooth grinding, and tail slapping; territory advertisement by scent mounding and scolding; dominance by castoreum deposition; submission by lying flat; and fighting by biting. Sexual messages include sexual advertisement by castoreum deposition; courtship by fighting; and bonding by dancing, talking, and allogrooming.

4
RABBITS
AND PIKAS

I really dare not name to you
the awful things that rabbits do.

ANONYMOUS

Rabbits and pikas belong to the order Lagomorpha, which gets its name from two Greek words meaning "hare-shaped." Like rodents, lagomorphs have large, beveled, continuously growing incisors and a *diastema* instead of canines. Unlike rodents, however, they have four incisors in each jaw, and these incisors are enameled on both faces. A difference more relevant to communication is that, unlike rodents, lagomorphs are induced ovulators (Orr, 1977).

THE EUROPEAN RABBIT, *ORYCTOLAGUS CUNICULUS*

Rabbits have been symbols of fecundity for centuries, and with good reason. They can breed year round in mild climates, and in severe ones they abstain only in late fall. Gestation is only about a month, so a female can have as many as eight litters a year, each with three to nine young. As though this were not enough, each birth is celebrated by a short post-partum *estrus,* during which conception of the next litter often ensues. The young must wait only 6 months for their own opportunity to make more rabbits. It is perhaps no wonder that their average life span in Australia is only about a year and a half (Mykytowycz, 1968).

Rabbits' fecundity has earned them a special place in our culture, maintained in part by purveyors of paraphernalia based on the myth of the egg-laying Easter rabbit. This myth probably had its origins in the ancient Teutonic saga of Otara, a goddess who once changed a bird into a rabbit. The rabbit expressed its gratitude by laying eggs on her feast day, about the same time of year as our Easter. This feast gets its name from the same root as *estrus,* and is thus a suitable name for a celebration of birth and rebirth. Why the rabbit was grateful for being changed is not explained in the legend.

Gratitude does not seem in order, for rabbits have been one of the most thoroughly exploited species on the planet. Rabbits' meat, which is lean and delicately flavored, and their fur, which is soft and easily sewn, have been used by humans for centuries. The Romans introduced rabbits to several Atlantic islands, which they both proceeded to colonize with enthusiasm and thoroughness. Probably as a result of such an introduction, rabbits flourished in England by the time of the Normans.

Colonists of a later empire introduced them to Australia, where their powers of proliferation became notorious, supposedly after a flood destroyed a number of rabbit pens, releasing their inmates. Within 50 years, rabbits were dense enough to be considered a problem throughout the continent (G. G. Goodwin, 1975).

Rabbits seem an appropriate choice for exploitation, not only because of their fertility and usefulness, but because of their appearance. They display in every aspect the innocent helplessness of the professional victim. Full-grown *Oryctolagus* adults are only about 30 centimeters long, somewhat smaller than their close domestic relatives. Females (does) are larger than males (bucks). Rabbit faces have the short snouts characteristic of the mammalian infant, and their quivering noses and large pupils suggest mounting panic. Their ears, only slightly shorter than their heads, seem designed to provide predators with a handle. The short tail, brown on top and white below, is the finishing touch on a creature seemingly evolved to provide a model for cuddly toys.

This impression of vulnerability is an illusion. It was not simply the rarity of predators that allowed the success of the rabbit in Australia. Rabbits have a nearly hemispherical zone of vision with binocular overlap in front and behind and above. Their ears may seem ludicrous but are also a major factor in their evolutionary success. Not only are the ears effective predator detectors, but they are also important temperature regulators (Bourlière, 1964). They are highly vascular, and the inside surface is nearly hairless. Thus, when erected, the ears can emit heat, and when folded down, they can conserve it. Effective heat control is important for rabbits, which avoid predators by either remaining motionless or sprinting, attaining a speed of more than 50 kilometers per hour in one or two hops and maintaining it for up to 70 meters (G. G. Goodwin, 1976).

Rabbits of the genus *Oryctolagus* are common throughout southern and central Europe, England, and Australia. They prefer areas where cover is good but generally avoid forest interiors. They can be found in and near fields, prairies, and brushland (Kirkpatrick, 1971). They dig complex systems of interconnecting burrows called warrens. This habit provided the basis for their generic name, which means "digging hare." Warrens are connected to feeding areas by networks of well-worn paths, some evidently used for several generations. Southern (1940) estimated that a warren with about 150 rabbits covered about .2 hectares and that the ranges of these animals covered about .8 hectares.

Rabbits adapt their diet to available resources. In spring and summer, they devour salads of clover, grass, and domestic crops. In the winter they eat twigs, bark, and fallen fruit (Kirkpatrick, 1971). They reingest their first, soft, fecal pellets, often taking them directly from the anus. By this process they extract vitamin B and nutrients in a manner analogous to ruminants' cud chewing, but without the use of multiple stomachs.

Rabbits are active mainly from dusk to dawn and hide in brush piles, burrows, or concavities in vegetation called forms during much of the day. The rabbit's main enemies are people, foxes, stoats, weasels, polecats, martens, badgers, eagles, and hawks. Speed, camouflage coloration, and knowledge of hiding places are their only defenses— only does with litters actively defend themselves.

Unlike hares and most other members of the *leporid* family, European rabbits are highly social. A population at a warren near Oxford University studied by Southern (1948) sometimes contained more than 100 rabbits. Using marked rabbits in .8-hectare enclosures, Australian investigators (Myers & Mykytowycz, 1958) have shown that such populations are divided into groups of eight to ten, dominant members of each group defending a well-defined territory. Groups containing the most dominant rabbits in such a population have the largest territories with the best resources. Ranges of individuals within each territory overlap, vary in size with rank, and are not defended against other members of the group. Subordinates, especially does and youngsters, are less confined to the territory and less involved in its defense.

Within each group there are separate dominance hierarchies for males and females. These hierarchies are established (or re-established) at the beginning of each breeding season, which in Australia is about 7 months long. High male rank is expressed by moving freely within the group territory, defending it, marking it by chasing subordinates, and monopolizing females. As a result of this monopolizing, high-ranking males father most of the young, called kits. Because dominant females have most of the litters, their movements are confined, and they have the smallest ranges within the group territory. They do, however, defend it. Much of their reproductive success is due to their exclusive access to breeding chambers within the warren. They chase lower-ranking females away from these relatively secure chambers, forcing them to bear young in "stops," shallow holes dug 10–50 meters from the warren. Even pregnant low-ranking does sometimes do not give birth but resorb an entire litter into the wall of the uterus. Subordinates of either sex are chased, have highly restricted ranges, do not defend the territory, and reproduce themselves less frequently than the dominants (Mykytowycz, 1968).

Southern (1948) cited Lincke (1943), who suggested that bucks and does may pair for the breeding season. Although he is too much a gentleman to point out that unmarked animals, like those studied by Lincke, are difficult to identify, Southern's own data suggest polygyny. Only the few high-ranking rabbits in a group "took active part in the affairs of the warren" (p. 191), and most of these made repeated sexual approaches to more than one female during each breeding season.

Like most lagomorphs other than the pika, rabbits generally remain silent, thereby reducing the probability of detection by predators. Their communication system is dominated by olfactory and tactile modes that are well adapted to cryptic, short-range transmission and that work well at night and in burrows.

NEONATAL COMMUNICATION

As parturition approaches, the doe lines the breeding chamber with fur from her breast. The kits are born naked and with eyes and ears closed. Development is rapid. After about 10 days the eyes and ears open. The kits are furred and hopping in 2 weeks and self-supporting in 4.

During most of the kits' first month, the doe visits and nurses them at 24-hour intervals. She then leaves the burrow, sometimes sealing its entrance with soil and then marking it with urine and feces. These marks seem to repel conspecifics (Deutsch, 1957; Mykytowycz, 1968; Mykytowycz & Dudzinski, 1972). Absenteeism of this sort is presumed to reduce the probability that the mother will lead a predator to the nest. Its disadvantages are that the mother cannot defend the nest should it be discovered and the young cannot share their mother's warmth (Martin, 1968). These disadvantages are considerably ameliorated when the young are hidden in a burrow, which retains both their odors and their body heat.

Zarrow, Denenberg, and Anderson (1965) have experimentally investigated the effects of this maternal absenteeism on the growth of a domestic strain of *Oryctolagus*. One group of mothers was given free access to their kits. Another group had access for a short period once every 24 hours, and a third group had access briefly twice every 24 hours. The weight-gain curves for all three groups were virtually identical. Evidently, does and kits are well adapted to the absentee regimen.

The absentee system of neonatal rearing ensures that the only odors presented to newborn rabbits are those of their mothers. It would be expected that such restricted experience would have profound effects on social development. Indeed, nestlings given a choice of artificial nests anointed with either rabbit odor or some other odor preferred the former (Gambale & Dudzinski, unpublished, cited in Mykytowycz, 1974).

Mykytowycz and Ward (1971) found that head-stretch orientations of nestlings were greater toward swabs containing their mother's anal and *inguinal* (groin) odors than toward corresponding odors from strangers. Breathing rate also increased. Maternal urine had a similar effect, but the largest differences were obtained with odors from anal sacs. Mykytowycz goes so far as to call these preferences a form of olfactory imprinting. When the kits are old enough to leave the nest, they enter a world permeated with their mother's odor. According to Mykytowycz and Dudzinski (1972), the nest area in the central region of the group territory is so well marked that intruders are rare and their aggression inhibited.

Neonatal olfactory messages also go in the other direction—odors from the young facilitate recognition by their mothers. Mykytowycz and Dudzinski found that does reacted aggressively to the kits of strangers but that this aggression was not shown to a doe's own kits or to the kits of other does in the group. This difference in aggressiveness occurred even if the does were blindfolded, which suggests that recognition was made with the aid of olfactory cues. These investigators further determined that mothers would attack their own kits if the young were smeared with inguinal secretions from strange adults. Similar treatment with secretions from anal sacs of strange females had no such effect.

Since the skin glands of neonates are usually relatively inactive, some of the newborns' odor probably comes from contact with the mother's inguinal region during nursing. Males also apply odors to kits. Mykytowycz and his colleagues found that bucks "chin" and lick kits in their group. Females chin kits only rarely, and then only kits less than three weeks old. Chinning is a behavior in which a rabbit rubs submandibular glands on the lower surface of the chin onto another rabbit or an object, transferring a yellow secretion (odorless to humans) containing proteins, hydrocarbons, and glycerides. These substances evidently allow parents to recognize both their own young and those of other group members.

Licking may also transfer odors to young rabbits, because saliva can contain odorous substances (Melrose, Reed, & Patterson, 1971). Goodrich and Mykytowycz (1972) suggest an intimate relationship between chinning and licking: since rabbits customarily investigate by moving their mouths over the investigated object, they may be able to taste chemicals applied by these two behaviors.

Odors from urine are probably even more important than those from saliva and submandibular glands in promoting recognition of group membership. Mykytowycz (1968) discovered that adults urinate on young rabbits in their group and that kits treated with urine from strangers were attacked.

INTEGRATIVE MESSAGES

Application of urine to a conspecific, or enuration, is also practiced among adults, who so anoint each other during aggressive and sexual encounters (Southern, 1948). Thus, all members of a group share many of the same odors. These odors are probably responsible for rabbits' ability to distinguish fellow group members and their offspring from strangers (Mykytowycz & Dudzinski, 1972).

Sexual identity may also be transmitted with the aid of odors from submandibular and anal glands. Volatile hydrocarbon fractions of these glands, examined by Goodrich and Mykytowycz (1972), differed in males and females.

Odors may familiarize rabbits with their home range as well as with one another. The scents of glandular secretions and urine may facilitate use and choice of routes. A rabbit's territory is a network of more-or-less well-worn paths connecting the warren with feeding areas. Mykytowycz (1970) mentions the possibility that these trails are marked with secretions from glands under the eyes. When grooming, a rabbit transfers these secretions to the paws, which then apply them to the ground. This process may provide an olfactory mechanism allowing rabbits to follow particular trails. Such a mechanism would be particularly important at night, and it would be advantageous because habitual use of a route would provide knowledge of areas of concealment and exposure.

Potential olfactory landmarks are provided by dunghills, small piles of fecal pellets. There are usually about 30 of these piles along trails around a warren, and they are situated strategically so as to ensure frequent encounters by any rabbit, resident or intruder, that moves through the territory. Some of the dunghills are small piles produced by mothers at the entrances to their nests. These are seasoned with urine and probably help mothers find their way back to their young. The urine of a recently parturient mother is likely to contain distinctive substances that may enable her to distinguish her dunghill from the others in the territory. These other dunghills contain pellets from all the adults in the group, but most of the dunghills' odor comes from a few dominant males that visit the dunghills most frequently and contribute most of the pellets. The effective odor of a dunghill comes not from the pellets themselves but from secretions of the anal sacs. These glands are situated just inside the rectum and can coat each pellet as it passes through (Mykytowycz, 1968).

Apart from the confidence and ease of movement that come from being at home, what might these odors mean to rabbits? Schalken (1976) has shown that domestic bucks can distinguish odors of their own chin glands and feces from those of strangers. They also can tell fresh feces from old ones. Rabbits thus have available information about the time course of their own movements and perhaps those of other rabbits. Since trespasses are common, this information might be of use to an intruder in determining the likelihood of discovery. It might also be of use to subordinate rabbits in avoiding dominant members of their own group.

Sniffing and licking are rarely seen in young rabbits. Southern (1948), during 162 hours of observation, only once saw two 1-month-old rabbits of unknown sex "nuzzling and licking" each other's head. He concluded that "these mild comforts may take place between young of the same sex" (p. 184). In any case, affiliative behavior, if it occurs at all, is very rare and seen only in young rabbits. An interesting exception to this general aloofness was reported by Mykytowycz (1958). In one of his experimental colonies, all members of a group would disregard individual territories and gather briefly around the main warren. This "general assembly" occurred daily.

Patterns of play are consistent with this extremely low frequency of affiliative interactions. Most of the play observed by Southern was solitary and engaged in by youngsters, which were often seen leaping into the air for no apparent reason. The play of adult rabbits seems synchronized rather than social and is quite rare. When adults do play, however, they make up in zaniness what they lack in frequency. Southern provides us with this delightful account:

> On one or two occasions, when the warren was watched just before a storm, a kind of craziness seemed to permeate the whole population; young rabbits would frisk and leap in the air, while others would hurtle headlong round the enclosure,

scattering various groups to right and left; still others would be rolling on their backs in the sandy earth and kicking their legs in the air like playful dogs. This kind of playful leaping is often accompanied by a shake or twist made by the animal kicking with its hind legs while in the air [p. 181]. *

Even during such antics, rabbits remain mute, reserving sonic signals for dangerous situations. In such cases, as when spotting a predator, they signal alarm with a rapid series of thumps given by the hind legs. Ewer (1968) interprets this signal as a ritualized flight intention, and, indeed, it often precedes flight. Thumping may have a tactile component as well as an auditory one. A rabbit sufficiently close to the thumper might well be able to detect the thumps through its skin. Rabbits are said to be able to detect silent human footfalls in this way, so there is some plausibility to this conjecture. It is possible that the white scut of a bounding rabbit may alert others to the presence of danger, but at present there is no evidence for this notion. Both these hypotheses could easily be tested with experimentally deafened and darkened rabbits.

The distress call of the rabbit is a loud, high-pitched shriek, sometimes repeated, given only *in extremis,* as when being seized by a human. Rabbits also release urine in such situations (Southern, 1948).

AGONISTIC MESSAGES

The dunghills that provide residents with information about their previous movements in the territory also inform intruders that they are trespassing. For more than two decades, Mykytowycz and his colleagues in Canberra and Albury have been exploring the territorial-marking system of *Oryctolagus.* They provide us with the most complete description of the operation of this system ever developed for any mammal. When a rabbit passes into a neighboring territory, its demeanor changes from confident to wary. Instead of feeding, it stretches its neck and sniffs continuously. If challenged by the approach of a resident, it withdraws immediately (Mykytowycz, 1968). The dunghills that mothers place outside their nest burrows have a similar effect and prevent interference with the nest.

* From "Sexual and Aggressive Behaviour in the Wild Rabbit," by H. N. Southern, *Behavior,* 1948, *1,* 173–194. This and all other quotations from this source are reprinted by permission of N. V. Boekhandel & Drukkerij Voorheen E. J. Brill, Leiden, Netherlands.

One of the Australians' most important results was a demonstration of a difference in odor between pellets that are used as territorial advertisements and those that are excreted merely for elimination of wastes. Hesterman and Mykytowycz (1968) presented samples of fecal pellets to human observers in a systematic and carefully controlled procedure, much like that used in industry to test foods and perfumes. They found that pellets produced during encounters with strangers, during the breeding season, or by intact bucks had a stronger, more "rabbity" odor than those produced when alone, outside the breeding season, or by does or castrated bucks.

In a related study Mykytowycz (1968) presented rabbits with artificial dunghills composed of turf sprinkled with pellets that were obtained either from dunghills or from scattered (nonmarking) eliminations. He found no particular response to turf alone or to turf decorated with nonmarking pellets. Pellets from dunghills, on the other hand, elicited intense sniffing and production of fecal pellets. Both these pellets and the marking pellets used as a stimulus were judged by a panel of 30 human observers to be stronger smelling than the scattered pellets.

Since the intensity of the odor of marking pellets varies directly with the size of the anal glands of the rabbit that produced them, Goodrich and Mykytowycz (1972) performed chemical analyses on anal-gland secretions. Using thin-layer and gas-column chromatography, they found that the hydrocarbon fractions of these secretions had a stronger "rabbity" odor than the other fractions. The composition of the hydrocarbon fractions varied between males and females in accord with the sex difference in odor of marking pellets.

Encountering a dunghill of marking pellets is probably responsible for the abrupt change in a rabbit's behavior when it enters a foreign territory. In large (.8-hectare) enclosures that did not seem to influence social relations, the Australian investigators found that dunghills were placed at points where intruders would be most likely to encounter them. They observed a similar phenomenon in 3-meter by 4-meter holding pens. Rabbits kept in these pens deposited most marking pellets along the sides bordering on other rabbit pens (Mykytowycz, 1958).

Rabbits have a second system of territory advertisement, which allows them to take advantage of visually prominent objects such as logs and tufts of grass. The configurations of these objects, which make them good visual landmarks, make them difficult to mark with feces. Rabbits solve this problem by chinning them as they do with kits. They also chin their pellets and those of intruders, but they never chin outside their own territory.

An ambience of familiar odors from anal and chin glands increases a rabbit's confidence in dealing with members of its group as well as with intruders. The effect of familiar odors on dominance was documented in a series of experiments by Mykytowycz (1972, 1973). He staged encounters between pairs of rabbits in an enclosure containing fecal pellets or chin-gland secretions from one of them. The rabbit whose odor was present dominated the other about two-thirds of the time. Odors from a close associate produced a similar but less pronounced effect, the rabbit with the familiar odor generally getting the upper hand. Urine was not as effective, and inguinal secretions had no effect at all. In these experiments, dominance was displayed by attacking first and then winning the fight.

Odors of chin and anal glands not only promote dominance but express it as well. Dominant males not only contribute more pellets to dunghills than subordinates but also contribute stronger-smelling ones, because they have the largest and strongest-smelling anal sacs. Dominants also chin more frequently and have larger chin glands than subordinates. Since dominants of either sex are most highly territorial, the size of the chin gland varies with the degree to which the animal is concerned with territorial defense.

In his observations of free-ranging rabbits, Southern (1948) observed behavior that parallels the results of Mykytowycz's experiments. Southern noted that at the end of the breeding season does were particularly intolerant of youngsters. A doe would frequently chase younger rabbits away from her burrow or favorite resting place, where she was presumably surrounded by familiar odors.

Chin-gland secretions have no odor to humans (Hesterman & Mykytowycz, 1968; Mykytowycz, 1968). Nevertheless, it is likely that odors from anal and chin glands are perceived both directly and through scent marks. Since strong anal odor is one expression of dominance, it is not surprising to find a display that propagates this odor. By analyzing motion pictures of agonistic encounters, Mykytowycz (1974) found that the winner invariably erected its tail, displaying its white lower surface, and extruded its anus. Although this display has a visual component, it must also serve to disseminate the odor of anal glands.

A similar combination of visual and olfactory modes is found in several of the threats emitted by both antagonists in an aggressive interaction. Southern (1948) describes a ritual of mutual threat in which two rabbits circle each other with their haunches raised. Mykytowycz noted that although the dominant raised its scut higher than the subordinate, each antagonist could be seen to erect its tail and extrude its anus, so this display can be considered an expression of threat as well as of dominance. Both opponents were often seen to chin and defecate. They also engaged in scratching, which may apply odor from the feet to the ground and which has an auditory component.

The enuration that sometimes occurs in conflicts can also be considered an attempt to threaten by applying an odor. In at least one and probably two cases observed by Southern, a doe urinated on a buck that was "pestering her" and that ordinarily would be dominant.

To the human observer, the most striking demonstration of submission is the deferential avoidance of dominants. This avoidance creates an invisible bubble of "individual distance" (Hediger, 1955) around each dominant rabbit. Whenever it approaches, subordinates withdraw. The precipitate flight of a subordinate when chased or charged can be regarded as a special case of this form of obeisance. In encounters like those arranged by Mykytowycz, where escape was impossible, submission is demonstrated by a display that is the antithesis of threat: the tail is clamped down over the anogenital and inguinal regions (Mykytowycz, 1974). This display can be seen and also restricts propagation of odors.

Submission usually succeeds, but sometimes even flight fails and the subordinate is overtaken. In such cases, according to Southern (1948), "the ensuing assault took the form of a vigorous leap with both front paws extended. On several occasions the fur flew as a result of such an attack, so it is probable that there was a downward striking motion of the paws involved" (p. 180).

In a second form of combat, which Southern calls jousting, each rabbit leaps into the air, and they collide about a meter above the ground. Each lands where its opponent took off, and the procedure is repeated. The impact of their bodies is sometimes audible to a human several meters away. One bout lasted about 3 minutes and was punctuated by episodes of "false" (displacement) feeding. After each such episode the champions would turn simultaneously and rush at each other again.

Rabbits sometimes engage in a third form of stylized combat, similar to the boxing described in several other mammals. Both antagonists rear up on their hind legs and strike at each other with their forelegs. As in the other styles of fighting, a victor emerges and the fight is ended when one of the combatants withdraws (Grzimek, 1955).

When artificially crowded, rabbits increase their rates of aggression. Males become more likely to attack members of other groups, and females become more likely to attack members of their own group. Dudzinski and Mykytowycz (1960) and Myers, Hale, Mykytowycz, and Hughes (1971) have shown physiological and behavioral impairment of adults born under crowded conditions. These investigators suggest that density-induced stress is responsible for the impairment and that similar phenomena may occur in natural populations.

SEXUAL MESSAGES

One clue that a doe will be receptive to sexual advances comes from a familiar source—the anal glands. During the breeding season females' anal sacs grow larger and become more active, producing an "index of potential odor production" 3.5 to 70 times higher than otherwise (Hesterman & Mykytowycz, 1968, p. 79).

Synchronization of litters among the females in a group is probably related to their advertisement of sexual condition. Myers and Mykytowycz (1958) ascribe this phenomenon to "some kind of mutual stimulation" (p. 1516). They point out that, once established, synchronization is maintained by the post-partum estrus. They suggest that spread of paternity is an important effect of estrous synchronization. When all the females come into estrus at the same time, there are too many for the dominant male to monopolize them, so subordinate bucks have a chance to breed.

The most complete description of courtship in wild rabbits is provided by Southern (1948), and the following account is drawn from his article. The most dramatic form of courtship is "fast and furious" chasing of the doe by the buck. This chasing becomes "half-hearted," with frequent interruptions when the doe runs out of the warren into neighboring fields, probably because rabbits are more vulnerable when away from their burrows. Courtship chases are easily distinguished from aggressive ones, which invariably culminate in attack if the pursued is overtaken. Courtship chases, on the other hand, are punctuated by other forms of courtship, episodes of displacement feeding or looking around, or the female's diving into a burrow.

One of the most common interruptions in a chase is provided by "tail flagging" by the buck. When the doe stops, the buck walks around her with his legs stiffly extended. Since his rear legs are longer than the front ones, his rear is elevated. His tail is erected so that it lies flat along his back and shows its white underside. In one variant of this display, which Southern calls the parade, the buck circles the doe at a distance of about 2 meters with his hindquarters twisted toward her. In another version, "false retreat," he repeatedly walks stiffly away and returns. Both variations present the doe with the salient sights and smells of his posterior.

Both versions of tail flagging put the male in a position to deliver a stream of urine directed backward at the doe. This enuration is performed with remarkable accuracy. His precision is even more notable when, as is often the case, the urine is squirted in the course of gymnastic maneuvers. For example, Southern writes that "a very common method is for the buck to run past the doe about a yard away from her and to twist himself toward her with a kind of skid as he draws level, so that she is in line of fire from his hind quarters" (p. 177). A less common but even more impressive feat is a kind of strafing in which the buck urinates on the doe in midair as he leaps over her.

A doe's responses to these ministrations are quite variable. Some remain unmoved, while others shake themselves and retreat. Sometimes, however, the display has an exhilarating effect, and the doe becomes excited, jumping and frisking about while shaking her head. On occasion she takes the offensive, striking at her suitor with her forepaws or employing a stream of urine of her own. In such cases the buck, torn between lust and fear, vigorously scratches the ground—evidently, like false feeding, a displacement reaction. Myers and Poole (1961) concluded that enuration itself is an incidental effect of the male's high arousal.

Eventually the doe, either succumbing or simply tiring, calms down, and the buck approaches her from the rear, often stealthily. If she does not run off, he may rest his head on her side or back or nuzzle her ears or rear. Sometimes she actively solicits attention and approaches the buck with her neck stretched out. This reduces her apparent size as she approaches and may thus be a form of appeasement. She then lies down next to her partner, and he licks and grooms her.

Attempts to mount are sometimes, but not always, preceded by the rituals just described. Copulation by wild rabbits is rarely observed, but it has been described by Rowley and Mollison (1955). The male mounts from the rear, grasping the female's flanks with his forelegs. He then oscillates his pelvis quickly and vigorously, evidently to place the penis in the vagina. He then makes several deep thrusts, arching his back so vigorously that his tail disappears beneath the doe. Mounts usually last less than 15 seconds and are usually terminated by the female. She is usually pursued by the male, and sometimes she orients toward his rear. The result is a whirling chase in a tight circle. Rowley and Mollison found that copulation is impossible without the doe's cooperation—if she remains completely passive, the buck soon loses interest.

Descriptions of copulation in domestic strains reveal refinements of technique not reported in wild rabbits. The domestic doe usually displays lordosis before the buck mounts. If she does not, he seduces her by licking the bases of her ears. If the doe responds appropriately, the buck mounts, and at the moment of intromission the doe gives a cry. The buck then thrusts violently. The final thrust is so strong that his hind legs leave the ground and he loses his balance and falls over backward or sideways, giving a cry similar to the female's. Since this clumsy disengagement is probably as painful to the buck as intromission is to the doe, Blount (1945) regards the copulatory cries at these moments as expressions of pain rather than delight. The differences in wild and domestic copulatory behavior may, of course, be due to the relative ease with which domestic rabbits can be observed. Blount's observations were made with the aid of slow-motion films.

The vigor of rabbit courtship and copulation is evidently necessary to induce the doe to ovulate. Ovulation can be induced in the laboratory by manipulation and, according to Hafez (1969), may occur when contact with other females results in an "orgasm."

SUMMARY

Rabbits display 18 different types of messages, primarily olfactory, postural, and gestural. Neonatal types include neonatal olfactory imprinting and parental scent marking to identify young. Integrative messages include olfactory familiarization with the home range, contact with and identification of group members, traces of affiliation in the form of nose touches and social gatherings, play, alarm thumps with hind legs, and a distress vocalization. Agonistic messages include territory advertisement and expression of dominance by scent marking, threat and submission by postures and possibly by odors, and stylized fighting. Sexual messages include sexual advertisement and synchronization of estrus, frantic courtship that induces ovulation, and a male copulatory cry.

PIKAS, GENUS *OCHOTONA*

The word *pika* comes from the Tungu *puka* (Orr, 1977). Pikas look like tailless baby cottontails with small, round ears (see Figure 4-1). Though frail in appearance, they reside at higher altitudes than any other mammal. Members of the 1921 expedition to Mount Everest found them at 6126 meters above sea level. Pikas of different species are found in a variety of habitats, including rocky slopes near sea level in British Columbia. This account concentrates on the species O. *princeps,* found in the Rocky Mountains of North America on rock slides and adjacent meadows.

Pikas thrive in this forbidding environment with the aid of several behavioral and morphological adaptations. In July and August, a patient observer can see them scurrying over the talus gathering vegetation from the meadow and placing it on or under rocks to dry for winter food. Those piles are the focus of summer and autumn life, for they are the pika's only source of nutrition during the winter. They are vigorously but not perfectly defended, for pikas occasionally steal food from piles collected by others. Like rabbits, pikas reingest their feces to gain extra nourishment, especially vitamin B.

Their scurrying is facilitated by dense, stiff fur on their soles, which gives such excellent traction that pikas can run right up nearly vertical surfaces. One observer (Markham, 1975) saw an escaped captive run 2.5 meters up the smooth walls of his laboratory.

The spaces between the rocks provide pikas with shelter from wind and predators. The rocks themselves provide vantage points for "musing," in which pikas bask motionless for minutes at a time, nearly invisible against the jumbled background. Pikas are diurnal and are most active in the morning and late afternoon. The major predators that threaten pikas are weasels, martens, and birds of prey, including ravens.

Pikas' vision is sensitive to movement but not to form. They respond to movements of birds and observers, but when Broadbrooks

FIGURE 4-1   A pika. Photograph by Preston Somers.

(1965) placed a stuffed pika atop several precious and laboriously constructed hay piles, the owners ignored it, and one actually bumped into it.

In some areas, pikas live in male/female pairs (Kawamichi, 1971), but Barash (1973) reports separate, often adjacent, territories for males and females except during the breeding season, when they share a common area. Kawamichi (1976) describes significant overlap between male and female ranges even in the autumn in the Rocky Mountains of Colorado.

Each pika, or pair of pikas, has more-or-less exclusive use from year to year of a small (1300 square meters or less) area surrounding a main burrow. This area includes one or more hay piles, musing points, and routes connecting them. The major defended resources are the hay piles, which are sometimes plundered by invaders, and, in the spring, the estrous female (Kawamichi, 1976).

In early spring, males begin to invade the territories of neighboring estrous females, presumably in order to mate. Sometimes their wanderings take them into territories held by other males, and fights between intruders and residents result. In June the family phase begins as young are born. This phase lasts into July as the mother cares for the newborn young. In July and August, the fast-maturing young wander off, leaving the parents in their territory. Agonistic interactions among pikas, usually adults chasing young, are frequently seen during this period (Kawamichi, 1971).

The boundaries of these territories are well defined. Kawamichi (1976) often saw an escaping intruder stop as soon as it had crossed back into its own territory. The pursuer usually stopped near the boundary, but if it went too far, a counterchase sometimes ensued. Estimates of the sizes of pika territories vary. Some representative areas are as follows: 2000 square meters (Kilham, 1958) and 400 square meters (Lutton, 1975) in Colorado and 900 square meters in Montana (Barash, 1973) and Japan (Haga, 1960).

A dominance system is superimposed on the territory system. Kawamichi (1976) found that some individuals were consistently more successful in repelling intruders than others. These successful defenders were also the ones that most successfully invaded the territories of others.

The social system of the pika requires that some messages be sent over relatively long distances. Pikas' small size, camouflage coloration, and poor pattern vision exclude visual display as an effective channel. It is not surprising, therefore, that their communication is confined to auditory and olfactory channels. In open alpine landscapes, sounds travel surprisingly well, and pikas are the most vocal of lagomorphs. Although odors do not travel very far, they are long-lasting and can thus effectively cover a pika's range.

NEONATAL COMMUNICATION

What little is known about neonatal interactions comes from captive colonies. After a gestation of 31 days (Severeid, 1950), a litter of two to four lightly furred young is born in a burrow hidden by rocks and often by snow. Like rabbits, pikas have a postpartum estrus and sometimes have two litters each season, the first in May or June and the second in July (D. R. Johnson, 1967). Development is extremely rapid. Japanese pikas (*O. yesoensis*) reach their full adult weight of 120 grams when they are only 48 days old. They can walk, albeit unsteadily, when they are 8 days old. At about this time they begin to emit a call that Haga (1960) transcribes as "chii, chii." Krear (1965) reports that they greet adults approaching the nest with a rapidly repeated call similar to peeping by young birds.

INTEGRATIVE MESSAGES

The first clue to the presence of pikas is often a high-pitched "ank" call, very birdlike, and resembling a short segment of the sound made by a "mama" doll. These "short calls" often occur when pikas are outside their territories or when they are musing without a discernible stimulus. Sometimes another pika replies; the first caller looks toward the second, and they both resume their musing or gathering. They seem simply to be reminding each other of their presence.

In early July, when young pikas begin to move out of their natal territories, they begin to produce the "ank" call, but with a characteristic high pitch. When resident adults hear these calls, they become very agitated and search actively until they find the young intruder, then chase it away (Barash, 1973).

The same short call, emitted spontaneously as a contact signal, is also a response to the appearance of large birds, aircraft (Markham & Whicker, 1973), bears, lynx, and humans (Barash, 1973). This call is repeated not only by other pikas but by nearby marmots and ground squirrels. Upon hearing this call, pikas out on nearby meadows run to the talus and often hide beneath rocks. Pikas that are already hidden peep out from their hiding places (Barash, 1973).

Pikas have regional dialects in their short calls. Somers (1973) has shown that pikas in the Rockies north of the Colorado River produce a longer, lower note than those to the south. Pikas with the northern dialect have calls with a fundamental averaging 460 hertz and a mean duration of .35 seconds, while those with the southern dialect have a fundamental averaging 1105 hertz and a mean duration of .18 seconds. Figure 4-2 shows sonagrams of these calls.

Pikas that have been injured, or are about to be, sometimes give a high-pitched shriek like that of a wounded rabbit. Other pikas respond to this scream with "anks" (Somers, 1979). Long (1938) describes an episode in which a weasel disappeared into the space between some rocks. Almost immediately there was a pika distress scream, and a pika ran from a crevice near where the weasel had disappeared. The pika screamed once more as it disappeared into another crevice a short distance away.

Since pikas' form perception is not keen, visual landmarks are probably not particularly useful as a pika moves from hay pile to musing point or from the meadow back to its territory. Scent marks probably act as olfactory landmarks, allowing a pika to orient itself with respect to shelter, food, supplies, the burrow entrance, and the edge of its territory. These are the places a pika must reach without a lengthy

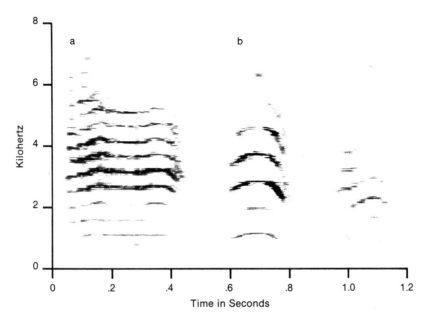

FIGURE 4-2   Sonagrams of northern (a) and
southern (b) dialects of pika short calls. Courtesy of
Preston Somers.

search, and these are exactly the places most likely to be marked with
cheek-gland secretions, urine, and feces. The first (cecal) pellets are
often posted on vertical rock faces, where they may act as marks before
they are reingested. Since some cheek-gland odor is probably applied
by the feet, pika territories are laced with odor trails connecting
concentrations of odor from cheek glands, urine, and feces. These
odor trails and landmarks must be of considerable assistance to these
dim-sighted creatures, which move about their steep, rocky habitat
with great alacrity and little meandering.

   At the rear of each of a pika's cheeks is a coiled apocrine-
gland complex that has been described in detail by Harvey and Rosenberg (1960). They found differences in size and chemistry of these
glands between sexually active and inactive pikas of both sexes. Pikas
often rub their cheeks with their paws. Since they use both front and
rear paws, it is unlikely that they are simply relieving irritation. Harvey
and Rosenberg suggest that paw rubbing may transfer apocrine secretions to the feet, whence they are transferred to the environment.

   Kilham (1958) noted that pikas often rub their cheeks on
rocks as they travel. Broadbrooks (1965) found that cheek rubs are

concentrated on prominent rocks and are most often at hay piles, musing points, and burrow entrances. These same locations are also marked with white splotches of urine, which pikas sometimes sniff before rubbing, and with small clusters of feces pasted on vertical surfaces.

Sharp (cited in Theissen & Rice, 1976) noted that juvenile males, which have had the least experience with their range, mark more frequently than any other sex or age class. Unlike adults, they mark most frequently in the areas they use the most. Both juveniles and adults, however, often rub on rocks outside their territories, especially along habitual routes to feeding and harvesting places away from the talus. These facts suggest that one function of cheek rubs and other scent marks is to familiarize pikas with the routes and locations they use most often.

AGONISTIC MESSAGES

Another, possibly more important function of cheek rubs is the advertisement of territory. Barash (1973) observed that 63 of 87 pika cheek rubs (72%) were performed along the perimeters of territories, where they would be most effective as territorial marks. A second kind of evidence for the territorial significance of cheek rubs is comparative: Kawamichi (1971) states that the Japanese pika is much less territorial than *princeps* and was not observed to cheek-rub.

When cheek rubs fail to deter trespassers, pikas often defend their territories by chasing the intruders away. Sometimes they save energy by advertising their presence with short calls. The short calls used in territorial defense sound like the "ank" alarm but differ in meaning from those that advertise presence or danger. The typical stimulus is the presence of another pika in the resident's territory, and the typical response is retreat of the intruder. Kilham (1958) saw a weasel hunting near several pikas, each of which stayed in the open but remained silent whenever the predator was in its territory. A few days later, at the same location, one pika invaded the territory of another. The resident ran toward the intruder, which retreated. The resident then advanced to the rock where the invader had perched, emitted several short calls, and returned to its hay pile. This pattern is typical in the San Juan Mountains of southwestern Colorado, but residents sometimes also give the short call as they approach the intruder. Further evidence that the short call sometimes has territorial significance is the observation that the calls are emitted less frequently when a pika is outside its territory. Finally, Himalayan pikas (O.

macrotis and O. royalei), which are not as territorial as other species, are much less vocal than their territorial cousins (Kawamichi, 1976).

Hay piles are the basis of pika territoriality, and pikas' dependence on these resources explains many unusual features of their territorial system. For example, female pikas are the only female herbivores known to possess their own territories. This anomaly is based on the female's use of her own hay piles. Furthermore, pikas display most territorial behavior, in the form of chasing, calling, and rubbing, in late summer, when they are building hay piles, rather than in the spring, when they breed. Kilham (1958) saw pikas urinate and defecate on their hay piles.

Perhaps because of these territorial advertisements, fighting among pikas is rare. Two observers (Barash, 1973; Broadbrooks, 1965) report seeing no fights in hundreds of hours of observation, but both Krear (1965) and Markham and Whicker (1973) reported occasional fights. Those seen by the latter authors, on Mt. Evans in Colorado, were brief and did not result in injury.

The fights that Markham and Whicker observed in captive animals, however, were often lethal. They illustrate the importance of aggression-limiting mechanisms such as territorial advertisement and flight, even among relatively puny animals like pikas. Most fights erupted when a dominant pika entered the area around the nest cage of another and tried to drive the resident away.

In the more serious fights both pikas stood on their hind legs, biting at each other's head and neck. They then fell over, each gripping the other's throat, and rolled around. The attacked pikas frequently attempted to escape but were pursued. Three fights lasted more than 30 minutes, including chases and searches by the aggressor. In all three of these long fights, the aggressor killed the other pika. It is not known whether such lethal fights ever occur in nature.

Pikas do not appear to have special postures for submission. Young pikas, which are the most common targets of aggression, typically avoid attack by fleeing, and the observation by Kilham (1958) described earlier suggests that adults do the same. Krear (1965) reports a call made by young pikas when attacked by adults. It is not clear, however, that this call inhibits aggression.

SEXUAL MESSAGES

During the pika's spring breeding season, a visitor to a rock slide inhabited by pikas can hear one of the most complex sonic productions made by any land mammal. Somers (1973) describes this long call as a song, and it does indeed meet many of Thorpe's (1961) criteria for classification as song, based on his work with birds. It is complex, has repeated elements, and has a strong fundamental pitch, usually around 700 hertz (see Figure 4-3). In Somer's words, a song is "a chattering vocalization made up of 30 to 40 calls each containing one to five notes . . ." (p. 126). Each call is about .2 second long. The first note in each call is slightly longer than the rest. As the song goes on, the notes get longer, as do the intervals between them, producing a noticeable *retard*. The song trails off toward the end, the number of notes per call decreasing gradually to one. This note is repeated, becoming a little longer with each repetition, until it becomes indistinguishable from the short "ank" call. Then, one after another, each from a different (usually invisible) source, other songs ring out, clearly audible even in a high wind. The calls are highly contagious, but when four or five different pikas sing, they rarely do so while another is performing (Somers, 1979). The excellent carrying qualities of the sound, contagion, and the tendency to avoid singing while another is singing make the song an effective means for males to advertise their presence to estrous females and to one another.

In his study of Japanese pikas, Kawamichi (1971) reported two characteristic male calls similar to the short calls and songs of North American pikas. These male calls, like songs, are most frequent in the breeding season. Females, often quite vocal at other times, remain mute when in estrus. Presumably, the male vocalizations provide females with information about the location of potential mates.

When a male and a female have found each other, perhaps with the aid of the male's song, they sniff each other's cheeks (Severeid, cited in Harvey & Rosenberg, 1960). Since these glands are largest and most active during the breeding season, this mutual investigation probably provides the couple with information about reproductive condition and may provide sexual stimulation. If satisfied with the results of their cheek sniffing, the couple disappears between the rocks, where their relationship is consummated.

Somers (1979) observed a copulation without preliminaries. The male simply mounted the female from behind, and there were several bouts of three to five pelvic thrusts in about 1 minute. The male dismounted, and the female resumed feeding.

SUMMARY

Pikas are known to display only ten different message types, all either auditory or olfactory. They include a neonatal greeting call; a contact signal, the "ank" short call; juvenile identity, by means of

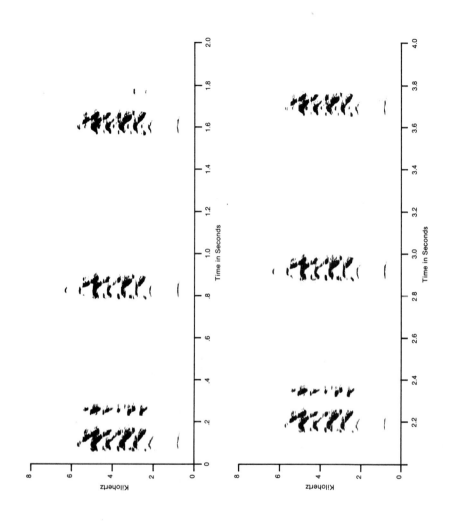

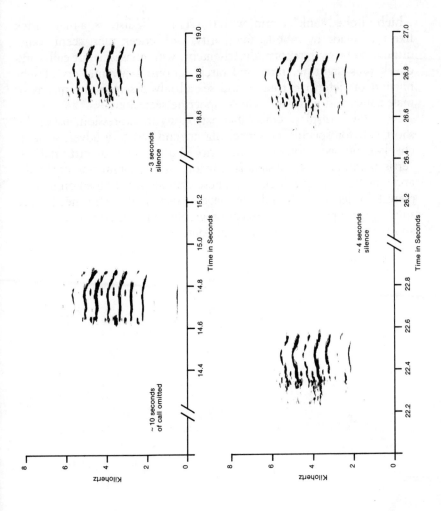

FIGURE 4-3   Sonagram of a
pika song. Courtesy of
Preston Somers.

a high-pitched "ank"; alarm, with the short call; distress, a high shriek like that made by rabbits; familiarity, with cheek rubs, scent trails, urine, and feces; territory advertisement, with chases, short calls, and cheek rubs; fighting, rare, and rarely serious; a juvenile cry of submission or defensive threat; and sexual advertisement by means of male songs and male and female apocrine secretions.

Pikas display remarkable parsimony in expression, using the short call for specific presence, alarm, and territory advertisement; cheek-gland secretions for familiarity and sexual and territorial advertisement; and special signals only for distress, submission or threat, and male sexual advertisement. These comments are based on incomplete knowledge of the pika message system. Particularly needed are studies, perhaps using infrared photography, on courtship, copulation, and parental communication with young.

5
WAPITI
AND DEER

We can show the marks he made
When 'gainst the oak his antlers frayed.

SIR WALTER SCOTT

Wapiti, commonly known as elk, and deer are members of the *cervid*, or deer, family of the order Artiodactyla. The name of the order means "even-toed" and refers to the fact that its members run on their third and fourth digits, which form a cloven hoof. Most *artiodactyls* have no upper incisors and use their lower incisors to rip vegetation from branch or ground. They swallow their food without much chewing, and in most species it passes directly into the first two chambers of a four-chambered stomach. When the animal has moved to a more secure location, the cud is regurgitated, chewed thoroughly, and swallowed again to pass into the third and fourth chambers of the stomach. Morris (1965) attributes much of the evolutionary success of the artiodactyls to this adaptation, which reduces the danger of predation during food getting, when herbivores are especially vulnerable.

Male cervids grow and cast their antlers every year. Each summer the antlers grow under a soft, vascular membrane called velvet (see Figure 5-1). With the exception of caribou or reindeer, female cervids do not have antlers. This sexual dimorphism is presumably an adaptation to a polygynous breeding system in which males compete for access to females.

Since artiodactyls' eyes are located on the sides of their heads, they can see other animals even when facing 90° away. Thus, they can use broadside displays, taking advantage of the fact that they seem larger when seen from the side than head-on (Walther, 1977).

WAPITI, *CERVUS ELAPHUS*

The wapiti was so named by the Shawnee long before settlers from Europe began to refer to them as elk. Thus, by convention, *wapiti* is the correct term for elk in North America. They are one of the largest members of the deer family, exceeded in size only by moose. Males weigh between 300 and 400 kilograms when fully grown, and adult females average about 250 kilograms. Wapiti are recognizable

106

FIGURE 5-1   This bull wapiti's antlers are in
velvet. Note that the tips of the antlers still seem to
be soft. Photograph by Roger Peters.

at a distance by their large white or beige rump patches, often the
only part visible to the naked eye. They can be distinguished from
deer by their size and by a gradient of brightness from light at the rear
to dark at the front. The best way to see them where they are common
is to drive along highways at dusk and carefully examine the far edges
of meadows.

The males, or bulls, unlike the females, or cows, have antlers. They often attain lengths of more than 1.5 meters. Each antler has a single main beam, often with as many as four projections, or tines. The antlers are used mainly to fight with other males (Bowyer, 1976).

Wapiti belong to the same species, *Cervus elaphus*, as the red deer of Europe and Asia, which they greatly resemble. In the Pleistocene, wapiti evolved in Asia and presumably crossed the land bridge into North America. They spread south through Oregon into California and the southwestern United States, and east as far as Massachusetts and North Carolina. Today they are found in large numbers only in the Rocky Mountains, northwestern California, and western Oregon and Washington.

Originally, wapiti were found in forested regions as well as in open plains (Nelson, 1930), but today they are most often seen in remote forest meadows. Even now, however, their habitat is remarkably varied and includes sagebrush grassland at lower elevations, pine and fir forest slightly higher up, aspen and montane forest at still-higher elevations, and finally high-altitude alpine meadow at and above timberline (Shelford, 1963). In the spring, wapiti follow the retreat of snow to higher elevations, where they spend the summer. In late fall they return to their winter range at lower elevations.

Wapiti subsist primarily on grasses, forbs, and woody browse. Some plants in each of these groups are eaten in all seasons, but grasses predominate in the spring and fall, forbs in the summer (Martinka, 1969), and browse in the winter (McCullough, 1969). Wapiti's dependence on grasses and forbs is probably a factor in their preference for meadows. Their diet demands that they consume large amounts of material in order to extract the necessary nutrition, and this in turn requires investment of large amounts of time. Martinka found that feeding was the major activity for elk when they were not disturbed. Even when resting, of course, elk are involved in feeding, for they, like other artiodactyls, chew their cud.

Coyotes and mountain lions are the major nonhuman predators of wapiti, and in their northern range grizzlies and wolves probably also take their toll. The most likely victims are calves and infirm adults, especially when the predator is the coyote.

Wapiti, like most other mammals, have keen ears and noses. They are, however, probably color-blind, and, though visually sensitive to motion, they will not respond to a motionless human in plain view only 15 meters away as long as the human is downwind.

Wapiti are gregarious throughout the year, but the size and composition of their groups vary. Although some associations, like those between cows and their calves, are stable throughout the year, other groupings are probably quite fluid. In late spring, as calving approaches, small "nursery" groups of females are seen. In southwestern Colorado the average membership of six such groups that we observed in 2 consecutive years was 25, with a range of four to 46. Martinka (1969) observed large numbers of single females and single yearling and adult males during this period. McCullough (1969) found a tendency for larger bulls to form separate "bachelor" groups throughout most of the year, but not in the breeding season.

The summer is a period of aggregation, with groups of adult males and loose associations of yearling males and adult females, some with young. The average group size during this period is 25; in southwestern Colorado, groups of 50 or more are not uncommon. These groups disperse in late summer as the breeding season approaches. Adult males are found with groups composed of females, young of that year, and male yearlings. Some yearling and adult males that do not compete successfully for females are seen as singles (Martinka, 1969).

In early fall the breeding season, or rut, begins, and large bulls chase other bulls and yearling males out of the female groups. A group of cows controlled by a bull is called a harem and may number up to 50 (Cahalane, 1947). The bull continually herds the cows in his harem, preventing them from straying off, moving them between feeding and resting sites, and warding off the approaches of other bulls. Bulls without harems are constantly in attendance, forming bachelor groups (McCullough, 1969) presumably waiting for a chance at one of the harem cows. According to Cahalane, there is some migration in and out of the harem as new cows are acquired and old ones wander off. The rut is an exhausting time for harem masters, for, in addition to herding and defense, they attempt to mount cows, sometimes unsuccessfully, but sometimes with two successes with different cows within 15 minutes of each other (McCullough, 1969). The proportion of younger males seen with harems declines as the rut reaches its peak, then increases toward the end of the rut (Struhsaker, 1967b).

At Jackson Hole, Wyoming (Martinka, 1969), and in southwestern Colorado, group size increases in winter as animals migrate down to their winter range at lower elevations, where the snow is light enough to allow them to find food. Herds in winter vary in size from nearly 100 to several thousand. The size of tule elk groups observed by McCullough in the Owens Valley of California, however, declined in the winter. Perhaps this difference is due to the milder climate prevailing in McCullough's area, where wapiti move to higher elevations for the winter.

Much of a wapiti's life is spent in a group (see Figure 5-2). Groups are adaptations for defense from predators, since membership

FIGURE 5-2   A herd of elk. Photograph by Roger Peters.

in a group lowers the probability of attack (W. D. Hamilton, 1971). Herding also facilitates way finding, since young animals can learn travel routes from adults (Scott, 1972). In Colorado, wapiti generally travel single file; this formation allows the leading animal to break trail for followers when traveling through heavy snow. Finally, the harem is a special group that benefits its master by reducing effort spent in locating estrous females because cows are always available as they come into heat (Davis & Golley, 1963). The harem system also tends to confine breeding to older males that have achieved high social rank.

According to Bowyer (1976) and McCullough (1969), each adult bull has a well-defined rank in a linear-dominance hierarchy. Dominance depends on size and antler development and is usually expressed by threat and successful sparring, after which the subordinate withdraws. Adult bulls dominate both cows and smaller males. Females also have a linear-dominance hierarchy, but female dominance is expressed much less violently than that of males. Subordinate cows simply give way to dominants (M. Altmann, 1952). Dominant cows

*Like deer and wapiti, bison, especially young, find
safety in a herd. Photograph by Roger Peters.*

not only outrank subordinate cows but also dominate males that are
under 3 years of age. Dominant bulls do most of the breeding (Bowyer,
1976; McCullough, 1969).

Herds of wapiti are not territorial in the strict sense, but they
do have home ranges, which are seemingly restricted in scope by habit
rather than physiography. Darling (1964) refers to three different types
of territory in red deer, but his "summer territories" include areas that
are used by several groups, and there is no evidence that his "winter
territories" are exclusive or defended in any way. There is agonistic
behavior associated with the "rutting territories," but his description
makes it clear that what is defended is not the territory but the harem
within it. The spatial organization of wapiti in southwestern Colorado
seems similar to that described by McCullough (1969) for tule elk—
that is, nonoverlapping, often non-contiguous herd ranges, without
intolerance of wapiti not of the herd. Claims of territoriality in Rocky
Mountain wapiti (Knight, 1970) and Roosevelt elk (Graf, 1955, cited
in Bowyer, 1976) are based on the mistaken assumption that the only
function of scent marking is territory advertisement. There is no quan-
titative evidence demonstrating territoriality in either subspecies
(Bowyer, 1976; Struhsaker, 1967b).

NEONATAL COMMUNICATION

The first sign that calving time is near is the cows' intolerance of their yearlings. In mid or late May, pregnant cows begin to chase, threaten, or even strike their yearlings with their front hooves, attempting to drive them away. About 2 weeks later, in late May or early June, the mothers-to-be show "marginal retreat" from the rest of the herd. They retire to cover but maintain contact by sight, sound, or scent (M. Altmann, 1963).

A birth observed by M. Altmann (1952) began at 5:30 in the morning. The cow lay in an aspen grove at the edge of a meadow in which several other wapiti grazed and rested. The cow licked her flank, then her genital area. The calf emerged 20 minutes later, and the cow, still lying down, licked it. According to Ewer (1968), the mother eats the placenta, but M. Altmann observed no such activity for the next 20 minutes. Then the calf stood up, the cow licked it, and it began to nurse, straining as it pushed against the cow. The cow observed by M. Altmann terminated the initial nursing session by nuzzling the calf and pushing it into dense underbrush.

The cow hides with her calf for much of the first day, nursing it at intervals that gradually increase from 20 minutes to several hours. Eventually the cow leaves the calf in order to feed or drink. Should it attempt to follow, she pushes it down with her nose or front feet. She returns to check on and nurse it at lengthening intervals and stays with it at night. This pattern continues for a few days to a few weeks, until the calf has the stamina to keep up with the daily movements of the herd.

The absentee system makes it necessary for cows to have some way of locating and recognizing their calves. Some writers (D. E. Johnson, 1951, for example) have concluded that the calf is practically odorless, but M. Altmann (1952) concluded that cows initially detect and recognize their calves by their odor. The licking and nosing associated with birth and nursing help the cow learn her calf's odor. Later on in the hiding period, auditory cues also play a part. Murie (1951) observed that a cow sometimes returns to the vicinity of her calf and then gives a loud, nasal whine. The calf returns this contact call, rises, approaches the mother, and begins to nurse. Ewer (1968) states that calves bunt the udder as they nurse, and Bowyer (1976) saw Roosevelt elk calves tug vigorously at their mother's teats. Such behaviors probably help stimulate the flow of milk.

Mothers defend their calves by pushing them down, barking in alarm, bugling in threat, and running away with a noisy, stiff-legged gait (M. Altmann, 1952; Murie, 1932). If cow and calf are in the open, the cow shields the calf with her body or pushes it away from the source of danger, often by striking the calf with her front hooves. Cows fiercely attack and persistently chase coyotes that get too close to their calves (M. Altmann, 1963).

M. Altmann (1963) found that mothers sometimes hide their calves in "pools" of from four to 70, with one or two cows remaining nearby as caretakers. Like mothers, caretakers defend their charges by running noisily away from the calves' hiding places. Calves generally remain in hiding unless the disturbance is within 10 meters, so the cows' flight may serve to lure predators away. Bowyer (1976), McCullough (1969), and Struhsaker (1967b) did not observe the "babysitting" described by M. Altmann.

Young mammals of many species have the ability to produce noises so unpleasant that they seem to compel succor. Young wapiti are no exception. Cahalane (1947) describes the distress call of wapiti calves as a high-pitched bleat. Murie (1951) transcribed it as "e-e-e-e-uh." D. E. Johnson (1951) calls it high-pitched and shrill, and McCullough calls it a squeal. The call is short and is repeated insistently at short intervals. The call is heard whenever calves are chased, handled, or in pain, and sometimes when they are separated from their mothers while in the herd, as during flight (D. E. Johnson, 1951).

In red deer (Bubenik, 1965), the typical response of any mother to the distress call of any calf is immediate approach, followed by attack on any potential predator in the immediate vicinity of the calf. In wapiti the bleating call of a lost or otherwise distressed calf is usually answered by several cows, which rush toward it and leave only after the mother arrives, sniffs it, and nurses it or leads it to safety. When a mother returns to the pool after feeding, she sniffs the ground along trails near where her calf is hidden. She may "test" several calves, sniffing their heads, necks, flanks, and posteriors before proceeding. Should a calf not her own attempt to nurse, she will reject it firmly, often striking it with her forelegs. During their first 2 weeks of life, calves frequently approach the wrong cows but soon, probably as a result of the punishment they receive, learn to recognize their mothers (M. Altmann, 1952, 1963).

When the calf is about 3 weeks old, it begins to follow its mother into a feeding group. The calf "heels," following its mother whenever she moves. The faster the mother goes, the closer the calf follows. During this period of integration into the nursery herd, the calf becomes increasingly sensitive to its mother's warning gait and freezing stance. When there is a disturbance, the lead cow runs toward it while the other cows and calves close ranks and run in the opposite direction (M. Altmann, 1963).

When the group moves, the cow and calf exchange contact calls. When produced by members of a large herd, these signals sound somewhat like the quiet bleats of grazing sheep and can be heard for several hundred meters. The calls continue throughout the summer. Darling (1937) heard similar vocalizations in nursery herds of red deer.

The contact call is sometimes used by mothers to summon their calves for nursing (Harper, Harn, Bentley, & Yocum, 1967). McCullough (1969) said that tule elk cows produce a signal that causes the calf to stop whatever it is doing and come to nurse, but he was unable to determine what the signal was. Perhaps it was a very quiet contact call. McCullough noted that nursing continues throughout the calf's first summer, even after it has begun to eat large amounts of vegetation. Nursing was especially common after rest periods. Bowyer (1976) noted that nursing was usually terminated by the cow's walking away.

Weaning is a gradual process accompanied by formation of attachments to the herd or to particular individuals within it. By midsummer, cows accept the company of their yearlings, and subgroups of cow, calf, and yearling are commonly seen throughout late summer. Female yearlings may accompany their mothers into harems in the fall, but males are driven away by the harem master (M. Altmann, 1963).

INTEGRATIVE MESSAGES

As calves are weaned, one class of behaviors that may facilitate the formation of attachments to other herd members is play. Social play, according to McCullough (1969) and Darling (1937), is seen in all sex and age classes but is especially common among calves. Darling felt that females were more playful than males, whose caprices he found to be dominated by mock combat. McCullough and Murie (1951) both noted that play was especially common in and around water. This generalization is supported by our observations in southern Colorado.

Bowyer (1976) found that Roosevelt elk calves initiated play by dipping their heads as they approached each other. The play that follows such a gesture includes elements from one or more of three categories of playful behavior. The first, described by McCullough as cavorting, involves bucking, spinning, jumping, and swimming. More highly structured play includes follow the leader and king of the mountain (Darling, 1937). Agonistic play, performed mainly by males, includes mock sparring, chasing, and dodging. Darling noted that chases often assumed the form of a game of tag, with roles of pursuer and pursued reversing as soon as one made contact with the other. Bowyer noted that in agonistic play Roosevelt elk often used exaggerated threats, but in only one of 17 cases did these displays lead to physical contact. Darling mentions "erotic" play among females during the breeding season. By this circumlocution he probably means mounting, which will be described later as a display of dominance.

The basic requirement of herd life is that members maintain contact with one another—that is, that they receive information about the location of other herd members. Wapiti do so by continually producing visual, auditory, and olfactory stimuli. The light-colored rump patch is a striking example of a contact signal. When watching elk at a distance, especially in low light levels, I have often been alerted to the presence of others by the flash of the rump patch as an animal turns or moves between trees. Although elk have poor pattern vision, they can detect movement of an observer's face in dim light at more than 200 meters and are probably even more sensitive to movement of rump patches, which are larger than a human face.

A class of tonically produced noises, according to McCullough (1969), serves to inform tule elk of the position of herd mates and thereby to coordinate activities. The first of these is the creaking of foot bones, which he describes as similar to the sounds produced by a leather saddle. When tule elk run, this sound can be heard up to 70 meters away. I have heard a similar sound produced by a single wapiti walking slowly at a distance of about 20 meters.

Another tonic sound is produced as elk rip and chew vegetation. McCullough was at first puzzled by the ability of the elk he studied to distinguish their own noises from similar noises he made but then realized that the elk were aware of one another's position and responded with alarm to sounds from unfamiliar locations. The stomach of an elk rumbles loudly enough to be heard by a human at distances of up to 10 meters away, even in the somewhat noisy environment of an urban zoo. McCullough is convinced that these sounds, too, have integrative significance and states that for animals whose hearing is as keen as that of elk, the creaking of feet, the ripping and chewing of vegetation, and the rumbling of stomachs must be a continuous din rendering them constantly aware of one another's presence.

The cohesion call used by cows and calves also serves to keep adults together. Bowyer (1976) noted that this call was used by members of all sex and age classes. Bowyer also described a "scenting posture" that Roosevelt elk used when they detected another animal. While in an alert stance, they moved their noses up and down in short arcs, sometimes licking their nostrils. This behavior suggests that olfaction is often an important channel for maintaining contact.

In wapiti there is considerable evidence that the leader of herd movements under several different circumstances is often a cow. In 12 cases of clear leadership observed by Struhsaker (1967b), the leader was an adult female even when bulls were present. In summer the movements of nursery herds, consisting of cows, calves, and young bulls, from feeding to resting areas and back are usually initiated and directed by older females (Lechleitner, 1969). Bolten (1970) states that the gradual migration of nursery herds to higher elevations is generally led by the oldest cow with the largest number of offspring. J. P. Scott (1972) suggests that in these migrations the herd follows the same paths year after year, so it is reasonable that an older animal, in a nursery herd necessarily a female, leads the way.

An observer in southwestern Colorado (Finstad, 1976) found little leadership in early spring, when herds are organized into nursery and bull subgroups. On two occasions when such a herd was disturbed, the bull subgroups and the nursery subgroup fled in different directions. During periods of intensive browsing, however, members of a nursery subgroup tended to follow the movements of one particular cow. This same cow led the subgroup to cover at the end of the morning browse periods. Darling (1937) noted that when alarmed, the red deer herd follows one particular female, usually older and larger than the rest.

Leadership among elk is evidently quite variable. However, large bulls are rarely leaders, as they tend to run in the middle of a group whether it is composed only of males or mixed. McCullough (1969) found that in cow or mixed groups an older cow was usually leader but that there were frequent exceptions and that leadership depends primarily on being the first to move, not on age or rank.

Leadership integrates the herd only if elk are disposed to follow. Allogrooming may supplement the effects of mere exposure by creating attractions among herd members. Allogrooming in wapiti consists of licking and nibbling body parts that an individual cannot reach itself, primarily on the neck and head. Most allogrooming occurs between cow and calf, especially during weaning, but adults sometimes groom each other, though never simultaneously (Bowyer, 1976; Struhsaker, 1967b).

McCullough (1969) believed that tule elk could identify one another by scent but presented no evidence for this hypothesis. This hypothesis should be investigated, perhaps with the aid of captive wapiti.

Marking of vegetation with antlers and incisors will be discussed as a male sexual advertisement, but Graf (1956) noted that females mark trees with their incisors, then rub the marks with their heads and necks, then rub their heads on their flanks. He observed that they particularly do so near their own "loafing grounds"—favorite areas for lying and chewing cud during the warmest parts of the day. It is thus possible that rubs serve as visual or olfactory landmarks.

Geist (1971) watched a cow rip bark off a pine tree, then rub her head and neck in the sap-filled wound. This action occurred in April, when sap flow is highest. This account brings to mind Darling's (1937) conjecture that rubbing of this sort serves to apply aromatic oils from the pine to the body. Geist cites experiments by Grundlach (1961) showing that this behavior is in fact often stimulated by pine oils.

Darling's hypothesis that rubbing serves to apply odor to the rubber for protection against insect pests leads to the prediction that rubbing would be most frequent before rest periods, when wapiti would be most vulnerable. The association of rubs with loafing grounds thus provides no way of deciding between the marking and the protective hypothesis.

Three pieces of evidence support the marking hypothesis. First, young wapiti, which are presumably as bothered by insects as adults, do not rub. Second, adult males mark only during the rut, when insect populations are probably no more bothersome than otherwise. Third, cows, which, according to Darling, are much more consistent than bulls in their use of particular locations, mark throughout the year. Of course, the two explanations, marking and protection, are not mutually exclusive, and rubbing may serve both purposes.

Whatever the explanation for rubbing, it is evidently pleasurable. Darling describes a female red deer rubbing with an expression similar to that of a horse being groomed—"half-closed eyes and drooped lower lip" (p. 72). This expression is seen in many mammals and is called the consummatory face because it is typically associated with the satisfaction of a drive, usually for physical contact. Whether this expression has significance for other wapiti is, however, unknown.

Wapiti express sudden fright through both auditory and visual channels. Murie (1932) describes a loud, harsh, alarm bark by both sexes over 1 year of age as a response to unidentified motion or sound, and Struhsaker (1967b) has developed sound spectrographs of this call. According to McCullough (1969), it is produced by a violent contraction of the abdomen and movement of the diaphragm. The bark is repeated until the source of the disturbance has been identified, and then it stops. Experiments in southwestern Colorado with realistic human imitations of the alarm bark produced results similar to McCullough's observation that investigation, not flight, is the typical response by other elk. The typical response in our area was a general increase in restlessness; feeding stopped, most animals looked around, and a few took a few steps toward or away from our position. Only occasionally were we able to delay flight by imitating the alarm bark, as McCullough was able to do.

According to Darling (1937), only females bark. Our observations in Colorado, like McCullough's in California, include several cases of barking by males. Barking is associated with a characteristic alert posture with head up and ears out to the sides, the ears becoming progressively more cupped as arousal increases. If the elk's investigation does not reassure it, it will, according to McCullough, raise its head above the horizontal, move its ears back, and start to move, turning its head from side to side. When not alarmed enough to flee or reassured enough to resume feeding or resting, elk continue to walk with a stiff-legged gait, head high and ears back. McCullough suggests that this posture helps the animal to see, scent, or hear the threat. Sometimes they hold their mouths open, which McCullough considers a signal of intention to flee ritualized from running with the mouth open. This signal is often combined with rump-patch erection, urination, and defecation, signs of sympathetic autonomic arousal. Similar behavior has been observed in Roosevelt elk by Bowyer (1976).

Another auditory alarm signal is produced tonically, as the sound of running elk alarms others that cannot see them. This sound is often quite loud. In open meadows the thudding of hooves can be heard at least 300 meters away even by a human, and in forest or brush the sounds of crashing vegetation are even louder.

Rocky Mountain wapiti sometimes amplify the sound of running by *stotting,* bouncing along with all four legs stiffened and moving in unison. Struhsaker (1967b) observed this behavior on six occasions, including both alarm and play situations.

Bowyer (1976) notes that the light-colored area around the Roosevelt elk's metatarsal gland, just above the lower joint of the leg, is clearly visible in the stiff-legged alarm gait and speculates that the coloration may serve as a visual supplement to an olfactory alarm. He points out, however, that there is no physiological evidence for the existence of a metatarsal alarm pheromone in elk. This suggestion should be investigated histologically and behaviorally.

Although adults are not known to use the high-intensity distress call used by calves, they do express a milder form of distress. Both Bowyer and McCullough (1969) noted a head-up twist with the ears flapping. This gesture was used when an elk was harassed by flies.

Illustrating Darwin's (1872) principle of serviceable associated habits, elk also display this behavior when bothered by conspecifics. For example, McCullough and Bowyer saw head-up twists when a cow was importuned by a bull and when a calf was prevented from nursing. The observers interpret head twists in this context as ritualized fly chasing.

AGONISTIC MESSAGES

Antlers can be formidable weapons, but only males have them, and they have them for only half the year. A display that is harmless during spring or summer, when antlers have not yet fully developed, can become a deadly attack in fall or winter, when the antlers sometimes extend the length of a bull's back. Although they are not very altruistic, wapiti have evolved two systems of threat—one used by females and antlerless males and the other used by males with antlers. Darling's (1937) description of threat in red deer is not nearly as complete as McCullough's (1969) but seems to be concordant. In particular, Darling describes stags with new or velvety antlers as behaving like the antlerless females.

According to McCullough, antlerless threats form a continuum from the mildest and most frequent displays to more severe and rare ones, which fall just short of fighting. The mildest ones appear first in agonistic sequences that can be terminated at any point by retreat of the receiver. The mildest form of antlerless threat is tooth grinding, which produces a creaking sound audible for only a few meters. This sound can be heard on still days when one wapiti approaches another that blocks its path. Tooth grinding is often associated with vertical retraction of the upper lip, which exposes the upper canines, contrasting sharply with black patches on the lower lip. McCullough was the first to report this threat, and he interprets the presence of upper canines as an adaptation to this display. This interpretation is reasonable because there are no lower canines and the upper ones evidently play a minor role in ripping and chewing vegetation. The black patches on the lower lip certainly seem well adapted to making the canines more salient. The threatening grimace often accompanies other threats.

The next-most-intense antlerless threat is the charge, in which the aggressor lowers its head, extends its neck, lays back (pins) its ears, thereby protecting them, and approaches to within a few meters of the animal it is threatening.

A graded system of postural threats by cows, calves, and spikes is described by Struhsaker (1967b). In this system the lowest level of threat is conveyed by pinning the ears and raising the head (see Figure 5-3). Intermediate levels are expressed by exaggerating these gestures, rising on the hind legs (see Figure 5-4), or kicking with a stiffened foreleg (not done by spikes, young bulls whose antlers lack tines). At the highest level of threat, the aggressor rears up on its hind legs and waves its forelegs rapidly up and down. McCullough (1969) refers to continued rearing and foreleg waving as boxing. This is still threat, not fighting, because contact even on forelegs is rarely made. In more than 130 encounters involving boxing, McCullough saw only one case in which a blow landed anywhere but on the forelegs, and this seemed accidental because it occurred on an upswing. The outcome of these encounters is not determined by the vigor of rearing and boxing but by dominance, as revealed by displays discussed later.

In addition to tooth grinding and canine display, antlered wapiti have many of their own auditory and visual threats. Darling (1937) describes a "roar" typically produced by a charging stag. McCullough does not mention the roar in tule elk, but Cahalane (1947) describes a bark that he transcribes "Eeough!" This vocalization was produced by an aggressive wapiti shortly before a fight.

Bowyer (1976) saw several homosexual mounts by cows and calves in agonistic situations. Cahalane describes mane erection, jaw

FIGURE 5-3   The two elk in the foreground raise their heads with their ears back. Photograph by Roger Peters.

FIGURE 5-4   The elk at the left rise on their hind
legs. Photograph by Roger Peters.

snapping, and "muscle display" in agonistic encounters in ways that
suggest that these signals are used as threats.

Darling (1937), citing Hingston's (1933) contention that dark
body parts often serve as threat signals, suggests that wallowing by red
deer stags, in addition to cooling them, acts to make them more
frightening to other males. Evidence for this hypothesis is slim, but,
according to Darling, females prefer less muddy water for their wallows.

The most complete description of threat by antlered wapiti
is McCullough's (1969). In the following account, drawn from his
study of the tule elk, it is clear that the highly complex and gradually
escalating series of threats provides many opportunities for disengagement, thus reducing the possibility of the deadly consequence of fighting with antlers. A bull with antlers typically charges with his head
higher than one without antlers. This posture tilts the antlers back

where they cannot strike the opponent. Typically, the antlered charge is performed with a stiff-legged gait. Charges are often followed by withdrawal by the receiver, but when this does not happen, according to McCullough, a fight is likely to ensue.

There are, however, five more threats that may precede or prevent fighting. First is a display that McCullough terms *parallel walk*, in which the opponents pace side by side about 4 meters apart. They bugle (see the discussion of sexual advertisements) and twist their heads so that their antlers move toward each other. All at once, one whirls to face the other and brings his head down so that his antlers point at the opponent, but instead of attacking, he brings the antlers even lower and scrapes at the ground with them. The opponent does the same, and sometimes they back away from each other, still tearing at the ground with their antlers. McCullough calls this display "thrash-thrash." Sometimes thrash-thrash is performed with a figure-eight motion, the antlers striking the ground at the center of the eight.

Mock sparring is similar to thrash-thrash but with the antlers directed at those of the opponent rather than at the ground. In mock sparring, elk act exactly as though their antlers were really engaged, twisting them from side to side and swiveling their bodies around as though exerting great force. Sometimes bulls mock-spar with a bush between them, leaving the bush much the worse for wear.

If neither retreats, fighting may begin, but there remains one last safety valve. Sometimes they charge but stop just short of contact between the antlers. At this point, one may run away, or both may go back into parallel walking. These threats are often associated with a rattling respiration not heard when elk are merely winded. McCullough states that the significance of this sound is not known, but it does seem to be associated with threat.

There are several forms of threat used by both antlered and antlerless wapiti. Bugling, the most frequent and far-reaching form of threat, is also the most impressive, particularly in high passes above timberline, where echoes sometimes seem louder than the bugles themselves. Their haunting, ethereal music can thrill the hunter just as it inspires the naturalist. M. Altmann (1952, 1963) mentions bugling by mothers as they defend their calves. According to Bowyer (1976), however, the use of the bugle as a threat is secondary and evolved from its primary use as a sexual signal. Bowyer also describes a "yelp" used by bulls as they chase spikes and a hiss from the herding posture performed only by nonspike bulls.

The mildest form of fighting is sparring, in which two bulls face each other, carefully put their antlers together, and then push

A bull elk bugling. Photograph by Ronald Gaines.

at each other. Although the opponents push at each other quite hard, they are easily distracted and frequently pause to chase flies away or look around. Sometimes they simply bump their antlers together. Sparring and mock sparring typically occur between bulls of approximately the same size and rank (Struhsaker, 1967b). After sparring, bulls often face each other but turn their forequarters away. Bowyer (1976) suggests that this posture allows each to assess the size of the other's antlers.

Combat with serious physical contact and injury is common only during the rut, but it may be sufficiently frequent then to be a significant contributor to mortality. Of more than 200 dead red deer stags found in the Woronesh sanctuary in Russia, 13% died as a result of injuries sustained in fighting (Heptner, Nasimovitsch, & Bannikov, 1961, cited in Geist, 1971). Because of the difficulty of separating the effects of injury from those of the general exhaustion prevalent among bulls by the end of the rut, however, it is hard to arrive at a good estimate of the extent to which fighting affects survival. According to McCullough (1969), a large proportion of older bulls engage in serious fights and are struck by antlers. Darling (1937), in his chapter on reproduction, states that in red deer the effects of fights are negligible and fatalities rare.

Fighting with antlers occurs only during the rut and is usually a competition for cows (Bowyer, 1976). After an exchange of threats, two bulls run toward each other, circle, then clash their antlers together. The clash is followed by neck twisting as each attempts to unbalance his opponent or deflect his horns. Darling once saw a bull receive a quick jab to the ribs following 5 minutes of antler wrestling. The recipient retreated immediately.

Bowyer found that cows, calves, spikes, and bachelors, but not masters, sometimes bite their opponents, often pulling hair from their heads or necks. Biters pinned their ears as they lunged. These same sex and age classes also kicked opponents with a stiffened foreleg. De Vos, Brokx, and Geist (1967) found that antlerless wapiti may also rear up and strike opponents with both front legs. Intraspecific fighting apparently differs from defense against predators. Nelson (1930) states that in the latter wapiti strike out and down with their forefeet, sometimes rearing up to use both, and that a blow delivered in such a manner can break a wolf's back.

As a result of occasional fights and frequent threats, hierarchies develop among the bulls and among the cows. In encounters between animals of approximately equal rank, relative position is established when the subordinate retreats. When there is disparity in rank, the dominant gives signals that assert its position.

In herds of wapiti in southwestern Colorado, the most frequently observed expression of dominance is supplantation, in which the dominant elk walks straight toward the subordinate as though it did not exist and the subordinate gives way. McCullough (1969) observed dominant tule elk displacing subordinates from their beds and noted that sometimes the subordinate would in turn displace an even lower-ranking animal from its bed. Struhsaker (1967b) observed the same sort of behavior in elk in the northern Rocky Mountains.

In a dominance display frequently seen in southwestern Colorado, a dominant animal, usually an adult, pushes the subordinate, usually a calf or yearling, with a foreleg. McCullough observed similar behavior in California and noted that sometimes the pushing was done with the nose.

A third expression of dominance is the foreleg stamp. The dominant elk walks up to the subordinate, places its muzzle over or alongside the subordinate's flank and stamps a foreleg, but does not direct the stamp toward any part of the subordinate's body. McCullough states that there is never any contact made by the leg and interprets this display as a ritualized blow by the foreleg. Perhaps the stamp is a ritualized form of foreleg pushing. Bowyer (1976) and Struhsaker (1967b) saw cows, calves, and bulls place their chins over the rumps of clear subordinates, which sometimes submitted by lowering their heads. Bowyer interprets this display as a ritualized mount. Walther (1977) states that dominant elk often assert themselves with a broadside display with a sideward tilt to the head. This posture is not described by Bowyer or McCullough.

Unequivocal messages are sometimes embodied in subtle signals. For example, dominance contests are not settled by vigor or amplitude of rearing and boxing. Rather, they are decided by staring, an expression so undramatic that it eludes all but the most careful observer. McCullough found that the victor in contests among antlerless wapiti was always the one that maintained a direct gaze at the opponent, whose gaze would eventually waver, then be directed elsewhere. Looking away was tantamount to giving up. Bowyer found that dominant Roosevelt elk often stared at subordinates, which invariably looked away.

All of the dominance displays mentioned so far are performed by both males and females. McCullough describes a display, antler threat, that is performed only by bulls and directed only to bulls of considerably lower rank. Antler threat consists of pinning the ears and lowering the head quickly into attack position. The subordinate moves off with alacrity.

At the bottom of the wapiti hierarchy, even below the lowest-ranking calf, are cripples. Their position is analogous to that of the omega rat, and, if McCullough's observations are typical, they are driven out of the herd by a kind of continuous displacement repeated upon each attempt by the cripple to rejoin the herd. All the crippled elk observed by McCullough were solitary, but one, at least, survived in good condition for more than a year. These animals may avoid attack with signs of submission. The most common expression of submission is often too subtle to be noticed. It consists of avoiding situations in which a less subtle expression might be required. As a large bull joins a grazing herd, there is often readjustment of position superimposed upon the movements necessary to find food. Some of these readjustments are performed by low-ranking wapiti, and their movements tend to take them out of the course of the newcomer.

Since staring is a threat, its antithesis, looking away, has become an act of submission. Both Bowyer (1976) and Struhsaker (1967b) noted that cows and calves sometimes lowered their heads and turned them away from dominants. Adult males did not display this posture. Bowyer points out, however, that interpretation of this gesture is difficult because once the submissive animal's head was down, it always began to graze.

Struhsaker saw cows rapidly open and close their mouths when approached by bulls. He concluded that this "jawing" is a submissive gesture that inhibits further interaction. Geist (1966) describes a submissive posture used by cows as they withdraw from courting bulls. The head is tilted down, the ears out to the sides. Sometimes the neck moves sinuously or is twisted sideways, so that one cheek is up.

Subordinate wapiti sometimes cautiously approach a dominant with their ears back and then touch noses with the dominant. McCullough, in a personal communication to Bowyer (1976), interprets this gesture as appeasement. Bowyer found that Roosevelt elk calves and cows seemed to graze closer to dominants with which they had touched noses than to other dominants.

While sparring with masters, bachelors often give short, high-pitched squeaks that vary in volume with the intensity of the master's efforts. Bowyer interprets these squeaks as appeasements that act to reduce the aggression of the dominant. Sometimes these signals do not suffice, and what McCullough (1969) calls decamping occurs; the subordinate abruptly shies away from the approaching dominant.

The ultimate gesture of submission is lying down. According to Burckhardt (1958), cited in Ewer (1968), red deer use lying down as an expression of submission. Walther (1966) suggests that this posture is derived from infantile concealment, which it resembles in that the head is extended and pressed flat against the ground.

SEXUAL MESSAGES

During most of the year, the tempo of elk social interaction is sedate. Apart from the nearly continuous background of contact sounds, the interactions that do occur are predominantly agonistic but are rare and brief. In early fall, however, there is a rapid acceleration in the rate at which messages are sent. Much of this increase is attributable to the maintenance and defense of the harem. The rest is due to calls, odors, marks, and displays that advertise reproductive maturity and readiness.

Bulls maintain harems with contact calls, by herding cows with charges identical to those described as offensive threats, and with a special herding posture, usually used behind the cows, in which the head is stretched forward, the nose raised, and the antlers tipped back over the body. Bulls sometimes hit cows with their antlers (Bowyer, 1976). McCullough (1969) observed that cows did not always act as though impressed. They sometimes responded by moving back into the group only when the charge was supplemented by tooth grinding.

His description leaves little doubt that herding is a full-time job. Not only do the cows intermittently wander off, but rival males appear in groups, occasionally while the master is involved with a receptive cow. As he chases his rivals away with the full repertory of threats, the cows again unobtrusively meander away. It is no wonder that harem masters lose weight during the rut.

The most common vocalization by bulls in rut is bugling. It is variable but usually sounds like a metallic whistle. It was first described in the scientific literature by Murie (1932). Sonagrams were developed by Struhsaker (1967b). The wapiti of the northern Rockies bugle with a fundamental at about 700 hertz, rising abruptly to about 1000 hertz. There are usually several harmonics. Bugles usually last between 2 and 5 seconds. Sometimes the bugle ends with a rapid, descending glissando to a pitch near that of the original note. At close range, a coda in the form of a nontonal moan can sometimes be heard. Bowyer states that Roosevelt elk bugles rarely carry more than 70 meters. In southern Colorado, however, they can often be heard at distances of several hundred meters. McCullough describes a roaring component to the bugle that cannot be heard beyond 60 meters. Darling (1937) describes a roar in red deer that late in the season becomes much more like a moan, but he does not mention the bugle.

Bulls adopt a characteristic posture while bugling. The head is stretched out horizontally, the ears laid back, and the mouth puckered into an "O." Bowyer points out the similarity between this posture and the posture used in herding. Struhsaker also observed this posture and further noted that the region around the penis sometimes shook as the bull bugled. On occasion, this shaking was accompanied by penile erection and emission of a clear fluid. Struhsaker speculated that this fluid may be semen, but as we shall see, it is urine.

Bugling is characteristically performed by mature bulls, and its frequency of occurrence roughly parallels the course of reproductive development, beginning in late summer, reaching a peak at the height of the breeding season, and then dropping off (Bowyer, 1976; Struhsaker, 1967b). During any short period within the rut, the frequency of bugling varies with several factors. One factor is time of day, the highest frequencies in McCullough's (1969) area occurring between 11 A.M. and 3 P.M., when copulations are most likely. Another factor is the number of cows in the harem. The highest rates of bugling in McCullough's area occurred in large harems. McCullough also noted that bugling occurred more frequently when one of the cows in the harem was in full estrus. Bowyer's data for Roosevelt elk are generally in accord with McCullough's for tule elk, showing that most bugling is performed by harem masters. On the other hand, Struhsaker's data suggest that solitary bulls whistled more than harem masters and that the least bugling occurred at midday.

Bugling occurs in many different contexts. Struhsaker often heard bulls bugle when cows withdrew from them, after herding and mounting, and before and after thrashing and digging with their antlers. Bowyer found that almost half the bugles performed by harem masters occurred during herding. The remaining bugles whose context he could determine (about 20%) occurred during agonistic interactions with other bulls.

McCullough (1969) observed cases in which a challenging bull's bugling stopped as soon as he was defeated by the harem master. In cases in which a group of males challenged the harem master, McCullough noted that only the dominant male of the cohort replied to the master's bugles and that low-ranking members of the encroaching bachelor herd remained silent. These factors suggest that bugling means not only that there is a mature bull present but that he is willing and able to defend his harem. Thus, bugling, though sexual in context, expresses both sexual readiness and defensive threat.

The harem system solves one problem, finding a mate, but it exacerbates another—defense against rivals. Nikol'skii (1975), in a study of red deer in Turkestan, found modifications of the male's mating call related to these problems. One type, probably Darling's (1937) "roar," was of lower frequency and was given at the appearance of rival males. Another type was similar but of higher pitch and was given when cows were present but inaccessible. This second type is probably the "grunt" of Murie (1932) or the "yelp" of Bowyer (1976), which occurred at the end of bugles or alone during herding.

Besides bugling in reply, other bulls often respond by approaching the bugler. Presumably, the bulls that approach are without harems of their own (Struhsaker, 1967b). Struhsaker observed no response to bugling by cows or calves. On the other hand, Bowyer found that cows ran toward the bull in 15% of the bugles he observed. He also found that bulls did not respond to recordings of their own bugles but bugled and responded aggressively to recordings of the bugles of other bulls. Bowyer inferred an ability to discriminate among bugles of different individuals. Thus, bugling and roaring, its analogue in red deer, present a puzzle for the student of wapiti communication. They are long-range signals associated with strong sexual motivation, but as Darwin (1872) noted, bugling does not seem to function to attract cows that are already gathered into harems. Bugling does, however, attract rival bulls and so is seemingly counterproductive.

Citing the similarities between the bugling and herding postures and between components of the bugle and the bull's cohesion
call, Bowyer suggests that bugling acts primarily to draw the cows'
attention to the bull. He hypothesizes that its use as a threat or
dominance display is secondary.

There are two other ways in which bugling might benefit the
transmitter. First, when copulating, a bull is least able to defend other
members of his harem from approach by rivals, so it would make sense
for him to produce a signal that will elicit a reply from nearby males
when he is about to mate. If there is no reply, he has some assurance
that he can proceed with impunity. If there is a reply, he learns that
it may be dangerous to copulate. Second, since it is possible that the
rivals will have determined by smell that there are females in the
vicinity, he is adding to the message "female present" the message
"mature male, too," thus warning potential rivals that they approach
only at some risk. He thereby may save himself the time and energy
needed to chase rivals that are not prepared to fight.

Bull elk have a strong body odor, and it is likely that this odor
is strongest during the rut and serves as an advertisement of reproductive condition. One reason that body odor is probably stronger
during the rut is heightened endocrine activity, which typically increases sebaceous-gland and apocrine-gland output (Strauss & Ebling,
1970). In addition to undifferentiated sebaceous and apocrine glands,
which cover much of the body surface, wapiti have at least three
special gland complexes. At least one of them, consisting of the
preorbital glands, is especially active during the rut (Darling, 1937).
These glands are located just forward and below each eye and connect
with the orbital apertures (D. E. Johnson, 1951). Darling observed
that in the breeding season of deer, they "secrete a musky, amber-
colored, waxy fluid which trickles down the face and dries on the hair"
(p. 174). Walther (1977) states that red deer open these glands when
they roar.

Two other glands, the metatarsal gland and the tail gland, are
also possible contributors to rutting odor. The metatarsal gland has
a tuft of short, fine hairs surrounded by longer hairs (D. E. Johnson,
1951). The special urination posture that in mule deer causes urine
to flow over this gland has not been described in wapiti, but its location
places it in contact with the ground whenever an elk lies down. It
may, therefore, produce a scent mark that could serve as a signpost
indicating the marker's proximity. The tail gland is located on the
dorsal surface of the tail at its base (Schaffer, 1940), where it may rub
against overhanging branches.

Wapiti exhibit several behaviors that supplement heightened exocrine-gland activity in producing strong body odor. One of these has been described in detail by McCullough (1969), who terms it *thrash-urinate:*

> The bull rakes the ground and low herbage with his antlers while releasing spurts of urine upon the long dark hair of the brisket and neck. The penis is projected and the muscles of the sheath contract, causing the penis to swirl from side to side, spattering the urine all over the underparts. The orifice of the urethra is directed upwards, further assuring that the urine is directed onto the body [p. 82].

As McCullough points out, it was undoubtedly the movements of the penis that led Darling (1937) and others (M. Altmann, 1952; Struhsaker, 1967b) to refer to this behavior as masturbation, but the liquid emitted is not semen but urine. McCullough indicates that the act is "virtually identical" in all races of *C. elaphus.* Struhsaker noted that Rocky Mountain wapiti frequently paw the ground with their forelegs, then lie or rub in the freshly turned earth.

Like bugling, thrash-urination seems to be a sexual advertisement whose audience includes other males. McCullough suggests that this display serves to inform rivals of the physiological condition of the displayer. When a male is in good condition and draws energy from the metabolism of food and fat, his urine almost certainly smells different from urine produced when, because of the stress of maintaining his harem, he approaches starvation and begins to metabolize muscle tissue. The best behavioral evidence for this suggestion is that some bulls are not challenged at all until late in the rut, when the challenge has a high probability of succeeding, which implies effective communication of general physical or psychological condition.

Bulls sometimes thrash-urinate as they wallow. Wallowing is much more common in the moist highlands of the Rockies than in the arid Owens Valley, where McCullough worked, but his description of wallowing by tule elk applies to Rocky Mountain wapiti. A bull digs into the mud with his antlers, throws mud into the air with a quick upward movement of the head, splashes with the forefeet, thrash-urinates, lies down and rolls in the mud, then repeats one or more of these elements. McCullough suggests that wallowing, like thrash-urination, acts mainly to coat the body with urine.

In the Rockies, wallows are often near springs and old beaver ponds. The odor of the mud in such cases is quite powerful and may supplement the odors from glands and urine (Struhsaker, 1967b). The dark mud may, as Darling (1937) suggested, render the bull more threatening, but McCullough found no relationship between muddiness and breeding success. It would be interesting to see whether this is also the case in the Rockies, where wallowing is much more common. In any case, wallowing probably also has some noncommunicative functions, such as cooling or parasite control (Murie, 1951).

Fresh wallows, reeking of organic decomposition and urine, are themselves a form of sexual advertisement informing humans and probably wapiti as well of the proximity of a rutting male. Another signal that may convey a similar message is the scent mark, a highly visible sign left by antlers on trees and fence posts. In the Rockies of Colorado, the trees that are rubbed are usually aspen or conifers, both of which emit rather strong odors when bark is removed. Graf (1956) noted that Roosevelt elk rubbed their faces on their antler marks, which could deposit preorbital-gland secretions, and that they then rubbed their cheeks and chins on their flanks. Bowyer (1976) mentions that they often licked and sniffed their marks. McCullough saw no such behaviors, however. Darling noted that red deer sometimes rub their bodies on antler-rubbed conifers and suggested that the piny odors thereby transferred to the body might provide some protection from insects. Bowyer observed many such scratches. It is possible that these marks, however striking, are an incidental effect of rubbing in order to remove velvet from or to polish antlers. This is unlikely, however, because McCullough and others have noted that rubbing with the antlers was combined with scraping with the lower incisors. All these behaviors are quite different but have a common effect— the production of a lasting, salient, olfactory and visual mark. Furthermore, frequency of scraping is not correlated with the presence of velvet (Graf, 1956).

McCullough describes bulls rubbing the antlers all the way up and down both sides, so that the entire surface is brought into contact with the tree. Although he does not mention an optical effect, this practice would produce in Owens Valley wapiti an effect often observed in Colorado—brightly flashing antler tips, visible for much greater distances than the elk itself.

Antler rubbing may sometimes have significance as a display as well as a means of making a mark. Graf (1956) noted that rival males, upon sighting each other, often engaged in intense antler rubbing, with face and flank rubs. Thus, like bugling and thrash-urination, antler rubbing is often addressed to other males and expresses threat as well as sexual readiness. Graf noted that bulls also made the display when approaching groups of females and juveniles in the absence of other mature males. Since bulls do not ordinarily attack cows or their young, the close association of antler rubbing with sexual behavior justifies grouping it with acts of male sexual advertisement as well as with threats.

Because the harem system assures that there will always be at least one male nearby as each cow comes into estrus, cows have no need for a long-distance sexual advertisement. Short-range olfactory cues suffice to inform bulls of a nearby cow's reproductive status. The vulva of a cow elk, like that of other mammals, is folded. Within the folds are concentrations of apocrine glands whose odors can be transmitted directly through the air or indirectly through urine, which may contain odorous chemicals related to reproductive status.

The best evidence that apocrine glands, urine, or both may advertise sexual readiness is a peculiar pattern of olfactory investigation performed by males called "flehmen." In flehmen, which McCullough (1969) calls "grimacing," the head is held in a raised position and the lips are drawn back, exposing the teeth and gums. Struhsaker (1967b) noted that the preorbital glands are open during flehmen. Knappe (1964) believes that the function of flehmen is to bring odorous substances, especially those contained in female urine, into contact with the vomeronasal organ, which is similar in microanatomy and neural connections to the nasal olfactory epithelium, the receptor surface for the sense of smell. This view has been questioned by Dagg and Taub (1970), who suggest that flehmen facilitates olfaction through the nasal epithelium, not the vomeronasal organ. However, Bowyer (1976) observed that the nostrils of Roosevelt elk were closed during flehmen. In any case, the function of flehmen seems to be to increase the sensitivity of the sense of smell, and, according to McCullough, it is characteristically performed by rutting bulls downwind of cows. Bowyer observed Roosevelt bulls grimacing in response to female urine. After flehmen a bull wapiti often approaches a cow from behind, rapidly lapping with his tongue. Usually the cow retreats, but sometimes the bull briefly sniffs and licks her anogenital region before she moves off.

The anogenital region is not the only source of sexually interesting odors. McCullough often saw bulls approach bedded-down cows and sniff their shoulders and the region around their tail glands. Struhsaker observed nose-to-nose contact, with the bull flicking his tongue. The cow's response to these forms of investigation provides an index of her sexual readiness. When fully receptive, a cow stands and allows investigation and mounting. Before and after she is ready, however, a bull's approaches are liable to be met with several apparently ritualized responses. One is a stereotyped, evidently unaimed kick with a hind leg. Another is the head twist already described as a sign of mild distress. It is often combined with jaw snapping, probably a defensive threat. These displays frequently precede flight, and Mc-Cullough suggests that bulls recognize this fact, for they usually desist immediately. A running female will usually be followed by the other females in the group, and bulls cannot always control a running harem.

Another form of investigation is described by Bowyer (1976), who observed bulls rapidly lapping their tongues as they trotted toward an alert cow. Since the bulls licked their consorts immediately after this gesture, Bowyer considers it a sign of intention to lick, informing the cow that the approach is sexual, not aggressive. The licks, usually directed to the *perianal* area but sometimes to the neck, ears, or back, probably inform the bull about the cow's receptiveness, by either her flavor or her response. Murie (1951) and Struhsaker (1967b) describe a "tapping" or "popping" sound emitted by the bull as he approached a cow. The significance of these sounds is unknown.

Once a bull has determined by sniffing, licking, and the absence of signals of intention to flee that a cow is receptive, he approaches from the rear and stands motionless about 1 meter away. The cow stands still, often licking one of her flanks, then the other. The bull walks slowly up to the cow and places his chin on her back from the rear. He then raises his forequarters and slides them onto her back. The cow generally stands still but occasionally takes a few short steps to maintain her balance. McCullough (1969) did not observe the lordosis reported by Morrison (1960) in Rocky Mountain wapiti, and Bowyer does not mention it in Roosevelt elk. The bull continues to slide forward, his penis extruded. The penis contacts the vagina, and the bull makes a few treading steps with his hind legs in an attempt to insert it. Usually the cow terminates the mount by walking forward, but occasionally the bull slides off first. The cow licks her flanks, and the sequence is repeated. In the 20 copulations observed by McCullough, the number of mounts before ejaculation varied from one to 33, with a mean of about six.

Darling (1937) recorded the following copulatory sequence in red deer: Bull approaches cow. Cow withdraws. Cow approaches bull, rubs whole flank against bull's. Cow attempts to mount bull. Bull attempts to mount cow. Cow kicks at bull with hind leg as bull dismounts. Bull approaches cow, which withdraws. Cow approaches bull, licks bull's muzzle and penis sheath. Cow attempts to mount bull. Bull mounts cow, roars while mounted. Cow withdraws. Bull withdraws, lies down.

Morrison (1960) and Struhsaker (1967b) reported that ejaculation occurs at the first successful intromission. Signs of ejaculation are, according to McCullough (1969), unequivocal. The bull makes a single, violent thrust upward, sometimes knocking the cow off balance: on one occasion, the bull fell over onto his side. The cow maintains a humpbacked posture, occasionally licking her flanks. The bull wanders off, paying her no more attention. Neither McCullough nor Bowyer (1976) observed the elaborate postcopulatory rituals of cows reported by Morrison. In fact, a striking feature of McCullough's observations is the low level of interest shown by bulls in cows that have recently been bred. Morrison reported captive cows copulating more than once during a 17-hour estrus period, but McCullough never saw a cow bred twice a day. He suggests that odors on a bull's chest from thrash-urination may olfactorily mark mounted females, as domestic ewes are visually marked by chalk placed on rams' chests.

SUMMARY

Life in herds has imposed special demands on the communication system of the wapiti. In addition to the basic mammalian messages found in solitary or colonial species, they need signals to coordinate travel, feeding, resting, and escape from predators. It is thus not surprising to find great variety and complexity in the content and form of the messages exchanged by these highly social animals. This variety and complexity is not expressed by a greater number of message types than are seen in other mammals but by a diversity of forms within each type.

Wapiti have 19 different types of messages. Neonatal messages include maternal recognition of calves by odor and, later, possibly by voice, as well as contact and distress calls. Integrative messages include three forms of play; contact through visual, auditory, and olfactory channels; female leadership; affiliative allogrooming; environmental familiarity by visual and olfactory marks; an alarm call and gait; and a gesture of mild distress. Agonistic messages include a tremendous variety of threat and dominance displays, submission, and fighting. Sexual messages include a variety of male vocal, visual, and olfactory advertisements; female advertisement by scent; and relatively simple courtship.

DEER, GENUS *ODOCOILEUS*

Two species of North American deer comprise the genus *Odocoileus*: *O. hemionus*, the mule deer, and *O. virginianus*, the whitetail deer. Although these two species differ somewhat in appearance, way of life, and habitat, their social and message systems are similar enough to be discussed together.

A glance at an alert mule deer explains its name—ears comparable in size to those of its namesake form the arms of a "Y" with a black dot of a nose at the bottom (see Figure 5-5). The ears of a whitetail are slightly smaller than those of the mule deer. The whitetail's namesake is about twice the size of the mule deer's black or black-tipped scut.

Mule deer bucks weigh between 60 and 120 kilograms. Whitetail bucks are smaller, on the average, but the biggest bucks of each species are similar in weight. Does of both species are one-half to three-quarters the size of bucks. In fall and winter, bucks of both species have antlers. Whitetail antlers grow from a single beam, like those of wapiti, whereas mule deer antlers branch. Both types grow under velvet.

Both mule deer and whitetails supplement their basic diet of shrubs and twigs with grasses, forbs, acorns, and fruit as each comes into season. Fawns of both species are preyed on by coyotes and lynx. In the Canadian portion of both species' range, wolves and cougars contribute to both adult and fawn mortality. In most areas human hunters and drivers also take their toll.

Mule deer thrive in a variety of habitats. In southwestern Colorado their range extends from semiarid lowlands thinly covered with sage and scrub oak, through shrubbed grasslands and coniferous forest, to timberline. In many areas they have adapted to the encroachment of humans and can, for example, be found on campus at a number of western Colorado colleges. Mule deer are not often found in the same habitat as whitetails, which prefer forest, swamp, and

FIGURE 5-5 A mule deer buck. Photograph by
Thomas Kucera.

brush. Only whitetails are found east of the Mississippi River, and the
blacktail variety of mule deer inhabits moist forest in much of the
Pacific Northwest.

Members of the genus *Odocoileus* munch and meander in
bunches throughout the year. Some, especially old bucks, wander
alone, but even they usually join others sooner or later. These aggre-
gations foster fitness in three ways, each involving communication.
First, aggregation provides protection from predation by decreasing
the probability that any individual will be taken (W. D. Hamilton,
1971). This advantage is enhanced if individuals scan more or less
independently and alert one another when they detect a potential
predator. Second, aggregation is a matrix for imitation. Even after
they leave their mothers, young deer probably learn locations of water,
browse, danger, security, and the routes connecting them by following
others. Third, aggregation allows allogrooming, which removes some

parasites and exposes deer to one another, thus increasing mutual attractiveness (Zajonc, 1971).

Linsdale and Tomich (1953) observed five different types of mule deer groups in California's Santa Lucia Mountains. Their description also applies to mule deer groups in southwestern Colorado. B. A. Brown (1974) and Hirth (1973) described analogues of most of these groups in whitetails. In early summer, groups consisting of a doe with her fawns are frequently seen. Linsdale and Tomich call them milk groups. Does with newborn fawns are usually intolerant of their yearlings, but after a few weeks this intolerance wanes, and maternal groups consisting of a doe with one or two fawns and yearlings are seen until spring, when fawning time approaches. Brown's "family groups" of whitetails correspond to maternal and milk groups of mule deer.

In the fall several maternal groups often gather at particular feeding or drinking sites. During the winter these maternal groups gradually merge to form a feeding group containing 10 to 20 deer. These groups are seen until spring, when pregnant does become aloof. Feeding groups of whitetails are frequently seen in winter around abandoned orchards in southern Michigan. It is not known whether membership in these groups is stable.

During the fall, or breeding season, deer display a tending bond, which means that a male, usually a dominant, closely follows one doe at a time and does not attempt to herd harems. Except during this period, bucks tend to travel in groups of fewer than six, which Linsdale and Tomich (1953) call fraternal groups. Their observations suggest that these groups are stable for at least 3 weeks at a time. During the rut they break up as each buck is attracted to a succession of females and larger bucks become intolerant of other males. In whitetails older bucks form the nucleus of fraternal groups united by strong social bonds (B. A. Brown, 1974). Both the stability of these groups and the role of older bucks have been questioned by Bowyer (1978).

Several different kinds of temporary associations can be lumped together as "experimental groups." The description of these groups by Linsdale and Tomich suggests some social attraction, but some experimental groups are probably produced by common attraction to a resource. Brown reports similar temporary associations among whitetails.

Dominance hierarchies have been described in whitetails (B. A. Brown, 1974) and in blacktails (Müller-Schwarze, 1972). In both species, ranks are defined by outcomes of competitions for food, water, beds, mates, rights of way, and displays resulting in the withdrawal of the subordinate. In both species, males dominate females, which in turn dominate yearlings and fawns. Neither species is territorial, but there is a tendency for does to use a well-defined range. Males of both species show dominance at particular locations but usually do not defend them (Moore & Marchinton, 1974; Müller-Schwarze, 1972).

In terms of both the types of associations and the dominance patterns, the social systems of the two species of *Odocoileus* seem highly similar. This similarity, combined with morphological and ecological similarities, leads us to expect few differences in their message systems.

NEONATAL COMMUNICATION

As spring unfolds, pregnant does become increasingly aloof and eventually retire to whatever cover is at hand to give birth. Pura, who provided Linsdale and Tomich (1953) with his observations of mule deer born in captivity, noted that newborn fawns could crawl but not walk. Voss (1965) observed twins born with the mother standing. Since birth takes place in seclusion, naturalistic observations of the first doe/fawn interactions are rare. They include maternal licking, nursing, and the rapid formation of a mutual attraction between fawn and mother. Voss noted that a mule deer mother began to lick her fawn as soon as it was born.

Pura indicated that the first nursing takes place immediately after birth; Voss gives the interval between birth and first nursing as 26.5 minutes. Linsdale and Tomich observed that does licked the anogenital regions of their young fawns as they nursed and noted that this licking did not seem to induce elimination, as it does in many other mammals. Probably this licking facilitates recognition of young, for Müller-Schwarze (1974) states that blacktail mothers "recognize their young by sniffing the anal and tail region" (p. 320).

Müller-Schwarze (1977) obtained experimental evidence that a kind of olfactory imprinting may be implicated in the formation of a fawn's strong attachment to its mother. He raised two young male and two female blacktails with a surrogate mother anointed with rump-gland secretion from a pronghorn antelope. He found that the fawns preferred the company of a deer surrogate scented with antelope odor to that of a deer surrogate scented with female odor at least until they were 33 days old. At 2½ months the fawns preferred the company of female pronghorns to that of female deer.

Both species of *Odocoileus* are hiders of their young, and in late spring and early summer the doe frequently leaves her fawn and begins to search for food. A mule deer fawn sometimes bleats in distress at the mother's departure. As distress increases, the bleats become two-syllabled, with accent on the second syllable. Eventually the fawn lies down and waits for her. Upon her return to the vicinity of the fawn's hiding place, the slowness of her movements alerts the infant, which rises, approaches her with its tail wigwagging, and immediately begins to nurse. Linsdale and Tomich (1953) suggest that the wig-wagging may serve to alert a twin of the mother's arrival. Single fawns, however, exhibit the same behavior (Bowyer, 1978), so it may simply be a contact signal.

As a fawn approaches a doe, several types of recognition sniffing occur. Both Linsdale and Tomich and Müller-Schwarze (1971) report that fawns sniff the tarsal area of does. In a case reported by the latter, when the doe was the fawn's mother, nasonasal contact followed, and in three other cases, when the doe was not the mother, the fawn withdrew. In a case reported by Linsdale and Tomich, a fawn sniffed its mother's tarsal gland, then began to nurse. Müller-Schwarze also saw a doe sniff the tarsal area of her fawn. These observations suggest that tarsal sniffing allows does and their fawns to recognize each other. Bowyer notes that fawns initially have trouble recognizing their mothers and sometimes try to nurse from males.

The physiology of tarsal scent is a lesson in the complexity often found in olfactory communication. The tarsal organs are crater-like depressions on the inner surface of the rear hocks. A central area of short, stiff hairs is surrounded by a tuft of long, erectile ones (Linsdale & Tomich, 1953). Lipids secreted by sebaceous glands under the skin are trapped on microscopic scales and ridges on the short, central hairs (Müller-Schwarze, 1977). According to Müller-Schwarze, scent is applied to the tarsal organ by rub-urination, an olfactory display performed by fawns and adults of both sexes. A deer rub-urinates by hunching slightly and urinating on the tarsal organs while rubbing them together and then licks the organs. Rub-urination has been described in mule deer (Cowan, 1936), blacktails (Müller-Schwarze, 1977), and whitetails (Haugen, 1959).

Brownlee, Silverstein, Müller-Schwarze, and Singer (1969) showed that a major socially active component of tarsal scent is cis-4-hydroxydodec-6-enoic acid. This lactone was not found in sebum freshly expressed from the tarsal glands but does occur in blacktail urine (Claesson & Silverstein, 1977). By an adroit bioassay Müller-Schwarze (1977) was able to demonstrate that in rub-urination chemicals pass from the urine to the lipid secretion by a process known as enfleurage, which is used by perfume chemists to extract odors from flowers by squeezing them into nonodorous oils or fats. Müller-Schwarze collected urine both upstream and downstream from the tarsal organs of a male and a female blacktail as they rub-urinated. Gas/liquid chromatographic analysis and comparison of this urine showed that volatile components in the upstream samples were missing in the downstream samples. Evidently, they had been pressed into the lipids by rubbing and held on the scales and ridges of the central hairs. Licking the organs afterward covers the short hairs with the outer, long ones, preserving the odor until the outer hairs are erected. Thus, some odor may be saved for later use, while some is emitted directly.

Rub-urination by captive blacktail fawns attracts does. In one case a fawn's rub-urination attracted not only its mother but three other deer as well (Müller-Schwarze, 1971). Captive blacktail fawns consistently rub-urinate upon separation from their human caretaker, but the display by fawns is rare in the wild. Linsdale and Tomich (1953) reported a case of rub-urination by a free-ranging separated fawn. The odor produced is quite strong and carries up to approximately 20 meters (Müller-Schwarze, 1977). Linsdale and Tomich describe it as an "overpowering dry heavy musk" (p. 112) similar to that of the poison sanicle.

Nasonasal contact is frequently seen when a doe and her fawn meet and is frequently preceded by sniffing and licking of the fawn's shoulders and neck and followed by nursing (Linsdale & Tomich, 1953). Bowyer (1978) suggests that nasonasal contact is an appeasement gesture.

Nursing periods are almost always less than 5 minutes long and are repeated at intervals of about 4½ hours. Twins nurse simultaneously. Nursing provides a basis for later maternal leadership, for a mother sometimes leads her fawn to a new resting place by walking as the infant nurses (Linsdale & Tomich, 1953).

During the summer the duration of nursing bouts shortens and the intervals between them lengthen, so weaning is a gradual process. Lent (1974) states that bunting, in which the fawn strikes the udder with upward movements of its forehead to stimulate milk flow, is involved in weaning. He points out that mothers of many ungulate species respond aggressively to bunting.

As weaning progresses, the attachment of the fawn to its mother remains strong. During short movements from one grazing area to another, the fawn follows closely, perhaps aided by its mother's white rump patch and the white underside of her tail. Mutual grooming probably helps maintain the attractions between does and their fawns. Mothers groom the entire bodies of their young except for the underparts, which, perhaps because of their thinner coat, may need it less. These areas are consistently neglected when deer groom themselves.

Does and young fawns have a complex of signals and reactions to danger not seen among adults. At any major disturbance the mother moves off, leaving the fawn in hiding. Her rump patch may serve to attract predators to her and away from her fawn. She can exercise this ruse with impunity because she can easily outrun predators under most conditions and can successfully defend herself and her young against their attacks. In general, she will not attempt to defend her fawn from humans but can successfully drive away coyotes and bobcats (Linsdale & Tomich, 1953). Linsdale and Tomich once saw a mule deer mother strike her fawn with her hind leg, apparently to warn it away from a man. A more common alarm signal is a maternal bleat, at which the fawn lies prone and remains still (Seton, 1929). Prevention of following by maternal vocalization and contact has been observed in whitetail fawns (Rue, 1962, cited in Lent, 1974). Downing and McGinnes (1969) saw whitetail fawns give the prone response to bounding by their mothers.

When the danger has passed, doe and fawn locate each other by exchanging bleats similar to those uttered by the fawn at the mother's departure (Linsdale & Tomich, 1953). According to Nichol (1938), mule deer mothers erect their tarsal hairs to warn or attract their fawns.

INTEGRATIVE MESSAGES

During their first year, fawns interact in cavortings seldom seen in older deer. Pregnant females sometimes frolic with their fawns, but according to Linsdale and Tomich (1953), adult males do not play. Males in their second year sometimes butt each other playfully, but these interactions are often related to the serious business of establishing dominance.

Play, say Linsdale and Tomich, is often most frequent in morning and evening, especially after nursing. A common context is female intolerance toward a fawn in a feeding group or the movement of the feeding group out into an open area. Linsdale and Tomich state that fawns "often deliberately initiate play" (p. 182), and there seem to be several signals that may convey nonserious excitement. These include a light, bounding gait, dashes past another deer, and a bristled tail held vertically and curled slightly forward at the tip.

Three types of play and their characteristic course can be abstracted from Linsdale and Tomich's descriptions and our observations in the San Juan National Forest of southwestern Colorado. In cavorting, a doe, usually the fawn's mother, serves as a focus for erratic orbiting, frenzied passes, and sudden course changes, punctuated by pauses of several seconds during which the fawn looks around and regains its wind. In tag, one fawn dashes past another or feints until the other begins pursuit. Roles of pursuer and pursued can change suddenly when the pursuer pauses to rest. The third type of play consists of head butting. This behavior differs from the serious dominance contests of older males mainly in its brief duration and frequent interruption.

Especially when several fawns are involved, a play sequence intensifies to a peak that is reached within a minute or so. It is usually terminated within 3 minutes in one of four ways: the players seem to get winded, the group takes flight, the nonparticipants move on to feed elsewhere, or the adults actively intervene by threatening the participants.

Play in deer probably performs several functions. Linsdale and Tomich suggest that play, involving "more avoidance than engagement" (p. 183), is primarily practice for evading predators. As Ewer (1968) points out, play bonds participants into an integrated group by providing practice in social response and motivation for association. These social functions seem appropriate to mule deer play, which often involves leadership, dominance, threat, and attention to movements of the group.

Groups are integrated by contact signals that reveal their location. Mule deer contact signals include visual, auditory, and olfactory forms. Mule deer vision is much more sensitive to motion than to form, as is shown by their tendency to ignore an unconcealed human observer as long as the observer is completely motionless. Even when they stand in the open, mule deer are often nearly invisible to a human's excellent form vision because of the highly effective camouflage. It is thus not surprising that several visual contact signals gain effectiveness from motion. The large, white rump patch may be an effective advertisement, at least to humans, but it becomes especially salient when a deer bounds away or moves through vegetation that intermittently hides it. On one occasion near dusk, a student and I were watching a group of deer in a valley when our attention was suddenly caught by a flash nearly 90° off to our left at the very edge of vision. The flashing continued and moved along a ridge just under the brow. Only with binoculars were we able to determine that the flashing came from the rump patch of a deer walking slowly through light oak brush. There is little doubt that such a signal would often be as apparent to the keen motion vision of deer as it was to us. Linsdale and Tomich describe a similar effect caused by the flashing of the inner white surface of a mule deer's legs as it walked in the open approximately 900 meters away.

The effectiveness of the rump patch as a medium-distance contact signal is sometimes enhanced by tail switching as the dark dorsal surface of the tail quickly uncovers and recovers it. Tail switching itself may act as a short-distance contact signal, for deer seldom go long without switching their tails. Our observations show that tail switching is frequent even at times of the year when there are no flying insects, so it is probably not merely an antiparasite behavior. Linsdale and Tomich mention several tail switches that preceded transitions from looking to grazing and suggest that a switch "conveys to the watcher advance notice of some activity about to take place" (p. 57). Our observations of mule deer in southwestern Colorado fail to confirm this hypothesis. Even in early spring, when no flying insects were detectable, tail switches occurred independently of changes in activity. Linsdale and Tomich themselves mention tail switches that occurred without changes in activity and changes in activity that occurred without tail switching.

Evidence that tail switches maintain contact among mule deer is indirect. Sudden motion of the tail can only rarely escape the keen, wide-angle motion vision of nearby conspecifics. According to Linsdale and Tomich, deer "seem generally not to recognize or even to see immobile animals" (p. 120). We have frequently seen deer move downwind toward others whose only discernible movement was tail flicking, so it is likely that in these cases it was the tail flick that informed the approachers of the others' presence.

Hirth and McCullough (1977) argue persuasively that tail flagging, a wigwagging of the erect tail given by running whitetails, is a signal that maintains group cohesion during flight. They cite the advantages accruing to individuals that flee in a group and reject explanation on the basis of kin selection. They point out that does with fawns are no more likely to flag than males.

Even when out of one another's sight, deer maintain their associations by following habitual routes, especially those recently used by others. They do this by frequently sniffing the trail for olfactory traces of their associates. Much of this odor comes from interdigital glands just above each hoof. These pouched sebaceous concentrations connect via a hair-lined channel to a groove running down the middle of the hoof. The pouch and channel are coated with a musky, sour, greaselike exudate that flows onto the hollow underside of the hoof and is applied to the ground at each step (Linsdale & Tomich, 1953).

Does probably maintain contact with their fawns with the aid of the fawns' powerful interdigital odor. Linsdale and Tomich maintain that the longer pedal hairs of fawns are more effective than the shorter hairs of the adults in carrying odors to the ground. Since fawns' movements are more restricted than those of adults, their pedal odor should be more concentrated. This concentration should also aid in detection by the mother but would aid predators as well.

Other major sources of odor along deer trails are urine and feces. When moving along heavily used deer trails in the San Juan Mountains in any season, one is rarely out of sight of fresh pellets, and in winter urine is seen to be equally densely distributed. The distinctive sagey or piny odor of deer urine is readily detectable by the human nose at more than 1 meter's distance.

Deer maintain olfactory contact not only through interdigital and eliminative scents applied to the environment but also by exposure to odors from one another's dermal glands, the strongest smelling of which are the tarsal glands. Other odors detectable by humans emanate from the preocular and metatarsal glands. Each of these odors has its own special significance, but each must contribute to a deer's general responsiveness to the presence of conspecifics.

Mule deer do not seem to have special auditory contact signals, but Linsdale and Tomich (1953) noted that the clicking of hooves and the ripping of vegetation attract the attention of nearby deer.

Deer often associate simply because of the common attraction to a feeding or watering site. That they often escape and travel together, however, suggests some form of attraction to one another. Presumably, infantile and continued exposure to conspecific odor plays a part in this attraction. For example, allogrooming may provide tactile as well as olfactory channels for the maintenance of attraction. Although deer sometimes groom themselves by scratching, they groom one another only with their lips and teeth, bringing their noses close to the areas groomed. These areas are mainly the neck and shoulders, which are hard for a deer to reach by itself. Simultaneous mutual grooming is common, and during such grooming Linsdale and Tomich saw biting, which they interpreted as crushing of ectoparasites.

Mutual licking is another expression of a social bond. Licking is performed mainly by close associates: doe and fawn, siblings, or captive animals raised together. A deer invites licking by approaching another from the front, standing close in front of its head, and lowering its own head. Most licking is directed at the head and neck (Müller-Schwarze, 1972).

Linsdale and Tomich (1953) noted that mule deer often touch muzzles upon meeting and consider this contact to be a form of recognition sniffing. However, they also report nasonasal contact after mutual grooming, when recognition is probably not required. This suggests that nasonasal contact may function among adults as it does between a doe and her fawn, as an expression of affiliation, appeasement, or identity. The last two messages are especially likely to be useful when strangers meet. Linsdale and Tomich discovered that strange deer are especially likely to touch muzzles upon meeting. Since these authors do not present figures on the relative frequencies of nose touching among strangers and familiars, the full significance of this behavior needs further investigation.

A model for such an investigation is the report by Müller-Schwarze (1971) on another form of individual recognition, tarsal sniffing. He found that an experimental group of three captive black-tailed deer sniffed one another's tarsal organs at a rate of 1.5 times per individual per hour. When an unfamiliar female was introduced, this rate doubled, and when an unfamiliar male was introduced, the base rate increased by a factor of about 7. Müller-Schwarze also showed that in a group of six blacktail fawns the rate of tarsal sniffing doubled after dark. Both of these findings are in accord with the hypothesis that tarsal sniffing assists in the recognition of conspecifics. Tarsal sniffing of strangers is often followed by chasing or striking, so it has an agonistic component in such contexts.

Two informal experiments, also by Müller-Schwarze (1971), suggest that tarsal scent may convey information about sexual identity. On both occasions when he applied tarsal scent from a female fawn to the rear leg of a young male, another deer, upon detecting the odor, approached a female fawn also present. Müller-Schwarze concluded that "the age and the sex of the tarsal scent [were] recognized" (p. 146). When a tarsal tuft from a slain male was rubbed on the hocks of live deer, others, particularly females, approached, followed, and licked the treated legs. Both bucks and does showed lingual extrusion and yawning, which, according to Müller-Schwarze, are considered to be signs of fear. In both of these experiments, deer seemed to respond on the basis of an expectancy about the sex of the odor's donor. Brownlee and his colleagues (1969) have shown that a major difference in the tarsal scents of bucks and does is the presence of a lactone in the male scent. Müller-Schwarze, Silverstein, Müller-Schwarze, Singer, and Volkman (1976) have shown that blacktails can discriminate among isomers of this lactone.

Volkman, Zewanek, and Müller-Schwarze (in press, cited in Müller-Schwarze, 1977) examined the roles of antorbital-gland (a sac just above each eye, secreting an odorous waxy substance) and forehead-gland secretions in recognition of age classes. Fawns and yearlings were able to discriminate age classes of donors only when both types of secretion were present. These secretions are applied to branches by yearlings and adult males (De Vos et al., 1967).

The hypothesis that herding in mule deer is primarily a defense against predators gains credence from an examination of their alarm system. In contrast to their two forms of affiliative expression, they have at least ten different ways of signaling varying levels of alarm, including visual, olfactory, and auditory modes. Mule deer sometimes seem to attend more to one another than to the alarming stimulus itself—a rifle shot will elicit only curiosity, but an ear twitch can galvanize the herd.

When seen from the side, the ears of an alert mule deer stick straight up. The large size of the ears is probably an adaptation for sensitivity. At high levels of curiosity and low levels of alarm, the ears cup forward toward the stimulus. Linsdale and Tomich (1953) observed that the cupped ears and erect neck of a mildly alarmed deer alerted others but that the others seemed unable to infer the direction of the disturbance from the position of the ears. Müller-Schwarze (1971), as previously mentioned, interprets yawning and extrusion of the tongue as symptoms of fear. However, in the absence of evidence about the response of other deer to these behaviors, their communicative significance remains uncertain. A similar situation exists with head bobbing, which Linsdale and Tomich describe as a typical response to an unidentified object or motion. These authors also state that bristling of the long hairs of the tail and edge of the rump patch "conveys a message" but qualify this opinion by pointing out that it often "occurs along with other accompaniments of excitement" (p. 102). In any case, caudal bristling is strongly associated with retreat from humans and other potentially dangerous situations. Some other customary visual accompaniments of mild alarm are a clamped-down or rapidly rotating tail, an erect stance, and a stiff-legged gait (Linsdale & Tomich, 1953).

The snort of mule deer and whitetail deer usually sounds like a cough but sometimes has a hissing or whistling component (Linsdale & Tomich, 1953). It is characteristic of low levels of alarm and often precedes slow retreat or cautious investigation. Deer sometimes snort every few minutes over a period of half an hour. Snorts are quite loud and thus often cause a startle response in concealed predators (or ethologists), thereby giving away their location. According to Linsdale and Tomich, common responses to snorts are looking in the direction of the snort, looking around, particularly in the direction of the snort, and, in the case of fawns, freezing followed by slow withdrawal. These reactions have been observed in deer almost 200 meters away from the snort. Snorting is sometimes accompanied by forefoot stamping, tail switching, and pacing. There is no evidence that these other actions by themselves affect other deer.

Mule deer *in extremis*, as when wounded, scream at high volume and pitch. According to Linsdale and Tomich, the sound can carry for more than 3 kilometers. There seems to be no special response to the distress scream, but it may serve to frighten an attacker or, in some cases, to attract other deer.

At high levels of alarm, a lying deer rises suddenly. Standing or walking deer bound rapidly away with a bouncing gait called *stotting* or spronking. Deer can travel quite quickly in this gait, but its major function seems to be to alert other deer. The vertical motion, especially of the white rump patch, is a striking visual signal that often evokes flight from any deer in a position to see it. Since whitetails usually bound through vegetation, they are not often seen to stott as strikingly as mule deer. Stotting also has an auditory component. At each stride a stotting deer lands with all four legs stiffened, producing a thud that under optimum conditions can be heard by a human well over 200 meters away. Müller-Schwarze (1971) reports that "only deep footprints of deer escaping from a dog, sheep, or other deer are thoroughly investigated" (p. 151) by other deer. This observation suggests that either the activity of interdigital glands increases during states of alarm or that stotting applies these odors to the ground more effectively than walking or running.

There are three other olfactory forms expressing high levels of alarm. The most important of these is discharge of tarsal scent, which, as already mentioned, can affect deer 20 meters away. I have detected tarsal scent released by a whitetail several minutes after it fled. Nichol (1938) observed deer of both sexes discharge tarsal scent by flaring the tarsal tuft into a rosette in response to dogs or other frightening stimuli. Linsdale and Tomich (1953) detected the characteristic dry, musky odor of tarsal glands as a fawn retreated from a truck and saw flaring of tufts on many occasions when deer were approached by a human. Apart from Nichol's statement that maternal tarsal scent "warns and calls" (p. 36) her fawns, we have no evidence about response to alarm-induced tarsal discharge.

The metatarsal glands on the outside of the hind legs just below the middle joint have been described in detail by Quay and Müller-Schwarze (1970). These glands, found in both mule deer and whitetails, are elongated concentrations of sebaceous and sweat, or sudoriferous, glands covered by a grooved, oval tuft of long hairs similar in color to the surrounding *pelage*. Müller-Schwarze (1971) has investigated stimulus and response contingencies relevant to the alarm function of metatarsal odor. He detected metatarsal odor, which he finds similar to that of garlic, when deer encountered a dog or were chased, cornered, or placed in an unfamiliar environment. To separate the effects of metatarsal odor from those of other alarm reactions, he applied metatarsal scent to a caretaker's hand and found that it inhibited licking the hand, the deer's customary greeting. He also found that metatarsal scent under a food bowl halved the frequency of eating from the bowl. Similar results were obtained using samples of the major chemical components of metatarsal-gland extract.

Müller-Schwarze found another source of alarm-induced odor in the enlarged sebaceous and sudoriferous glands of the tail. When a deer is excited, it is likely to sniff its own tail or that of an associate, and on such occasions an "almost fecal odor" (p. 151) can be detected in the tails of fawns and adult males.

Any salient and lasting olfactory stimulus may help a deer to recognize a location. The stimulus may be an odor from the deer's body, an odor produced by another deer's body, or an odor released by destruction of plant tissues. Recognition of a location may have several effects: attraction or aversion to the site or its environs, an increase or decrease in confidence, and possibly choice or avoidance of a route. The thrashes, rubs, scrapes, and urinations described in the next section as dominance displays are especially likely to operate in these ways. They are salient, interesting, and concentrated at nodes such as trail junctions and communal water holes, where they are likely to be encountered. De Vos and his associates (1967) and Linsdale and Tomich (1953) report consistent tendencies for whitetails and mule deer, respectively, to rub and thrash the same trees year after year. Kile and Marchinton (1977) found that whitetails prefer to rub species of trees with highly aromatic sap and wood. These preferences should increase the trees' value as landmarks even though the primary function of rubbing and thrashing is sexual advertisement. Tonically deposited odors from urine, feces, and interdigital glands also probably familiarize individuals with their environment.

According to Müller-Schwarze (1971), the sleeping site is "the most important fixed point in the home range of a blacktailed deer" (p. 151). It is a focus for forehead rubbing and antler thrashing. Both these behaviors leave lasting odors, either from sudoriferous glands on the forehead or from frayed plant tissues. Fawns frequently urinate on their beds. Müller-Schwarze claims that they will stop using their beds if the urine is disturbed or removed. Another source of odor in the bed is the metatarsal glands. According to Linsdale and Tomich, these sebaceous complexes are located so that one always touches the ground wherever a deer lies. They observed a deer sniffing a bed, possibly testing for the scent of these glands. These behaviors suggest a familiarization function for these various odors.

AGONISTIC MESSAGES

In *Odocoileus*, strife occurs mainly among males in rut, but it also occurs among females, especially in spring and early summer, when they are most intolerant of their yearlings. Even in encounters between rutting males in the presence of females, however, conflict is not inevitable. On several occasions Linsdale and Tomich (1953) saw one large buck allow another to approach a group of does and leave with one of them. Pugnacious harem defense like that of wapiti has not been reliably reported in *Odocoileus*, claims of sports magazines notwithstanding.

When conflict does arise, fighting sometimes occurs without preliminaries. Often, however, dominance displays precede fight or flight. When dominance is not clearly established, escalating mutual threats ensue and usually culminate in retreat by one of the antagonists. Ozoga (1972) found that among Michigan whitetails 58% of all aggressive encounters were without physical contact. Linsdale and Tomich report that most mule deer strife is "merely a bluff" (p. 506). Special submissive signals may occasionally avert attack, but the most common form of appeasement is withdrawal.

Although members of the genus *Odocoileus* are not territorial, they display three forms of marking, some of which have been viewed by Moore and Marchinton (1974) as "suggestive of territoriality" (p. 454). Forehead and antler rubs, thrashing, and scraping are, however, primarily expressions of dominance (Kucera, 1978), even though the marks produced by these displays may convey their message after the transmitter has left.

Müller-Schwarze (1972) has investigated forehead rubbing in captive blacktails. He often observed a buck or doe approach a low, dry branch, sniff the branch, rub its forehead on it with predominantly vertical movements, then sniff it again. Bucks often repeat the sequence, especially during the breeding season. Forehead rubbing is often associated with similar rubbing of the burrs or irregular protuberances, at the base of each antler, but this rubbing is usually directed at a larger limb or trunk (Linsdale & Tomich, 1953). Moore and

Marchinton describe a highly similar behavior in free-ranging white-tail bucks. Like forehead rubbing, antler rubbing alternates with sniffing the rubbed area. Moore and Marchinton's illustration of this behavior shows the forehead in contact with the branch being rubbed. Hair, possibly removed from the forehead by friction, is often found in bark rubbed by whitetails (De Vos et al., 1967).

In mule deer forehead rubbing frequently escalates into thrashing in which the antlers are violently pushed, twisted, and whipped against branches. This behavior, too, alternates with sniffing, and sometimes licking, the thrashed branches (Graf, 1956). Thrashing by a whitetail buck observed in Michigan by Pruitt (1954) was accompanied by "low grunts" (p. 129) and by scraping the ground with the forefeet. The sniffing associated with production of these closely related marks suggests that the behavior has olfactory as well as visual consequences. Geist in a personal communication to Walther (1977) noted that the sounds of thrashing often caused subordinate or young bucks to withdraw, so thrashing is sometimes an auditory signal as well.

In captive blacktails, adult males, which generally dominate females, rub eight times more frequently than females. Frequency of rubbing by desert mule deer correlates positively with dominance rank based on frequency of winning in agonistic encounters (Kucera, 1978; Müller-Schwarze, 1972). Moore and Marchinton (1974) report that in agonistic encounters between whitetail bucks it is the dominant that rubs his antlers. All but one of the cases of antler thrashing reported in mule deer by Linsdale and Tomich (1953) occurred when the thrasher was alone. In the single exception a large buck thrashed upon encountering a yearling female, which conceded his dominance by withdrawing. Mule deer have not been observed to scrape, but all of the scraping adult whitetails whose rank Moore and Marchinton could determine were dominants, at least in the area where they scraped. These observations suggest that these marks and displays are expressions of dominance.

The spatial distribution of these marks maximizes their efficiency as signals. All three types of marks are spatially clumped. In blacktails forehead rubs and antler thrashes occur near beds and at communal trails and water holes (Müller-Schwarze, 1972). Moore and Marchinton found that whitetails periodically return to scrapes and scrape again. They found rubs and scrapes concentrated at trail junctions, along roadbeds, and along the edges of open areas. There was no evident relationship between the locations of clusters of rubs and scrapes and the boundaries of a buck's range.

Temporal patterns also suggest that these behaviors are acts of communication. Forehead rubbing by blacktails occurs throughout the year. Males, however, rub more frequently during the rut, when dominance really matters, than they do in the spring, when access to does is not an issue. During the fall breeding season, bucks significantly increase their rate of rubbing after rain, as though replacing odors that have been washed away (Müller-Schwarze, 1972). Antler rubbing by mule deer and whitetails is similarly associated with the rut (Linsdale & Tomich, 1953).

The best evidence about the meaning of a message is the response of the receiver. Müller-Schwarze noted that when branches rubbed by resident blacktails were sniffed by newcomers, the newcomers exhibited "escape behavior." Moore and Marchinton (1974) also observed fearful responses in males that sniffed another's scrape. Müller-Schwarze also found that sites rubbed exclusively by high-ranking males were sniffed mainly by their "owners" and by low-ranking males. High-ranking bucks only rarely sniffed the rubbing sites that they shared with bucks of lower rank. Rubs and scrapes are frequently examined by females, especially during the breeding season.

Like bugling in elk and raised-leg urination in canids, these markings probably act as male sexual advertisements as well as dominance displays. They are often associated with one or more of five other dominance displays: rub-urination, head tossing, charging, foreleg striking, and mounting. The rub-urination of adults resembles that of fawns, but when performed by the former, it intimidates rather than attracts. When using rub-urination as a threat, the transmitter sometimes turns away, bringing the tarsal glands closer to the receiver's nose. Rub-urination in aggressive encounters is performed by adult blacktails of either sex (Müller-Schwarze, 1971) and by rutting whitetail bucks (Haugen, 1959). In blacktails the frequency of rub-urination increases during the rut, when aggressive encounters are most common. Moore and Marchinton hypothesize that rub-urination acts in similar ways in whitetails and mule deer; that is, it intimidates opponents. Several of the rub-urinations reported by Linsdale and Tomich (1953) were followed by withdrawal by the threatened deer. Nichol (1938) thought that the least fearful of an excited group of mule deer rub-urinated most often, but Linsdale and Tomich were unable to substantiate this opinion.

Kucera (1978), in his investigation of free-ranging Texas mule deer, found high correlations between antler size and social dominance. He found that antler size correlated with frequency of thrashing, rubbing, chasing, rub-urinating, and head tossing. The last display was often used by dominant bucks to prevent a subordinate from approaching a doe.

Charging is a very direct and clear display of dominance. In mule deer, charging invariably leads to retreat by the animal at which the charge is directed (Geist, 1971). The charger, which may be male or female, usually holds its bristled tail vertically, but sometimes it is held stiffly to the side. Sometimes the neck is held low, and usually the ears are laid back. Charges usually cover only 1 or 2 meters, but sometimes they turn into extended pursuits. The charger often increases the ferocity of its rush by bellowing (Linsdale & Tomich, 1953). This vocalization is sometimes also described as a bawl or a groan.

Does commonly assert their dominance over younger deer by striking them with a foreleg to drive them away. This tactic does not seem to intimidate them severely, for although they withdraw, they usually stay fairly close and sometimes return (Müller-Schwarze, 1971).

Müller-Schwarze mentions mounting as an aggressive behavior pattern in captive blacktails. This behavior has not been described as an aggressive act in *Odocoileus* elsewhere, but in most mammals in which it occurs in agonistic situations, mounting establishes the mounter's dominance.

There are 10 forms of agonistic display that do not generally establish dominance. Some of these threats operate in more than one sensory mode. In mule deer some threats typically occur early in an encounter. In whitetails Thomas, Robinson, and Marburger (1965) ranked visual threats in order of typical occurrence and increasing intensity. In both species, however, encounters may begin with any threat or other form of agonism.

Three threats that usually occur early in encounters among mule deer are tarsal sniffing, rub-urination, already discussed, and lateral display. Tarsal sniffing was the first interaction in two of three agonistic sequences described by Müller-Schwarze (1971). According to him, tarsal sniffing is "functionally and temporally" (p. 142) associated with chasing, striking, and other acts of aggression.

The third threat, lateral (broadside) display, usually occurs near the beginning of an aggressive sequence. Both mule and whitetail deer stand in front of an opponent and present their largest aspect (De Vos et al., 1967). Sometimes the display is mutual and becomes the sidle described later. Kucera (1978) found that in a typical agonistic encounter both opponents would thrash vegetation and then approach each other in a head-low crouch, with ears pinned back and their inner surfaces turned out. This posture is also described by Cowan and Geist (1961). The deer circle each other in this posture, and then one may turn away, ending the interaction. Sometimes, however, there is a sudden rush with lowered antlers.

Nose licking, snorting, and stamping may occur at any point in an agonistic sequence. These threats have been described in *O. hemionus* by Cowan and Geist. Front and rear legs are slightly flexed, the rear legs more so than the front. Muscles are tense, and hair all over the back bristles. The walk is slow and stilted. The tail quivers, and the whites of the eyes are partially exposed. The tongue flicks in and out constantly, each time from the alternate side but always contacting the nose. The impression is of pent-up energy about to explode. Soon there is an explosion in the form of a convulsive, nasal snort. Threatening snorts sometimes hiss and sometimes whistle, but they are always explosive. They are also more sibilant than alarm snorts (Bowyer, 1978). According to Cowan and Geist, deer snort by forcing air through closed nostrils. As they snort, they bulge their neck muscles, arch their backs, bristle their hair, and stiffen their tails. Snorting is often performed during lateral threat and is sometimes associated with flaring of the openings of the preorbital glands. De Vos and his colleagues state, however, that rutting whitetails do not snort or lick their noses. Moreover, stamping both forefeet is a rare form of threat in mule deer. Linsdale and Tomich (1953) observed it on only four occasions. In every case, stamping by an adult caused a fawn to withdraw rapidly.

Thomas and his coworkers (1965) describe a temporal sequence of threats of increasing intensity in whitetail bucks. Sometimes threats that usually occur early in the sequence are omitted, but an earlier, less intense threat is eardropping, in which the ears are laid back along the neck. This display is also described as a threat in mule deer by Linsdale and Tomich. Bowyer suggests that ear dropping signifies an intention to rear and flail. Linsdale and Tomich claim that dominants lower their ears more than subordinates, but I have not been able to find any support for this statement either in print or in the field. If ear dropping does not inspire withdrawal, the threatener goes on to the "hard look." He lowers and extends his head and neck, drops his ears, and stares fixedly at his opponent for 3 to 10 seconds. Sometimes his hair is bristled slightly. Geist (1971) mentions a similar threat in mule deer and states that it is used mainly by inferiors, indicating that it may be defensive. The opponent may retreat or reply in kind, in which case a third level of threat is reached as both antagonists begin to sidle. With heads erect, necks vertical, and bodies at about 30° to each other, they slowly approach and slowly circle each other.

The fourth level of male threat is termed antler threat. The head is lowered, directing the tines of the antlers at the other deer. This threat is common to all antlered deer, but antler threats are not used by whitetails in velvet (De Vos et al., 1967). Antler threats may be derived from charging and, if so, can be considered a charging intention.

Sometimes fights are interrupted by "profiling" (Kucera, 1978), in which the head is turned sideways, as though to display the antlers. Kucera found no relationship between dominance and profiling. Profiling was always performed by both opponents.

Aggressive whitetail does and fawns sometimes use ear dropping and sidling as mild threats. An intense form of threat for antlerless deer is a head-high posture, which often precedes or accompanies foreleg striking. Their most intense form of threat is flailing, in which they rear on their hind legs, striking out with both forelegs in a manner similar to the boxing of some other mammals. Thomas and his colleagues found that contact was rare during flailing. Rearing and flailing without contact are common threats made by antlerless mule deer (Linsdale & Tomich, 1953).

Adolescent, or spike, bucks use striking, flailing, and antler threat interchangeably, according to Thomas and his coauthors. These investigators describe staring as a threat. Geist in a personal communication to Walther (1977) found that it was typically used by inferiors addressing superiors. Thus, staring is best classified as a defensive threat.

Mule deer usually signify their lack of aggression toward a dominating or threatening conspecific by withdrawing. Sometimes the withdrawal is complete, and the submitting deer leaves the vicinity. More often, however, the withdrawal is token, and the animal walks one or two steps away, avoiding any semblance of a threatening posture.

Moore and Marchinton (1974) state that whitetail bucks can avoid attack by maintaining "subordinate postures" but do not describe these postures. Hirth and McCullough (1977) found that submitting whitetails tucked their tails down over the *perineum*, covering their rump patches. As these authors point out, this posture is to be expected on the basis of the principle of antithesis, since the tail is elevated in threat. Guthrie (1971) argues that display of the rump patch is a submissive signal in deer, as it is in bighorn sheep (*Ovis canadensis*). According to Hirth and McCullough, the evidence is against this hypothesis, since dominance mounting does not occur in *Odocoileus*. Their opinion must be questioned in the light of Müller-Schwarze's (1972) mention of dominance mounting in captive blacktails. Nevertheless, Hirth and McCullough argue convincingly that rump patches

*Two Père David's deer rear and flail. Photograph by
Leland LaFrance, courtesy of the Brookfield Zoo.*

in cervids function mainly to promote group cohesion, not to invite
dominance mounting. Hirth (1973) observed that subordinate whitetails greeted dominants by approaching and touching noses. The use
of this gesture as an appeasement in wapiti suggests the possibility that
it performs an analogous function in deer.

Combat between antlered deer can be mild or intense. In mild
combat two bucks face each other, lower their heads so that the antlers
point toward each other, and gently ease their heads together and
push, sometimes with twisting movements. Mutual pushing continues
until one retires or both pause for rest, in which case fighting often
resumes after a few minutes (Kucera, 1978; Linsdale & Tomich, 1953).

High-intensity fighting takes the form of a rush in which the
antlers are lowered as in threat and the opponents run directly at each
other. When their antlers meet, the impact can be heard through

heavy timber more than 300 meters away. Linsdale and Tomich saw no fatal injuries but were sure that they did occur. When Cowan and Geist (1961) were rushed, tines went right through their plywood shield.

These authors saw no differences in the rushes made by mule, blacktail, and whitetail deer. Charging deer often hold their bristling, stiffened tails up or to the side. Mule deer sometimes make grunting vocalizations that Linsdale and Tomich transcribed as "hu-hu-hu-hu," "hoo-pa-pa-pa-pa-pa," or "pa-pa-pa-pa-pa." These observers saw many mutual rushes culminating in prolonged pushing bouts, each buck twisting his neck in an attempt (sometimes successfully) to strike the other's neck or flank. Several such bouts are linked by interruptions for looking around, grazing, or resting. A series of bouts may last half an hour. Kucera (1978), on the other hand, found that fights lasted less than 2 minutes and were terminated by flight by one of the combatants.

Spikes replace the rush with a gentle engagement of the heads—perhaps because the stakes are not high, since spikes do not fight over estrous does (Bowyer, 1978). Fighting by does and young bucks is not as serious as that by mature males, but antlerless mule and whitetail deer sometimes contact each other as they rear and flail in threat. Linsdale and Tomich describe several such combats in which blows were parried and others in which one opponent was knocked to the ground. Thomas and his colleagues (1965) saw a doe knock a fawn to the ground by striking it on the back with both forefeet. Small bucks sometimes incite attack by touching larger bucks with their antlers and then running off. It seemed to Linsdale and Tomich that large bucks did not take such challenges very seriously.

SEXUAL MESSAGES

Odocoileus does come into estrus at approximately 4-week intervals (Cheatum & Morton, 1946, cited in Haugen, 1959; Linsdale & Tomich, 1953). Thus, during the fall breeding season, a doe will have at most two or three periods during which she can conceive. Each of these periods lasts only a few hours, so it is essential that a female have at least one male in attendance when her time comes. She accomplishes this mainly by the use of scented urine, which attracts males. Courtship begins when a rutting buck finds a doe. It ends with copulation and mutual withdrawal or withdrawal and defensive actions by the doe. Courtship patterns are very similar in mule and whitetail deer (De Vos et al., 1967). In neither species is there harem herding, as in wapiti, but in both species, males become intolerant of one another and "tend" estrous females. Unreceptive does allow males to approach but then jump away when the males begin to sniff at their rumps. Such retreats give rise to short chases, which become longer as the rut progresses. In these later chases, females are more wary and males more circumspect. For example, in later encounters bucks announce their intention to approach a doe by sharply lowering their heads as they step forward. They continue their approach with their heads held horizontally, often bobbing them up and down (Geist, 1966). According to Linsdale and Tomich (1953), the head-bobbing approach is directed only toward females.

By the time a male blacktail is 3 months old, he is already attracted to urinating females. When a female urinates, he stops whatever he is doing, walks over, and sniffs the urine (Müller-Schwarze, 1971). Moore and Marchinton (1974) noted that whitetail does often approached male scrapes and urinated on them. On many occasions these authors found small (doe-sized) tracks in scrapes. On seven occasions a buck detected female scent at a scrape and began making the grunts often associated with sexual pursuit. Does were always quickly found by the buck, usually within 200 meters of the scrape. Moore and Marchinton concluded that scrapes attract females, which urinate on them, alerting bucks to their presence. Bucks refrain from urinating on scrapes, making female urine easy to detect.

Linsdale and Tomich (1953) mention elevation and wrinkling of the muzzle as a common male response to fresh female urine. This flehmen response was described in detail in the section on wapiti sexual advertisement. It probably brings odors of estrous urine into the vomeronasal organ (Estes, 1972). Dagg and Taub (1970) state that flehmen occurs in all male ungulates except the pig, so it can be concluded that whitetail bucks also flehm. Of the flehmens recorded by Kucera (1978), 72% were during or immediately after urination by a doe. Does did not flehm.

Forehead and antler rubbing and thrashing have been discussed as threats, but, as is often the case, these threats act also as male sexual advertisements. Males rub most frequently during and around the rut. They rub much more frequently than females and sniff rubs no more frequently in the fall than in the spring. Females, on the other hand, sniff rubs more than twice as frequently in the fall as in the spring (Linsdale & Tomich, 1953). These relationships suggest that during the breeding season females are important receivers of messages encoded in rubs.

Antler rubbing and thrashing are formally and functionally related to forehead rubbing, which is also associated with the rut. No information is available on female responses to antler rubs or thrashes, but these behaviors do leave conspicuous signs of the presence of rutting males. Thus, females may make use of these marks as they seek a mate.

According to Seton (1929), whitetail bucks wallow like wapiti during the rut. The communicative significance of this behavior, if there is any, may be similar to that conjectured for elk—it may have an attractive odor, it may cloak the wallower in his own urine, advertising his physical condition, or it may render the animal more threatening by darkening his coloration. Urine, especially on the tarsal glands, is probably a major source of the attractive odor. As mentioned earlier, deer can discriminate the sex of a donor of tarsal scent. This odor is powerful enough to be detected by a human minutes after a deer has passed.

Bucks often rub-urinate just before approaching a doe. Moore and Marchinton (1974) suggest that this behavior, like rubbing and thrashing, acts as both a threat and a sexual attractant.

Bucks sometimes resort to the ancient mammalian strategy of infantile mimicry. They frequently bleat like fawns to attract does. Sometimes one approaches a doe from the rear and bunts her udder. Bowyer (1978) points out that does do not accept fawns as mates, so those tactics may act mainly to prevent the doe's withdrawal. These tactics, reported by Linsdale and Tomich (1953), have not been described elsewhere.

A more straightforward confession of sexual attraction is a series of low grunts, uttered as the male trots after the female. Sometimes grunts are uttered in rapid succession, creating a "low, loud bellow" (Linsdale & Tomich, 1953, p. 490). Browman and Hudson (1957) report lip smacking as an accompaniment of close pursuit. Kucera (1978) heard bucks bellow, grunt, moan, and click as they followed does. He noted that tending bucks were usually dominants, that a buck followed only one doe at a time, and that the buck never attempted to influence the direction of travel of doe groups. Kucera observed that tending bucks assumed a special posture, with the head lowered to shoulder level, the muzzle extended forward, the ears laid back against the head, and the tongue flicking in and out. The buck would lick and nuzzle the female's perineum whenever she stopped. Tending continued until copulation occurred or until the buck was supplanted by another of higher rank. The time of copulation was determined by the doe's willingness to stand for a mount.

The following description of a "typical" copulatory sequence is based on several accounts by Linsdale and Tomich. Such sequences are often incomplete, but they typically begin with the female standing rigidly, facing away from the male with her back arched and tail bristled and sticking out obliquely to the side, exposing the vulva. The buck approaches from behind and sniffs or nudges her tail and vulva with his muzzle. If she does not move away, he may test her readiness by striking her with a foreleg. If she responds to this blow by moving, the buck follows until she stops, and the sequence is repeated. Sometimes the pursuit is interrupted as the couple assumes a mounting posture but with roles reversed. According to Walther (1977), however, the foreleg kick does not occur in cervid courtship. I have been able to find no descriptions of such kicks except by Linsdale and Tomich.

Finally the doe stands with tail averted, and the male mounts. Occasionally she displays lordosis with rump raised, shoulders low, and forelegs stretched out at an angle of about 60° to the ground. She is usually mounted immediately after adopting this posture, which evidently expresses the height of sexual receptivity.

Kucera (1978) counted multiple mounts in the four copulations he observed. The number of mounts varied from three to six. Golley (1957) obtained similar figures from captive blacktails. A complete mount usually lasts less than 5 seconds and consists of a few slow thrusts followed by a quick, violent one. Haugen (1959) observed that this last thrust usually knocked the doe to her knees. It is assumed that the violent thrust is a concomitant of ejaculation. Haugen noted that it was sometimes associated with a flashing erection of the male's rump patch but does not describe any particular response to this action. He found that when such thrusts occurred, does arched their backs briefly. His whitetail does were sometimes serviced twice in 15 minutes. Linsdale and Tomich (1953) also report multiple mounts but did not obtain evidence of ejaculation. After copulation the pair sometimes seem to lose all interest in each other. On other occasions the doe moves off with the buck 10–25 meters behind. Kucera did not see repeated copulations.

SUMMARY

Like the message system of the wapiti, the message system of mule deer is characterized by large numbers of message forms, but the number of types, 20, is not much larger than that of most of the other mammals already discussed. The message system of the deer includes neonatal tonic and phasic interactions and several signals: identity of the mother and infant by odor, affiliative grooming, and imprinting; maternal alarms and infantile distress; and contact and submission. Integrative messages include several forms of play, visual and olfactory contact, solicitation and use of affiliative licking and grooming, olfactory identification, an alarm gait, a distress cry, and environmental familiarization by means of visual and olfactory marks. Agonistic messages include eight kinds of dominance displays, 15 offensive threats and one defensive threat, two forms of submission, and ritualized fighting. Sexual messages include male and female advertisements and courtship.

6
VERVET
MONKEYS

The monkeys winked too much
and were afraid of snakes.

MARIANNE MOORE

161

Vervets *(Cercopithecus aethiops)* are probably the most abundant species of monkeys in the world (Struhsaker, 1967e). They are equally at home in the trees and on the ground. Thus, their behavior may shed light on our evolutionary genesis, when our ancient ancestors left their arboreal Eden.

The vervet is exquisitely caparisoned with a black, hairless face set off by a yellow, furry ruff around its upper half. The eyelids are bright pink, and the ears are black and humanoid. Except for the face, ears, palms, and soles, the entire body is covered with fur. This fur is a grayed brownish-yellow, lightening toward the extremities and almost white on the ventral surface and the inside of the arms and legs. The tail sometimes has an orange tip.

Both hands and feet are *prehensile,* or adapted for grasping, but the tail is not. It is used mainly for balance and support. Leathery gray pads, called ischiadic callosities, are located on the posterior, where they provide protection and comfort while the monkey sits.

Vervets display moderate sexual dimorphism. Adult males are considerably larger and have broader shoulders than adult females. Males weigh about 7 kilograms, females about 5 kilograms. When standing erect, they are about .7 to .9 meters tall. Males have bright-red penises and powder-blue scrota, outlined by white perianal pelage. Females have dark, pendulous teats located close together at the center of their chests. The vervet's bright coloration is probably socially significant. Since vervets are *primates,* they presumably have excellent color vision.

Vervets walk with a diagonal gait, moving forelimbs and hind legs on opposite sides in unison. They gallop with forelimbs together, meeting the hind legs on every stride. Especially when traveling in open terrain, they frequently stand erect and look around (Struhsaker, 1967d). Observations by McGuire (1971) showed that vervet groups spend between 30% and 90% of their time in the trees, depending on the terrain and vegetation. Struhsaker often saw them jump from

162

A male vervet monkey. Drawing by Heidi Reynolds.

trees to the ground, sometimes from as far as 10 meters up without apparent injury.

Vervets live throughout the savannahs of Africa, especially along rivers. In central Africa they also abound in other environments from rain forest to dry brushland. They seem to prefer open spaces near water that are sparsely vegetated with low brush (Brain, 1965; Hill, 1966). This preferred terrain provides opportunities for both concealment and reconnaissance.

Vervets are opportunistic *omnivores*. Although they subsist mainly on plants, they often eat invertebrates, and they occasionally eat small vertebrates. In the Amboseli Reserve of Kenya, Struhsaker (1967e) saw them eat insects, snails, and birds and their eggs. Even when hungry, vervets fill their cheek pouches before they swallow. According to Brain, this behavior is an adaptation to competition for limited quantities of food. Vervets stuff their cheek pouches even when alone, so this habit may also allow an individual to obtain a supply of food quickly and then retreat to trees or cover where it is safe from predators.

There are probably at least 16 species of these predators. Baboons are a threat to vervets when there are no trees nearby, although young baboons and vervets commonly play together (S. A. Altman & J. Altman, 1970). Other mammalian predators include cheetahs, leopards and several other cats, hyenas, jackals, and Cape hunting dogs. At least three birds prey on vervets: the martial eagle, Verreau's eagle owl, and the crowned hawk eagle. Cobras and puff adders are also a serious threat, for these snakes can follow vervets into the trees. Humans also take vervets for food, fur, and research.

The rich social life of the vervets revolves around stable heterosexual groups called troops. Membership in a troop facilitates food finding, provides protection from predators, and, for some, a chance to reproduce. Struhsaker (1967d) found that troops ranged in size from seven to 53 individuals, with a mean of 24. In Zambia, Lancaster (1971) found that a troop of 55 contained seven adult males, 15 adult females, one subadult (less than 4 years old) male, three subadult females, 15 juveniles, and 14 infants. Studies on vervets in other areas indicate that these proportions of sex and age classes are typical.

Troops are organized into male and female dominance hierarchies. Ranking is determined by fighting. Males are larger than females, but Lancaster observed instances in which a female dominated some adult males, and Rowell (1971) found that females generally dominated males. Whatever their sex, subordinates wait while dominants feed, but dominants often feed peacefully side by side. A juvenile's rank is just below its mother's. The troop is led by the dominant, or alpha, male, especially during encounters with other troops. Adult males may sometimes jockey for the alpha position, but control is generally maintained by one male that has defeated all contenders (Struhsaker 1967e). There seems to be considerable variation, however, in the degree of dominance displayed by different vervet populations. Gartlan (1969), for example, found no fighting or dominance among vervets in Uganda, but he did note that roles are differentiated on the basis of age and sex.

A troop travels as a unit within a well-defined area in which they feed, breed, and rear young. DeMore and Steffens (1973) radio-tracked vervets in the Nsumu Game Reserve of South Africa. They concluded that troop ranges varied from .2 to .8 square kilometers. The sizes of these ranges did not seem to depend on the size of the troop but did seem to vary with the quality of habitat. Ranges in areas where appropriate vegetation for eating, sleeping, and concealment was plentiful were smaller than ranges in areas where these resources were less abundant. Similar correlations were obtained by Struhsaker (1967e) in Kenya and by McGuire (1974) on the Caribbean island of St. Kitts.

Although some overlap of home ranges was noted in South Africa, Kenya, and on St. Kitts, there seems to be considerable variability in the degree of tolerance among troops in different areas. Gartlan (1969) found no territoriality among vervets in Uganda. In fact, males frequently moved from one group to another. DeMore and Steffens maintain that intergroup tolerance in South African vervets varies with the season. During the dry season, when food is scarce, ranges overlap considerably, but in the rainy season, when food is plentiful, boundaries become distinct. And, finally, Struhsaker observed that the vervets of the Amboseli Reserve in Kenya are territorial throughout the year. As suggested by E. O. Wilson (1975), these differences may be due to genetic differences among geographically distinct populations. Alternatively, as suggested by the seasonal variation found by DeMore and Steffens, they may be due to ecological factors.

Struhsaker (1967d), from whose work much of this chapter is drawn, found that larger troops dominated smaller ones and forced them into less desirable habitats. Both males and females defend the troop territory, but both Poirier (1972) and Gartlan noted that males played a more active role.

Since Amboseli vervets must rear *altricial* young (that is, young that need extensive parental care), establish and maintain dominance relationships both within and between troops, avoid several different kinds of predators, and maintain the integrity of the troop in a variety of *biomes*, or regional types, we should expect their message system to be quite complex. As we shall see, this expectation is borne out by field investigations of their social behavior.

NEONATAL COMMUNICATION

Struhsaker (1967e) saw newborn vervets in Amboseli only during the rainy season, which lasts from October through March. He implies that birth during this period is designed to capitalize on the abundance of food available then. Probably because much of a vervet's safety lies in numbers, a female about to give birth stays close to her fellow troop members. According to Gartlan (1969), birth usually takes place at night or early in the morning. Vervets always give birth to a single infant. Once it is born, the mother eats the placenta, cleans the infant, and holds it close to her abdomen for the first few hours. The infant is able to cling to its mother's belly immediately after birth, even before the umbilical cord is broken. Newborn vervets continue to cling while their mothers walk, run, or climb. At first a mother that is about to move around scoops up her infant and presses it to her belly. Later she need only touch its back to get it to cling. At this stage it may sometimes ride on her back.

Should the infant become separated from its mother, it emits one or more of five distinctive vocalizations (Struhsaker, 1967a). If the distance between mother and infant is less than a few meters, the infant gives the "lost rrah" call. At greater distances the infant alternates "lost rrr" and "lost eee" calls. The latter two cries are also sometimes used independently of each other. Both are considerably louder than the "lost rrah." "Lost eee" calls are single, long, high-pitched screams, and "lost rrr" calls (termed *gargles* by Brain, 1965) are composed of several shorter and lower tones uttered in rapid succession. When uttered alternately, "rrr" and "eee" are very easy to localize because of their combination of high and low frequencies combined with sharp interruptions. Two other calls produced by separated infants are the high-pitched "lost squeal" and the even higher "weaning scream," both of which show a pitch variation shaped like an inverted "u." The latter vocalization is often used to express distress at the approach of an unfamiliar adult male. Each of these calls evokes rapid retrieval of the infant by its mother. Struhsaker found that when a mother retrieves a calling infant, it is always her own, suggesting that she can recognize her infant's voice. When reunited with its mother, the infant sometimes emits the short, nontonal "eh-eh" call. Struhsaker suggests that this call may strengthen the bond between mother and infant. Presumably, it works by expressing the infant's satisfaction at retrieval, thereby reinforcing the mother.

Upon detecting members of another troop, infants give the "wawoo" call. This call appears to be analogous to the "aarr" that adults use to announce the presence of another troop.

Young vervets continue to nurse until they are at least 9 months old. The mother's nipples are so close together that the infant usually sucks them both at the same time. Weaning is a long process. The mother may begin to resist her infant's attempts to nurse when it is only 3 weeks old. Infants produce three different calls during weaning, each expressing a different level of distress, frustration, or solicitation. At the lowest level of intensity, as when the mother is simply inaccessible, infants give the "weaning rrr" call, similar to the "lost rrr." When the mother is accessible but frustrates the infant's attempts to nurse, it gives a "weaning squeal" very similar to the "lost squeal." At higher levels of frustration, as when being nipped or pushed away, infants give the "weaning scream," which is very much like the scream given at the approach of a strange male. Since the calls given during separation and weaning are so similar, the three classes of vocalizations can be regarded as distress calls at three levels of intensity. In his description of these calls, Struhsaker (1967a) distinguished between similar signals when the contexts were different. He noted that when the calls were a response to separation, they elicited approach by the mother, but when they were a response to weaning behaviors, they elicited no particular maternal action.

As the infant grows older, the mother spends less and less time holding it. During the first week the infant is held up to 35% of the time, but by the fifth week this figure has been cut in half. As time spent with the mother decreases, time spent with other females, especially older siblings, increases proportionately (Chalmers, 1972). Sometimes the mother encourages such contacts by depositing the infant with another female before leaving to forage. Such baby-sitting chores are evidently quite attractive, for females without infants of their own compete for babies to sit with. The baby-sitters, or "aunties," display several forms of such *alloparental behavior*. When an infant squeals in distress, all the females in the troop may rush to its aid and may even be joined by males, which ordinarily pay little attention to infants. Similar protective behavior is also displayed by females as young as 3 months old. The aunties handle, groom, carry, and feed the infants entrusted to their care and, rarely, may even try to nurse them. In fact, Struhsaker noted that immature females groomed infants more than the infants' mothers did. Similarly, immature females were more likely than mothers to hug infants while carrying them. Even males sometimes showed interest in infants by what Struhsaker termed muzzling. The males muzzled the infants all over, while mothers concentrated their muzzling around the infant's perineum. Struhsaker speculates that this maternal muzzling may facilitate olfactory recognition of infants.

Alloparental care increases the cohesion of the troop by providing a common focus of attention. It may also play a role in the development of maternal skill. It is confined to young females without young of their own. Their performance in tasks like carrying displays considerable improvement as they care for others' offspring. Some of this improvement may be due to practice (Brain, 1965; Struhsaker, 1967d; E. O. Wilson, 1975).

The end of infancy is marked by striking changes in appearance. At about 3 months of age, the infant's pink face blackens, and the furry ruff above the face lightens. The emergence of the resulting contrast between face and ruff signals the beginning of the juvenile period. This period is characterized by a peak in weaning conflicts, refusal by the mother to carry her baby even in the face of danger, and the beginning of the baby's attachment to group members other than the mother (Lancaster, 1971). One of the major activities responsible for the formation of such attachments is play, with which we begin the next section.

INTEGRATIVE MESSAGES

Social play begins during the second week of life and continues throughout the juvenile period. Adults do not play as often as younger monkeys, and when they do play, it is always with youngsters (Struhsaker, 1967d). Young vervets have favorite playmates, usually of about their own age. They usually play in pairs, but sometimes five or six play together. Adults and subadults not only tolerate play but sometimes encourage it by inviting it and participating in it. They also threaten juvenile males when they get too rough for infants (Fedigan, 1972).

Invitations to play include a gamboling approach followed by rapid withdrawal and a sudden stop (Fedigan, 1972). The inviter invariably displays the "play face," in which the mouth is held open, exposing the cusps of the teeth. Play faces are often maintained by all participants throughout a play sequence. During playful wrestling, juveniles sometimes purr softly. Struhsaker (1967a) speculates that the purr may "enhance the play bond" (p. 292), but it is also possible that, like the play face, purring is a reminder that the contest is not serious. This conjecture is supported by the fact that wrestling, the vervet's most aggressive form of play, was the only form that Fedigan found to be invariably accompanied by the play face. Perhaps wrestling requires extra reminders to prevent escalation into real fighting.

There are several other forms of agonistic play besides wrestling. One common form involves mutual slapping with an open hand, often combined with hopping. Fedigan notes that this combination resembles human boxing. Tail pulling and fur pulling often escalate into "mock biting," in which a vervet tilts its head from side to side, snapping its jaws. Mock biting often alternates with gamboling and chasing, the latter with frequent reversals of the roles of pursuer and pursued.

Struhsaker (1967d) observed two more highly structured forms of play. In tug of war, two young vervets would pull from opposite directions at a piece of bark or blade of grass, using both hands and mouths, until one of them gave up. Hide and seek took place in the tall grass that flourished during the rainy season. The game usually began with bounding chases and counterchases. Then one of the participants would suddenly drop out of sight and remain still until one of the others found it or came very close. At this point, chasing or some other form of play would ensue. Agonistic play is especially common when vervets play with conspecifics from other groups. The games are similar, but the mood is "tense." Agonistic displays are common and sometimes escalate into fighting (Struhsaker, 1967d).

All nonagonistic play among vervets is sexual. Sexual play is both homosexual and heterosexual but is uncommon among females. Struhsaker saw playful pelvic thrusts performed by an infant only 1 month old. Pelvic thrusts do not seem to be combined with mounting until infants are about 4 months old. Playful mounting was often associated with other sexual behaviors, such as back nuzzling and hip grabbing. There was never any evidence of ejaculation during sexual play, but otherwise the postures of sexual play were identical to those of real sex. The nonserious character of sexual play was most commonly revealed by its frequent and sudden interruption for chasing and grooming.

Fedigan (1972) suggests that play is an important factor in the development of "social perception." She suggests that in play young vervets learn the difference between playful and serious bites, the meanings of various signals, and the facts of dominance. She found, for example, that in their play groups, young vervets assume the relative ranks of their mothers. Since vervets have preferred playmates, play may also create cohesion within the troop.

The bonds of attraction that develop through play and propinquity are maintained and expressed by a variety of affiliative behaviors that persist throughout a vervet's life. The most important and common of these affiliative behaviors is allogrooming. Allogrooming in a captive troop studied by Brain (1965) was independent of rank—low-ranking and high-ranking monkeys were equally likely to groom or be groomed. Brain found that females were much more likely to groom than males, but Struhsaker (1967d) mentions no such tendency.

When interspersed with play, grooming begins suddenly. At other times it is preceded by preliminary behaviors, which include alternate reciprocal mounting without pelvic thrusting, gentle mouthing of the groomee's neck fur, and mutual muzzle contact. Other preliminary behaviors are dominance and sexual displays and grooming solicitations. Potential groomees solicit by presenting the part of the body they wish to have groomed or, sometimes, by briefly holding the groomer's face between their hands. Monkeys that wish to groom also

Like vervet monkeys, baboons cement social relations with elaborate grooming rituals. Photograph by Roger Peters, courtesy of the Chicago Zoological Park.

display preliminary behaviors that seem to function as solicitation to be allowed to groom. These solicitations include placing one or both hands on the head, shoulders, or genitals of the monkey to be groomed; placing the face near the potential groomee's face or inguinal region; nuzzling its scrotum; embracing it from front or rear; straddling from behind; and, when possible, handling the groomee's infant. In this last case the attractiveness of infants makes it likely that grooming itself acts as a solicitation, providing the groomer with an excuse to make contact with the infant.

Vervets groom with their hands, carefully combing a small part of their partner's fur at a time. Every part of the body is attended to except the genitalia. The groomer removes foreign material either with its mouth or with its hands. In the latter case the foreign material is then placed in the groomer's mouth.

Male groomers and groomees often displayed erections. Groomees often displayed erections in which the penis jerked back and forth in the sagittal plane or rotated in a small circle (Struhsaker, 1967d).

According to both Struhsaker (1967a) and Brain (1965), grooming not only promotes hygiene but reduces tension and fosters cohesion. It may also reduce tension during intertroop encounters. In such cases juvenile males from one troop sometimes groom their counterparts from the other.

Another form of apparently affiliative contact is tail twining. Two vervets sometimes sit on the same branch with their tails entwined. Struhsaker mentions the hypothesis that the primary function of this form of contact is social but points out that it may also help vervets to keep their balance.

Affiliative contact is not limited to waking hours. Brain (1965) noted that his captive vervets were eager for physical contact at night. They slept in groups of three or four, in which the sleeping partners were of similar rank.

When a hungry adult or juvenile sees another monkey with food, it emits a "war-hor-hor" call, sometimes prolonged into a "gargle" similar to that used by an infant separated from its mother. The same call is sometimes used by females when they see another monkey holding an infant (Brain, 1965). In either case the lips are protruded and pursed into an "o," and the muscles around the eyes are contracted. This call appears to be a general solicitation that may be used for food, contact, or the privilege of holding an infant.

Brain heard a "squeal" of distress whenever vervets were in pain or severely frustrated. When produced by an adult male, this squeal resembled the infantile distress squeal but was lower in pitch. The squeals of adult females resembled those of infants. When a vervet squeals, its mouth is open, and its lips are retracted in a narrow grin. Although Brain states that the social significance of the squeal is "great," he does not describe a typical response. According to Huffman (1976), who studied free-ranging vervets on St. Kitts, the squeal sometimes elicits rapid approach by other members of the troop.

Vervets have an elaborate repertory of alarm calls, far more specific than those of any other nonhuman mammal. Struhsaker (1967a), in his study of free-ranging vervets in Kenya, distinguished six different alarm calls. The following description of these calls is based on his report.

The "snake chutter" is given mainly by juveniles and adult females when they detect an Egyptian cobra or a puff adder. Three other species of snakes (presumably not predators on vervets) did not elicit the snake chutter. When giving this call, vervets grimace, exposing unclenched teeth. The snake chutter is staccato, nontonal, and low in intensity and pitch, with energy concentrated below 2000 hertz but extending to about 7500 hertz. Units of about .1 seconds in duration form phrases lasting about .4 seconds. These phrases are repeated by several monkeys as they gather within 2 meters of the snake and move along beside it. The low intensity of these calls probably warns nearby vervets and delineates the snake's progress without attracting the attention of other potential predators.

A very different alarm call, the "uh!" call, is sounded at the appearance of hyenas, jackals, cheetahs, Cape hunting dogs, and humans. Because these creatures do not usually cause vervets to flee, Struhsaker considers them "minor mammalian predators" (p. 306). The "uh!" call, which is a single, quiet, short exhalation, may be given by any vervet in a troop when the predator is as far as 300 meters away. Upon hearing this call, other vervets look toward the predator. Only if the predator is moving toward them do they withdraw. Otherwise, they resume their activities. The "uh!" is clearly a low-intensity alarm that serves mainly to put nearby monkeys on guard.

A third alarm call, the "nyaow!" call, is stimulated not by any particular kind of animal but by sudden, close motion. A single, short vocalization, it is given by any monkey that perceives sudden movement nearby and does not know what caused it. Upon hearing a "nyaow!" call, other monkeys look toward the movement but do not flee unless they detect a predator moving toward them. Thus, the "nyaow!" call, like the "uh!" call, is probably a low-intensity alarm.

A short, single "rraup" call, on the other hand, elicits a rapid run for cover. It is given by juveniles and females as soon as they spot a major avian predator, especially a martial eagle or a crowned hawk eagle.

Adult males commonly give the "threat-alarm bark," a low, loud, nontonal call usually repeated as alternating exhalations and inhalations. Exhalation barks usually carry more energy in their higher frequencies than inhalation barks. In a typical threat-alarm bark composed of two phrases, each phrase had four inhalations and four exhalations, each of which lasted about .1 seconds. It seemed to Struhsaker that the closer the predator got, the louder, longer, and faster the barks became. He noted that the combination of low frequency and high amplitude makes the threat-alarm bark easy to localize over a long distance. It was given only when a major predator—bird or mammal—was quite close.

Under similar circumstances, juveniles and adult females give a loud series of "chirps," which are short, abrupt, low-frequency exhalations. When they hear a chirp, any vervets that are in the open run for cover. Several monkeys sometimes join in a chorus of clamoring that may last for hours.

As Struhsaker points out, this repertory of calls provides vervets with specific information about the location and type of predator.

Juveniles and adult females give a "rraup" when they first detect a major avian predator. If, however, the predator is a mammal, they give a chirp. Vervets make different responses to different calls. If a "rraup" signifies the approach of a predatory bird, they run into thickets, but if a chirp announces the approach of a predatory mammal, they run for the trees.

The effectiveness of these calls places a premium on troop cohesion. It is therefore imperative that there be means of keeping the troop together even while its members forage separately. One important means of maintaining contact may be provided by their distinctive coloration, which must make it easy for them to spot one another even in light cover. When vervets are in dense cover, however, visual signals are not very effective and must be supplemented by a contact call. From the age of 1 month on, vervets commonly produce a "war-hor-hor" vocalization. This call has been described as a solicitation, but it is also produced while the monkey feeds, travels, looks around, or engages in other routine activities. According to Brain (1965), this call signifies approval at the sight of food or another monkey in the troop. Its frequent repetition probably facilitates group cohesion by informing others of the caller's presence and location.

Since a vervet is especially liable to become separated from its companions when traveling, it is to an individual's advantage to signal its intention to move. Vervets do this with a short, low-pitched, guttural "progression grunt," given up to 15 minutes before locomotion begins. As soon as one monkey gives this call, it is answered by a chorus of grunts from its neighbors. The initiator often gives a second grunt and is answered again. Each of these call/answer sequences lasts less than 5 seconds. The troop almost always moves more than 50 meters within 15 minutes of a chorus of progression grunts.

The visual characteristics that convey a vervet's age class and sex have already been described. There are also vocalizations and other sounds characteristic of each sex. Only males, for example, produce barks, tooth chatters, and grunts. Only females and juveniles emit squeal-screams and "aarr-raughs" (submissive and distress calls discussed later). "Lost eee," "lost rrr," "eh-eh," and "wawoo" calls are given only by infants. Other vocalizations are given with tonal variations that could convey information about age. Progression grunts given by young monkeys, for example, are higher in pitch than those given by adults, and the "nyaow!" calls and squeals of older males are deeper in pitch than those of younger males and females (Brain, 1965; Struhsaker, 1967a). Older males also identify themselves as such by characteristic displays of dominance and other agonistic messages.

AGONISTIC MESSAGES

One of the dominance displays characteristic of adult males is the "red, white, and blue display," a satire of chauvinism in which the red is around the anus, the white on the perineum, and the blue on the scrotum. This tricolored emblem is proudly paraded by elevating the tail and strutting back and forth in front of the receiver, then pausing and directing the hindquarters in the receiver's direction. Alternatively, the colorbearer, which is always higher in rank than the receiver, may march in a circle around the subordinate, which continually turns so as to face the display.

The elevation of the tail seen in the red, white, and blue display may generalize to other situations. During his investigations of free-ranging vervets on St. Kitts, McGuire (1971, cited in Huffman, 1976) found that the higher a vervet carries his tail, the higher his rank. Struhsaker (1967c) noted a complementary tendency for low-ranking vervets to keep their tails low.

Several threats enhance the imposing demeanor of the vervet that shows the colors in the red, white, and blue display. The displayer, during the initial parading back and forth, keeps his body axis perpendicular to the receiver's direction of gaze, an orientation suggestive of the sideways display described later. Furthermore, the displayer sometimes interrupts his strutting to stare at the subordinate, either by looking back over his shoulder or by rising on his hind legs and gazing to the side. Finally, as though the red, white, and blue were not colorful enough, the dominant supplements those hues with a bright-pink penile erection, sometimes rigid, sometimes jerking back and forth.

The penis also plays a part in an unusual display that is most often directed toward a younger male. The form of the display and the ages of the participants suggest that the display may be an expression of dominance. Struhsaker (1967d), however, does not draw this conclusion and cautiously describes penile display under "miscellaneous social behavior." To perform a penile display, a male stands on his hind legs and places his hands on another, sitting male's shoulder, back, or head. This posture places the displayer's inguinal region directly in front of the receiver's face. In three of the seven cases of this display observed by Struhsaker, the displayer had an erection; in the remaining cases the condition of his penis could not be seen.

Sometimes a vervet's bearing suffices to convey its dominant status. In the gait that Struhsaker calls the confident walk, a dominant strides in an alert, but relaxed, manner and does not glance rapidly to the sides. The confident walk is often a prelude to supplantation, in which a dominant approaches a subordinate, which stops whatever it is doing and moves off. The dominant then occupies the position, eats the food, or grooms the companion left by the subordinate. Struhsaker noted that success in supplantation sometimes seemed to depend on the proximity of certain other monkeys with which the supplanter might form a coalition. Spatial supplantation was most common, and sometimes the dominant took over a real resource, such as food or a comfortable position. Usually, however, supplantation seemed to be a purely symbolic assertion of dominance.

A monkey that is being supplanted or that receives a red, white, and blue display typically gives several visual and auditory signs of submission. One of these signs of submission, the "crouched walk," seems to be the antithesis of the confident walk. Steps are short and rapid. The back is arched, the head lowered, and the limbs flexed. The monkey glances rapidly back over its shoulder in the direction of the dominant and gives a nontonal "woof-woof" call. When the dominant performs a red, white, and blue display, the subordinate sometimes hops backward, facing the display and remaining crouched even as it hops. The crouching hop is always accompanied by a grimace, which may be both a mild defensive threat and a sign of subordination.

Another response to the red, white, and blue display is the "false chase," in which the subordinate, maintaining a slight crouch, gallops slowly and hesitantly toward the displayer. Young male recipients of red, white, and blue displays often give the "waa" call, a long, tonal exhalation with a fundamental frequency of about 1000 hertz and harmonics up to 14,000 hertz. On occasion they combine "waa" calls with "woof-woof" calls to produce "woof-waa" calls. Females give a call similar and probably homologous to the "waa," which Struhsaker transcribes as "wa-waa." The dominant recipient of the false chase and associated vocalizations moves slowly away, gazing ahead and keeping its tail erect. The chaser never catches the dominant, and the false chase usually terminates the interaction.

According to Bolwig (1959), both male and female primates sometimes present their hindquarters as a quasi-sexual demonstration of submission. According to Wickler (1967), the male's anogenital coloration mimics that of the estrous female, and this mimicry enhances the effectiveness of the male display. Presumably, it works by replacing aggressive with sexual motivations.

Before and during grooming, both participants may display three types of signals that signify nonaggression rather than friendliness. Lip smacking and tooth chattering are performed by both dominants and subordinates as they groom or are groomed. In similar situations young vervets give long and short versions of the "raugh" call. These calls have both tonal and nontonal components, with peak energy at 500 and 5000 hertz. According to Struhsaker (1967a), these three signals permit proximity of vervets of different rank by signaling a nonaggressive mood.

Perhaps because of the high frequency of conflicts both within and between troops, vervets have an elaborate repertory of threats. Since Struhsaker (1967a, d), from whom the following descriptions are taken, found several differences between threats used within and between troops, intratroop and intertroop aggression will be discussed separately.

Struhsaker found that conflicts within troops mainly involved juveniles and adult females. He noted that serious conflicts were more common in smaller troops and suggested that this tendency might be explained by the more numerous opportunities for peaceful interaction in larger troops—in a small troop a monkey has fewer choices of a partner for grooming, play, and so on. Intratroop conflict often involves redirected aggression, in which a threatened or an attacked vervet threatens or attacks another, which may in turn redirect its aggression. Such chain reactions usually have the effect of placating the original attacker.

The mildest form of threat used within the troop is staring directly at the receiver for 3 to 5 seconds. Staring was often combined with eyelid exposure, with the brow retracted to reveal pink eyelids set off by the black facial skin. Both eyelid exposure and staring could be either defensive or offensive threats. Their meaning depended on whether the monkey crouched with lowered head and tail, in which case the threat was defensive, or jerked its forequarters up and down, in which case the threat was offensive. While crouching and staring, a defensive vervet often gaped widely, but without exposing its teeth.

Several vocalizations underscore these mild threats and seem to be both defensive threats and solicitations for aid. These calls include the "intragroup chutter," a staccato noise with almost all energy below 1000 hertz; the squeal, a short, 5000-hertz cry; the scream, a longer, nontonal call similar to the distress screams of infants; and two combinations, the "chutter-squeal" and the "squeal-scream." These calls were given only by juveniles and adult females when they were threatened or attacked.

The highest level of defensive threat is expressed when a monkey grabs and slaps at the aggressor, usually without making contact. This defensive behavior often serves to deter further aggression.

In addition to these mild and defensive threats, there are offensive threats that form sequences of increasing intensity. One of the mildest offensive threats is jerking the head back and forth while staring at the receiver. A more intense offensive threat is conveyed by jerking the forequarters through an arc whose length is proportional to the intensity of the threat. The length of this arc is determined by the monkey's posture—when the monkey is seated, with hands remaining on the ground, the arc is short; when the monkey stands, the arc is longer, and the monkey is likely to attack (see Figure 6-1).

Two other postural displays also serve as offensive threats but were rarely seen. In one of these displays, the transmitter places its hands on the ground and jerks its forequarters laterally while staring at the receiver. In the other display the threatener stands on all fours while presenting its flank and staring at the receiver. This display alternates with a head-on orientation sometimes combined with the forequarters jerk.

Another uncommon threat was a bark. This bark was shorter than the threat-alarm bark and unlike it; each phrase of the threat-bark consisted of a single, short exhalation. It was given only by older males, usually in response to a squabble among juveniles or females. The bark effectively quelled such disturbances.

The highest level of offensive threat is expressed by running at full speed toward the receiver, then grabbing and slapping at it without making contact. When they connect, these grabs and slaps often start a fight, but only about 15% of the intragroup conflicts observed by Struhsaker resulted in physical contact. Nevertheless, the list of wounds observed by Struhsaker and others is a dispatch from a combat zone—only the lower legs seemed to be exempt from laceration, presumably caused by bites. Healing, however, was remarkably rapid, even when large flaps of skin were torn loose, exposing the muscles. Wounds were concentrated on the tail and were relatively rare on the shoulders and lower back. Juvenile males seemed to receive more than their share of injuries.

Fighting is even rarer in conflicts between troops than in conflicts within them. Only about 5% of the intertroop encounters observed by Struhsaker involved physical contact. Since almost 90% of these encounters were agonistic and involved some degree of territorial defense, Struhsaker concluded that the vervet's system of territorial threat effectively minimizes violence. About 10% of the encounters between troops were relatively peaceful, with the residents virtually ignoring a troop of trespassers clearly visible less than 50 meters away. On occasion, such tolerance lasted several hours before the residents herded the intruders out of the territory.

As soon as females or young males detect another troop, they yell "aarr." There seems to be two forms of this call—long and short.

FIGURE 6-1 Vervet threat gestures: (a) jerking of
the head; (b) jerking of the forequarters while
sitting, with the hands remaining on the substratum
(ground or branch); (c) jerking of the forequarters
while sitting, with the hands alternately moving on
and off the substratum; (d) jerking of the
forequarters while standing quadrupedally; (e)
jerking of the forequarters by rapid oscillation
between a quadrupedal and a bipedal posture.
Redrawn by Heidi Reynolds from "Behavior of
Vervet Monkeys and Other Cercopithecines," by T.
T. Struhsaker, *Science*, 1967, *156*, 1197–1203.
Copyright 1967 by the American Association for
the Advancement of Science. Reprinted by
permission of the author and publisher.

The long form can be completely tonal, completely nontonal, or mixed. It can be pulsed or smooth, but it always lasted about .7 seconds. Most energy was concentrated below 1000 hertz. The short form is similar in pitch, but it was always pulsed, nontonal, and about .2 seconds long. Since these calls were given only when a foreign troop was close enough to be seen, they occurred only along territorial boundaries. Upon hearing these calls, other members of the troop looked toward the foreigners. After gazing toward the other troop for a few minutes, monkeys often exhibited "head flagging," rapidly nodding their heads toward and away from the foreigners. "Aarr" calls combined with head flagging indicate the proximity and direction of the foreign troop. There is no evidence that they act as threats.

Although juvenile males of different troops may sometimes interact peacefully, the mood of most intertroop encounters is aggressive. This mood is expressed by a pandemonium of vocal and gestural threats. In addition to the barks and other threats seen in agonism within the troop, there are several observed only when two troops meet. In such situations, males give the "intergroup grunt." Struhsaker was unable to determine the meaning of this call, but its similarity to the progression grunt may indicate a readiness to withdraw. The "intergroup chutter A" was given by females and younger males. It resembles the intragroup and snake chutters but is longer, has more high-frequency energy, and has more pulses per phrase. "Intergroup chutter B" is similar to the "A" type but is more variable, ranging in quality from a "woof" to a bark. Its energy is restricted to lower frequencies. It was given only by older animals during intense agonistic encounters characterized by chasing, wrestling, striking, and biting. Although some vervets give intergroup chutters, others, mainly females and juveniles, give the alarm chirp often heard as a response to major predators. It seemed to Struhsaker that these chutters and chirps acted both as threats and as solicitations for aid.

An interesting combination of calls is sometimes given by a young female when approached by members of a foreign troop. If there is a dominant male of her own troop nearby, she runs to him, crying "aarr-raugh," and then attempts to groom him. Struhsaker interprets this call as a combination of the "aarr" that announces the presence of another troop and the "rraugh" used as an appeasement to allow the caller to groom a dominant.

The cacophony of yells, grunts, chutters, barks, and chirps provides a score for a pageant of chases, postures, progressions, and coalitions. All members of both troops chase and counterchase one another back and forth across the border between their territories. Meanwhile, some males pose with tail held erect or with arms and legs extended, displaying white chests and erect penises (Poirier,

1972). Others ricochet from tree to tree, occasionally pausing to grab a branch and shake it. In this frenzy of territoriality, branch shakers grab and bite any monkey they encounter, friend or foe. Sometimes attacks on compatriots are accidental, but sometimes they appear intentional, with the aggression redirected from foreigners to fellows. Struhsaker suggests that such redirected aggression may be adaptive to the aggressor in two ways. First, aggression directed at a member of one's own troop may be less likely to result in injury than aggression directed at a foreigner. Second, such aggression might enhance group cohesion by driving the aggressor's group into tighter formation. Such formations sometimes include as many as 17 vervets in an area only 3 meters in diameter. Presumably, a monkey in the midst of such a cluster is practically immune to attack. Females and juveniles in such clusters often groom males. This grooming can be interpreted as a solicitation for protection, an appeasement against redirected aggression, or a displacement of nervous energy.

Encouraged and reassured by grooming and contact, members of a cluster often form a coalition and walk en masse toward the other troop. The result is a phalanx that, however unruly, is imposing enough to herd an intruding troop out of the area. Being on home ground seems to confer an advantage, for an adult male can sometimes herd a trespassing troop away without help from other residents.

Since encounters between troops are so stressful, individuals should benefit from behaviors that advertise the territorial boundaries and thus allow troops to avoid trespassing. According to Poirier (1972), who studied free-ranging vervets on St. Kitts, there are two types of territorial advertisement. The first is the powerful odor of feces that surrounds a troop's sleeping trees. The second is a visual display performed by dominant males. They often place themselves on bare treetops, so that the sunlight catches their white chests. Given the rather high frequency of close encounters between the troops observed by Struhsaker, it appears that in Africa, at least, these signals are often ineffective.

SEXUAL MESSAGES

Vervets become sexually mature during their fourth year (Lancaster, 1971). Females are receptive only during the dry season, between May and October. Presumably, this seasonality assures that births occur during the rainy season, when food is most abundant. Struhsaker (1967d) observed some females that remained sexually receptive for more than 2 months. Since prolonged receptivity is usually associated with menstruation rather than estrus, vervets should have a menstrual cycle. Data on the lengths of vervet menstrual cycles cited by Struhsaker suggest that they are highly variable. According to Rowell (1971), vervets do not show cyclicity of sexual behavior during their period of receptivity but may copulate at any time. Struhsaker found that they may even copulate while pregnant or lactating. He suggests a high frequency of heterosexual behavior shown by females immediately after giving birth. Sexual behavior occurs throughout the day, but Struhsaker found marked peaks in attempts to copulate in the morning and in the afternoon.

Struhsaker found a direct relation between a male's rank and the frequency with which he copulated. He noted that red, white, and blue displays were most common during the breeding season and suggested that they allow dominants to assert themselves, thereby increasing their chances to reproduce.

When females are not receptive, they respond to suitors with a variety of unambiguous rejections. They grab, slap, and lunge at males that follow too closely. When in a less aggressive mood, they simply crouch, making further advances impossible. If an unreceptive female is caught or cornered after a chase, she gives the "anticopulatory squeal-scream," a highly variable shriek somewhat similar to the squeal-scream described as a defensive threat. According to Struhsaker, this protest was usually given from a crouch, with a grimace exposing the unclenched teeth. It invariably quenched the male's ardor and usually led to his retreat.

The absence of anticopulatory defense is usually a good sign that the female is receptive. There are also several positive signs that she is willing. One of these may be olfactory. The role of odor in advertising receptivity in the vervet's close relative, the rhesus, has been documented by Michael and Bonsall (1977). Furthermore, Gartlan (1969) noted that male vervets often sniff females' vulvae before mounting. Huffman (1976), in his observations of free-ranging vervets on St. Kitts, noted that males often sniff and nuzzle the genitals of several females in succession, as though testing them for receptivity.

Females also advertise their receptivity by changes in anogenital coloration. Wickler (1967) was able to observe these changes by sitting below the trees in which vervets fed. He noted that the perineum of the unreceptive female is pale pink, the vulval slit nearly invisible, and the teats light gray. These timid tones become bolder when the female is receptive. The perineum and teats redden, and the bright-red clitoris protrudes from the blue edges of the vulval slit.

Although Struhsaker (1967d) states that there are no external changes like those described by Wickler, he did observe a behavior that he calls presenting, which is admirably suited to their display. A receptive female presents by standing close to a male with her posterior toward him. The tail is often raised or moved to the side, but sometimes hangs limp. While giving this display, she sometimes gazes fetchingly over her shoulder toward the male. Females also show their receptivity to an approaching male by adopting a quadrupedal posture without orientation away from him. Any doubt about her readiness is removed as she nuzzles his scrotum from behind as he stands on all fours.

Males often respond to these invitations by nuzzling the female's perineum, grabbing her hips, or embracing her from behind. The female then takes a few steps, dragging the male with her. Still maintaining his embrace, he may wrestle her to the ground by rolling to the side (Struhsaker, 1967d).

After this relatively perfunctory courtship the male attempts to mount, grasping the female's hips with his hands and her calves with his feet. Sometimes he adjusts his position by climbing her calves until his feet grasp her thighs. Once in position, he thrusts with his pelvis, first with short, rapid strokes, then with long, slow ones. Struhsaker speculates that short and long strokes occur, respectively, before and after intromission. In 25 copulations he observed, the number of thrusts varied from one to 25. The female stands with limbs slightly bent and her tail turned to the side. From time to time, she glances over her shoulder at the male. She may terminate the mount by taking a few steps forward, carrying the male with her. Otherwise, the male stops thrusting on his own and then may nuzzle her back.

By careful observation of the details of copulation, including the presence of semen around the vulva, Struhsaker determined the ejaculation is usually associated with a 2-second to 3-second pause between cessation of thrusting and dismounting. Pauses and thus, presumably, ejaculation, occurred in about 90% of the copulations he observed.

When the male dismounts, the female may suddenly become aggressive, threatening him with a stare, eyelid display, grimace, forequarters jerk, or attack. Usually, however, she merely investigates her vulva by touching it and then bringing her hand to her muzzle, where she can presumably smell or taste material from the vulva (Struhsaker, 1967d).

In about 10% of the copulations observed by Struhsaker, young males harassed the copulating pair. This badgering involved running, staring, grabbing, and slapping at them, especially the male. This interference was sometimes bothersome enough to cause the male to dismount or even to threaten, chase, or attack the harassers.

Males over 6 months of age are sexually active all year long. During the months when females are unreceptive, however, their sexual activities are restricted to masturbation and homosexuality. Of the 11 heterosexual behavior patterns described by Struhsaker, eight occur in interactions between males. Only ejaculation, its associated pause, and subsequent nuzzling of the back are absent from homosexual behavior. There are three patterns of homosexual behavior not seen in heterosexual encounters. These patterns are a hand-on-back gesture similar to the hands-on-shoulders pattern; putting the face in the inguinal region; and the genital touch. All three of these patterns sometimes precede grooming and may function as grooming solicitations. Their homosexual significance is inferred from their association with hip grabs, penile erections, and mounts. Thus, there is considerable overlap between homosexual and heterosexual repertories.

Wickler (1967) suggests that male anogenital coloration may have evolved to imitate that of the receptive female. The male perineum is red, like the female's, but the male's color comes from fur rather than skin. The male's scrotum is blue, like the female's vulva, and the male's penis is red, like the female's clitoris. According to Wickler, the similarities in color occur in nonhomologous structures and are thus the result of convergence rather than mere absence of dimorphism. Since the primate penis is *homologous* to the clitoris, Wickler's contention is not accurate in every detail. Nevertheless, it is interesting that the male's red, white, and blue display, in which these structures are shown off, is quite similar to female sexual presentation (Struhsaker, 1967d). The functional significance of this possible convergence of male and female anogenital coloration and its relation, if any, to male homosexuality should be investigated.

SUMMARY

Vervets have an elaborate repertory of messages, but one that is probably typical of the *cercopithecids*, the Old World monkeys, in its complexity (Struhsaker, 1967c). Struhsaker describes 46 different communicative gestures used by vervets. Their eight different messages are solicitation, low and high levels of defensive threat, low and high levels of offensive threat, fighting, dominance, submission, and male and female sexual advertisements. Struhsaker also describes 36 different sounds, which, after deletion of coughing, sneezing, and vomiting, whose primary significance is not communicative, transmit, according to my analysis, 19 different messages. Seven of these 19 messages are also transmitted by gestures. The remaining 12 messages are infant identity, distress, affiliation and satisfaction; play; contact; adult affiliation, alarms, identity, distress and courtship. Thus, in addition to tonic and phasic neonatal interactions, vervets can be considered to transmit at least 20 different messages—8 by gestures, 12 by sounds, and 7 by both sound and gesture. Addition of tactile and olfactory channels to this catalogue would probably not raise the number of messages but places the number of physically distinct signals in the low eighties.

7
BOTTLENOSED DOLPHINS

Another kind of lonely,
almost disembodied intelligence floating
in the wavering green Fairyland of the sea . . .

LOREN EISELEY

Delphi, the center of the ancient world, got its name when a dolphin, a manifestation of Apollo, leaped into a Cretan ship, brought it to the shore near Parnassus, and appropriated the local oracle. Dolphins are still immersed in myth 2500 years later. We know that more than 50 million years ago some primitive mammals returned to the sea, but we do not know what kind of mammals they were or why they went. Fossils feed only speculation, for the earliest *cetacean* remains are of serpentine creatures that had already made the marine transition. The teeth, skulls, and skeletons of these fossils suggest that the dolphin's ancestors were among the earliest *carnivores,* and that they went to sea for fish. Some anatomical and behavioral characteristics of modern dolphins, however, suggest that they evolved from early omnivorous ungulates, which also could have found food in the sea. At our present state of knowledge, however, either version is more myth than theory.

The recent phylogeny of the genus *Tursiops* is no clearer. Some taxonomists distinguish as many as four geographically distinct species: *truncatus,* most common in the Atlantic and Caribbean but also seen in the Mediterranean; *gilli,* in the Pacific; *nuuana,* in the North Pacific; and *aduncus,* in the Indian Ocean. Other scientists maintain that there are only two species, *truncatus* and *gilli,* while yet others regard them all as subspecies of *truncatus.*

Even *Tursiops's* common name is largely a matter of opinion. According to one tradition, a dolphin is any small, toothed whale with an elongated beak. Many American scientists, however, refer to the beaked *Tursiops* as a porpoise, reserving the term *dolphin* for fish of the genus *Coryphaena.* There is no good scientific reason for preferring one term over the other, so I have chosen *dolphin* because of its greater antiquity. *Bottlenosed* refers to the dolphin's distinctive beak, which is said to resemble an old style of gin bottle. *T. truncatus,* the familiar performing dolphin, gets its specific name from its 88 conical teeth, which become truncated with wear. This truncation provides a reliable estimate of a dolphin's age, which may be as great as 20 to

30 years. Teeth may also provide information about sex, for males' teeth are larger than females' (D. K. Caldwell & Caldwell, 1977).

The lower jaw is undershot, and the chin forms the tip of the beak. The lower jaw is used both as an auditory receptor and as a ramming weapon. Bottlenosed dolphins owe much of their popularity to the shape of their mouths, which when open appear to grin and when closed seem to smile. Evidence for ungulate ancestry is found in the eye, which resembles that of a goat. The pupil has straight sides and contracts vertically, as though by the lowering of a shade. The eyes are just behind the corners of the mouth and are extremely mobile. The blowhole, used both for breathing and for producing sounds (*phonation*), is directly on top of the head. The brain, large (about 1700 grams) and highly convoluted, lies just behind and below the blowhole. Although the head can be turned slightly, there is no visible neck to interrupt the nearly perfect streamlining. Even the penis, like that of many ruminants, is ordinarily retracted. It is carried within a longitudinal slit about midway along the ventral surface. Males have an anal slit just behind the penis. Females have a single urogenital slit, often bordered by darker skin, about two-thirds of the way back from the nose. The skin is soft, smooth, and resilient. It is a slate gray on the back and lighter toward the abdomen.

Two triangular flippers about a third of the way back along the lower flanks are used for steering and touching. A swept-back fin extends vertically from the center of the back. Damage to the thin, trailing edge of this fin produces distinct marks that allow identification of free-swimming individuals at considerable distances. The horizontal flukes at the tail are the primary organs of propulsion. (See Figure 7-1).

Both the streamlined shape and the resilient skin minimize turbulence, allowing 3-meter-long, 150-kilogram adults to swim at almost 40 kilometers per hour (Hertel, 1969). Although dolphins can dive to more than 600 meters, they usually stay within 50 meters of the surface. They can remain submerged for up to 7 minutes but usually surface to breath two or three times a minute. They do this even while sleeping, an ability that Wood (1973) compares to the ungulate ability to sleep while standing.

Dolphins have acute vision in both water and air, as is shown by their ability to leap out of the water to grab small objects several meters above the surface. The anatomy of the retina suggests that they may have color vision (Perez, Dawson, & Landon, 1972). They are most active during daylight hours, when vision is most useful. Their hearing is excellent between 200 hertz and 150 hertz (C. S. Johnson, 1967). They have no olfactory bulbs and hence no sense of

FIGURE 7-1   A dolphin leaping. Photograph by
Roger Peters.

smell. They may, however, have a sense of taste (Sokolov & Kuznetsov,
1971), which they may use much as terrestrial mammals use smell.
They often swim with their mouths open or tongues extruded, especially while following another dolphin that is eliminating. Their skin
is quite sensitive, and they frequently caress one another with flippers
or flukes or rub against one another's bodies. They also have a sense
that is totally alien to most of us. Only the blind, who may learn to

detect walls, curbs, and objects with the aid of echoes, have some intuitive apprehension of what this sense is like. With the aid of this sonar-like perceptual system, dolphins can distinguish nickels from dimes even when the coins lie in mud. They also can distinguish species of fish and avoid thin wires even when blindfolded (Kellogg, 1961). This sense is probably most useful for detecting prey over long distances or swimming in murky water.

Ecologically, *Tursiops* is a group hunter. Most of the creatures dolphins eat are much smaller than they are. These include mullet, shrimp, squid, and rays. When taking schooling fish, dolphins often swim rapidly in circles, stirring them up and grabbing them as they go by. Saayman, Tayler, and Bower (1973) call this behavior herding and found that it was often conducted in an organized manner. On several occasions they observed a large group of bottlenosed dolphins divide into two subgroups, with one group driving fish toward the other, thus enclosing the prey in pincers. Herding was usually accompanied by long leaps, which are also seen when they pursue flying fish. Another feeding technique was observed by Hoese (1971), who saw two bottlenosed dolphins swim rapidly toward a mudbank so that their "bow waves" flung small fish onto the shore. The dolphins came completely out of the water to grab the fish, then slid back down the bank with their catch.

Because of their tendency to follow warm water, with its abundance of fish, dolphins have large home ranges. Saayman and his colleagues regularly recognized an individual at two locations 46 kilometers apart. The Wursigs (1977) studied a group that returned to their study area after being sighted 300 kilometers away. The regularity of the movements of some groups is illustrated by the group of dolphins that swam past my house on the bay at Sarasota, Florida, almost every day for 3 months. They entered the bay every morning and swam out into the Gulf every afternoon. I was unable to identify individuals, but a steady decrease in flight distance suggested that this was indeed the same group.

Although dolphins have home ranges, there is no evidence that they defend territories. Several different groups are frequently seen together without noticeable conflict. They also associate peacefully with nonprey species, especially pilot whales and tuna.

This latter association results in hundreds of thousands of them being killed each year by commercial fishermen. It has been proposed that some of these dolphins might be spared if, before deploying their nets, tuna fishermen would transmit simulated killer whale calls to scare the dolphins away. The rationale for this proposal is that the killer whale, *Orcinus orca,* is the dolphin's major natural enemy. Although members of the dolphin family, killer whales are much larger than bottlenosed dolphins and can easily swallow them whole. One specimen was found with 13 undigested dolphins (and 14 seals) in its crop (Slijper, 1977).

Another enemy is the shark. Sharks often attack injured or newly born dolphins. In such cases, dolphins attack the shark, ramming it at high speed and often causing serious internal injury. According to Cousteau and Diole (1975), dolphins usually succeed in injuring sharks because several dolphins act in concert. Divers' lore has it that when dolphins are about, one need not fear sharks. That dolphins do sometimes attack sharks is the germ of truth in this belief, but both in captivity and in open water dolphins and sharks can be seen coexisting peacefully.

Concerted action directed against both prey and predators reveals a social capacity that transcends mere gregariousness. Indeed, our glimpses of dolphin society reveal a great deal of structure, both in patterns of association and in hierarchies of rank. Dolphins are almost never seen alone and, in waters off California and Florida, seldom in groups of fewer than four. In the Mediterranean Pilleri and Knuckey (1969) saw groups ranging in size from ten to 15. In the Indian Ocean Saayman and his coauthors (1973) rarely saw fewer than ten and often saw several subgroups of 25–50 *aduncus* organized into large schools numbering in the hundreds. The Wursigs' (1977) photographic data, gathered off the coast of Argentina, allowed precise determination of the size of one group of 53 *truncatus*. Within this group were stable units of five or six individuals and unstable subgroups numbering between eight and 22, with a mean of 15. Evans and Bastian (1969) note that *gilli* groups off California usually number fewer than 20 and are often divided into subgroups of at least two or three.

Some of the smaller stable units are probably mothers with the young of several years, which are known to associate with their mothers for 4 years or even longer (D. K. Caldwell & Caldwell, 1966). Larger stable units are probably composed of two or more closely associated females with young or of immature males, which, according to Evans and Bastian, also tend to swim together. A few of these units may temporarily combine to form the unstable subgroups of ten to 50, which may in turn combine to form large schools numbering in the hundreds. McBride and Kritzler (1951) frequently observed small groups of bottlenosed dolphins in estuaries along the coast of Florida. On the basis of dorsal-fin shape (and presumably body size), they concluded that the basic social group was an adult male, three to five adult females, and their young. McBride and Kritzler's attempts at breeding dolphins in captivity were most successful when they established a group with this composition.

Pilleri and Knuckey (1969) observed one such school swimming in an unusual formation in which the division into units of four and five was especially clear. These units all held the same course and formed an elongated ring. The usual formation is a tight disk, in which each dolphin surfaces and dives in a characteristic rolling arc.

M. C. Tavolga (1966) investigated dominance relations in a stable, self-perpetuating colony of 12 bottlenosed dolphins at Marineland studios in Florida. She found a well-defined pecking order based on threats and other aggressive acts. High-ranking dolphins initiated such acts, while low-ranking ones received them. The adult male was the largest and highest-ranking dolphin. During the breeding season, in the spring, he frequently swam with one or another of the four adult females. He was the sire of all the nine calves born to five different females during the 4-year study period. Outside the breeding season he usually swam alone, but sometimes he swam with a mature, multiparous female that was next in order of rank. She was often the first to investigate new objects introduced into the tank and was a leader at public performances.

Below these two dominant dolphins were three groups whose members were of approximately equal rank. The first of these groups consisted mainly of older females, which swam together and were occasionally joined by individuals from the other two groups, all of which were their offspring. The second group consisted of young males, which stayed together almost continually and engaged in homosexual behavior during the breeding season. The third group consisted of nurslings, whose excursions from their mothers' sides were frequent but brief. They were consistently submissive to all the others and, perhaps as a result, were never attacked. In another captive group, containing three males, three females, and one infant, McBride and Hebb (1948) found a well-defined male hierarchy in which rank varied with size, but they saw no evidence of rank among the females. Dolphins are clearly highly social, both in the sense that they live in groups and in the sense that these groups are highly structured. As we shall see later, messages that express membership and dominance comprise an important part of the message system of the dolphin.

One modern myth about dolphins is the popular notion that they have human or even superhuman levels of intelligence. Another is that they have a language. These two myths are related because, if either one were true, the other would seem much more plausible. In any case, our assumptions about intellectual capacity affect our interpretation of the meaning of messages. Thus, it makes sense to preface a discussion of dolphin communication with a brief treatment of dolphin intelligence.

Dolphins have large brains and long lives. They have an extraordinary ability to learn by imitation and an extended period of *parental investment* in which to use this ability. Thus, the notion of alien, yet somehow humanoid, minds just offshore is plausible as well as appealing. But we cannot have it both ways—if their intelligence resembles our own, we ought to be able to measure it; if it is alien, to compare it to ours is to be tangled in our own conceptual net. It is difficult—some would say impossible—to devise tests that permit comparisons of the intelligence of members of different human cultures. To do the same for a mammal of a different order, living in a different physical and sensory world, is obviously much harder. Nevertheless, there are some data that bear on the question of dolphin intelligence. On *reversal-shift discriminations*, which have been proposed as a kind of phylogenetic IQ test, one dolphin did about as well as an elephant (Kellogg & Rice, 1966) but not nearly as well as 9-year-old humans (White, 1966). Dolphins excel at sound production, so tests based on this ability should provide data for a fair estimate of dolphin intelligence. Unfortunately, such tests yield ambiguous results. According to Andrew (1962), dolphins' great ability to mimic sounds places their intelligence somewhere between that of a dog and that of a chimpanzee. On the other hand, their ability to modify vocalizations for a reward exceeds that of any nonhuman primate.

Although seen in many vertebrates, use of tools is often considered to be a sign of high intelligence, but one hardly expects to find such a sign in an animal virtually without limbs. Yet dolphins do use tools. D. H. Brown and Norris (1956) watched a dolphin dislodge a moray eel from a crevice by poking the eel with the poisonous dorsal spines of a scorpion fish. Similarly, M. C. Tavolga (1966) reports that a young dolphin used one (presumably dead) fish to lure another fish out of hiding. Tayler and Saayman (1973) saw a dolphin use a scraper to dislodge seaweed, which she then ate. When the scraper was removed, she resumed her attempts to obtain seaweed by using a broken piece of tile from the floor of her pool. Another dolphin was later seen performing in a similar manner. On the basis of this and other cases of observational learning, Tayler and Saayman concluded that dolphins are less "stimulus-bound" than many other mammals.

Another indirect approach to the issue of intelligence is through an analysis of emotional and motivational behavior. On the basis of the persistence of their fears, the strengths of their attachments, and the variety of their play and sexual behavior, McBride and Hebb (1948) concluded that the complexity of dolphins' motivational systems is "somewhere in the range of development of dog to chimpanzee" (p. 122).

Thus, observations of emotional and manipulative behavior, like direct tests of problem solving and sound production, provide no firm evidence of extraordinary intelligence. What, then, of the dolphin's large, highly convoluted brain? Granted that brain structure is not always a good index of intelligence, dolphins must use their brains for something. Tomilin (1957, cited in Wood, 1973) has suggested that the brain of the dolphin is an adaptation to life in a two-phase, three-dimensional world and, in particular, to perception by means of echolocation.

Some of the size of the dolphin's brain may also be attributable to the processing of acoustic information from other dolphins—to communication as well as perception. This hypothesis brings us to a second popular myth: the notion that dolphins have a language. It is true that in their ability to emancipate sound production from emotion and in their ability to modify vocalizations for a reward (Lilly & Miller, 1962) dolphins are more promising subjects than chimpanzees, which are capable of complex and abstract forms of communication (Gardner & Gardner, 1969). Nevertheless, there is at present no direct evidence of any unusual ability to transmit information. Bastian (1967) conducted an experiment to see if one dolphin could tell another whether or not a signal light was flashing. The apparently positive results of this experiment were widely publicized, but Bastian (1967) has interpreted his results as an artifact of training and not as evidence of any unusual ability to communicate. Neither this experiment nor others that have produced negative or ambiguous results are conclusive. It is in principle extremely difficult to prove that dolphins do *not* have language. It is possible, however, to compare the dolphin's communications to those of other mammals. Before doing so, however, it is necessary to describe the various kinds of sounds they make and some general findings about them. These sounds can be classified into three basic types: clicks, whistles, and a miscellaneous category variously referred to as quacks, squawks, blats, and barks.

Clicks are known to be used in echolocation but may also serve in communication. They may occur singly or in bursts that sound somewhat like a high-pitched creaking door. Clicks are produced intermittently when dolphins are in clear water or familiar surroundings. When they are in murky water, unfamiliar surroundings, or are blindfolded or presented with an unfamiliar object, the rate and volume of clicking increase. A variety of experiments have shown that by listening to echoes from these clicks, dolphins can detect and discriminate objects of different sizes and materials and even different species of fish (Kellogg, 1961). Each emission is a complex of sine waves whose frequencies vary within and between pulses. Energy is widely distributed between 1 hertz and at least 196,000 hertz, but most energy is below 30,000 hertz (Kellogg, 1961; Lilly & Miller, 1961a). In some pulses, however, energy is concentrated in the ultrasonic (above 20,000 hertz) region. These pulses are highly directional and are emitted straight ahead (Norris, Prescott, Asa-Dorian, & Perkins, 1961). Clicks may be as short as .1 milliseconds, and bursts may contain as many as 600 clicks per second and last from .5 seconds to 15 seconds (Lilly & Miller, 1961b). They originate in the laterally paired nasal sacs (Evans & Prescott, 1962), but exactly how they are produced is not understood. Lilly and Miller (1961b) found that when dolphins are separated by a barrier but can hear each other, they engage in "exchanges" in which each refrains from clicking while the other clicks. Although they do not say so, T. G. Lang and Smith (1965) present data that show a similar pattern. Lilly and Miller noted that exchange clicks are of lower frequency than echolocation clicks and that sonic and ultrasonic components can be emitted independently.

To the student of communication, whistles are dolphins' most interesting type of sonic production. Dolphins can whistle and click independently or simultaneously. According to Lilly (1962), simultaneous clicks and whistles are produced by whistling with the nasal structures on one side while clicking with those on the other. As is the case with clicks, the anatomy of whistle production is well-known, but the mechanism is not. Whistles range in frequency from 4000 to 18,000 hertz. They are sinusoidal in form, with a well-defined fundamental and several harmonic multiples of the fundamental. The frequency pattern of a whistle may remain stable while it is repeated at intensities varying by as much as 100 decibels. Whistles last from .1 to .4 seconds and are usually about .3 seconds long. They may occur singly or in groups of up to nine, but groups of two or three are most common (Lilly & Miller, 1961a).

The number of distinguishable whistles depends on the sensitivity of recording and spectrographic equipment as well as on the criteria used to discriminate one whistle from another. Some researchers have identified as many as 2000 different "phones" (Cousteau & Diole, 1975). Dreher (1961, 1966) has analyzed thousands of whistles on the basis of pitch variations in the fundamentals, which he calls contours. He has distinguished 24 different contours, 18 of which are used by Atlantic bottlenoses, 16 by Pacific bottlenoses, and nine by both. T. G. Lang and Smith (1965), using similar criteria but different subjects, distinguished six types of whistles. Simplified drawings of pitch changes characteristic of the whistle types distinguished by Dreher and by Lang and Smith are shown in Table 7-1. Each contour is a plot of the frequency of the fundamental against time. The average length of the whistle shown is about half a second. As shown in Table 7-1, Lang and Smith found a fairly good match between their Type A and Dreher's Contours 1 and 2; between their Types B and D and Dreher's 3, 5, and 7; between their Type C and Dreher's 3; and between their Type F and Dreher's 32. This correspondence is supported by concordance between the rank orders of frequency of occurrence of the types recorded in the two studies.

Dreher (1966) found that rank order of occurrence varied linearly with the log probability of occurrence. He pointed out that this probability structure is characteristic of information-bearing arrays including, for example, common English words. He concluded that the information structure of his array of contours "could be intelligence bearing" (p. 1800). A similar analysis of whistles produced by different dolphins in a different environment yielded a Spearman's rho of .17 between ranks of whistles common to the two situations. Dreher and Evans (1964) concluded that this correlation demonstrated a new organization of the array of whistles. When Dreher used Shannon's equation to calculate the information content of the array of whistles, he arrived at a figure of about 2.2 bits per symbol. By way of comparison, he points out that this figure is quite close to the approximately two bits of information for letters in written English. Thus, whistles' *potential* for information transmission is quite high. The emphasis here is important—comparisons with English should not be over-interpreted. Information is a statistical property of a frequency distribution, not a measure of meaningfulness.

There are several sorts of evidence that bear on the question of meaning. Some types of whistles correspond to the message types that dolphins share with other mammals and will be described later. Here we will deal with results that are not easily assimilated into the discussion of message types.

Dreher and Evans (1964) often heard a short, descending chirp, repeated rapidly, when they removed objects from the dolphins' tank. They inferred that this call signifies curiosity and anticipation. In another study Dreher (1966) made hydrophonic recordings of the whistles of four adult and two juvenile bottlenoses as they swam slowly about their tank. He used these recordings to determine base rates for emission of the six most common whistle contours. He then played the recordings back into the tank, one whistle at a time. After each playback he recorded all whistles and other relevant behavior for 5 minutes. Each stimulus produced a different pattern of responses.

TABLE 7-1
COMPARISON OF WHISTLE CONTOURS BY RANK
ORDER OF FREQUENCY OF OCCURRENCE

DREHER (1961)		LANG AND SMITH (1965)		
Contour number equal to rank	Contour	Rank	Type	Contour
1				
2		1	A	
3		2	C, B, E	
4				
5				
6				
7		3	D	
8				
9				
32		4	F	

Adapted from "Linguistic Considerations of Porpoise Sounds," by J. J. Dreher, *Journal of the Acoustic Society of America*, 1961, 33, 1799-1800; and from "Communication between Dolphins in Separate Tanks by Way of an Electronic Acoustic Link," by T. G. Lang and H. A. P. Smith, *Science*, 1965, 150, 1839-1843. Copyright 1965 by the American Association for the Advancement of Science. Used by permission.

However, there was evidently no attempt to counterbalance for order of presentation, so some of these differences in response pattern may have been a result of habituation or some other order effect. Moreover, Dreher had no way of determining which dolphin made which response, so some of his results may be attributable to the identity of the responder rather than to a particular state induced by the stimulus. Nevertheless, Dreher's results are extremely interesting. Responses to Contour 1, a long, upward sweep that dolphins often produced while searching, included, first, repetition of the stimulus, then a series of falling, "distress" whistles together with echolocation clicks. Responses to Contour 2, which Lang and Smith (1965) called a "contact" signal, included periods of silence interspersed with intense echolocation and an increase in the rate of production of the double-humped Contour 4, which at other times had been associated with "irritation." Other contours produced excitement, orientation to the loudspeaker, and sexual arousal. As intriguing as Dreher's results are, it is important to point out, as Dreher does himself, that these findings are based on a classification into contour types, or phonemes, that may or may not correspond to that used by dolphins.

Saprykin, Kovtuenko, Korolev, Dimitrieva, Ol'sjamslo, and Becker (1977), using a computerized Fourier algorithm, have analyzed whistles in a different way. They examined not the patterns in pitch of the fundamental but variations in levels of acoustic energy in various harmonics. They concluded that whistles encode information in terms of relative strengths of these harmonics. It is possible, therefore, that analysis into contour types will not reveal all, or even any, of the "morphemes" used by dolphins.

Quacks, also called barks, squawks, and blats, are the third type of dolphin phonation. They seem to be associated with emotional behavior, including play, courtship, or contact with a human. Quacks may be emitted alone, or together with whistles. Lilly and Miller (1961a) suggested that quacks may be rapidly repeated click trains. Dolphins can modify their rate of production of quacks to receive a reward and sometimes use them to get the attention of their trainers (D. K. Caldwell & Caldwell, 1972). They are thus not only totally reflexive, but as is the case with most whistles, their significance to dolphins is not known.

Having dealt with some of the facts and mysteries of dolphin phonation, we can turn to an analysis of their communication system.

NEONATAL COMMUNICATION

Females may bear their first young in the spring of their fifth year. Gestation is approximately 12 months (Essapian, 1963), which means that young are born in and around the breeding season, when males are most aggressive. At least in captivity, this coincidence of reproductive events leads pregnant females to begin their withdrawal from others as early as 4 months before term. McBride and Kritzler (1951) noted that as parturition approached, two pregnant dolphins avoided the other dolphins in their tank but stayed close to each other, even though they had not previously shown such an attraction. McBride and Kritzler observed five normal births. Except where otherwise stated, the following description of birth and related behavior is drawn from their description.

Several days before birth, the female's withdrawal from her conspecifics becomes nearly total. During this period her vulva is distended from time to time. As parturition begins, she floats in place as though in sleep. Abdominal contractions begin to accompany the distensions of her vulva, and she gapes from time to time. Within a few minutes of the appearance of abdominal contractions, a particularly violent contraction results in the appearance of the tail of the fetus, with the flukes folded ventrally. At about this time the tank is often filled with quacks quite unlike those heard at any other time, but the source of these quacks is unknown. Contractions continue at intervals of 30 seconds or fewer. At each contraction a little more of the tail appears, but some ground is lost at each relaxation. The intensity of the contractions increases as the trunk begins to pass. At the peak of intensity, the mother's tail flexes upward, almost at right angles to her forequarters. During these powerful contractions the entire trunk of the fetus slides in and out of the vagina. Finally, a particularly violent contraction expels the infant in a cloud of blood. The mother immediately whirls, snapping the umbilicus, and the infant rises to the surface for its first breath. Normally, the infant is able to rise unassisted, but the mother stands by should it need help. In several stillbirths observed by McBride and Hebb (1948), the mother attempted to lift her dead infant to the surface. The entire birth sequence lasts from 20 minutes to an hour (Essapian, 1963). The placenta is passed several hours later and is not eaten. From the moment of birth on, mother and calf whistle almost constantly. These whistles are difficult to hear, for, according to Essapian, delivery is celebrated by a clamor of whistles and other sounds produced by all the other dolphins in the tank, all of which are in attendance.

Within an hour or so after birth, the 1-meter-long, 12-kilogram infant can swim alongside its mother at her top speed. It is probably assisted by a "drafting" effect, for the preferred position is just behind and slightly to the side of the mother's dorsal fin. Attempts to nurse may begin as early as 9 minutes or as late as 4 hours after birth. In such attempts the infant is assisted by the delineation of the area around the nipple by contrast between the dark dorsal and light ventral surfaces (Evans & Bastian, 1969). Eventually the infant finds a nipple and grasps it between its grooved tongue and its palate. Underwater nursing is accomplished with the aid of muscular vesicles on either side of the mother's genital slit. Milk flows into the vesicles from the mammaries, and when the infant grasps a nipple, the mother expels the milk into its mouth. During the first 2 weeks, the infant nurses once or twice an hour. Later, nursing is less frequent. During early nursing sessions the mother rolls onto her side. It is not obvious how this behavior facilitates nursing because the infant can approach from below. Perhaps it is a vestige of the dolphin's terrestrial past.

For the first few weeks, mother and infant are always close to each other, even when sleeping. The infant sleeps just below the mother's tail, rising with her once or twice a minute to breathe. It solicits care and attention from her by nudging her genital region. Because the bulls are physically aggressive as well as sexually importunate, the mother must defend herself and her infant from them. She guides it by nudging it with her snout and attempts to keep it away from all other dolphins. In captivity she steers it away from the sides of the tank and all foreign objects. She guards it by chasing other dolphins, especially males, away.

Infants quickly become strongly attached to their mothers. McBride and Hebb (1948) observed a 1-week-old infant that was momentarily separated from its mother. It swam rapidly at the surface in a circle 2 meters in diameter, whistling continuously, until she rejoined it.

As the infant grows more independent, it sometimes resists her guidance by fleeing. In such cases the mother may discipline it in a highly terrestrial manner—she pins it to the bottom of the bank. The opposite treatment is also effective punishment. M. C. Tavolga (1966) saw a mother swim upside down, grasp her obstreperous offspring with her flippers, and hold it out of the water. Thereafter, the chastened infant remained at her side.

The mother is not alone in her attempts to protect her infant. Even before her labor begins, adult females repeatedly swim beneath her, turn on their sides, and inspect her belly. As soon as the newborn drops free, one of these females will often swim beside her as she rises to the surface with her infant. Within a few hours the infant swims between the mother and an escort, usually one of the attending females. Much of a mother's effort is spent retrieving her calf from these solicitous associates, which compete with her and one another for the privilege of escorting it. Even bulls, ordinarily guarded against, may become guards themselves. They sometimes share in baby sitting, swimming with the infant while the mother feeds (Essapian, 1963). In one case a calf repeatedly answered a solicitous bull's calls by joining him, and together they attempted to escape from the mother. Such alloparental care has also been described by M. C. Tavolga and Essapian (1957) and by Tavolga (1966). In some cases it may extend to virtual adoption, as when two 10-months' pregnant dolphins repeatedly slowed to assist a foundling in its (futile) attempts to nurse. Both females guarded the infant until it died.

The calf's teeth begin to erupt when it is about 6 months old, and at about the same time it begins to eat fish. Weaning is not complete until it is about 10 months old, when the mother actively repulses its attempts to nurse. By this time the calf is spending much of its time away from its mother, but it continues to join her to rest until it is at least 2 years old. If the mother becomes pregnant, she will reject all approaches with increasing vigor as her term approaches (Tavolga, 1966). Otherwise, the association can last much longer. Tavolga observed 4-year-old and 6-year-old dolphins that returned to their mothers' sides whenever they were tired or alarmed.

INTEGRATIVE MESSAGES

Young dolphins are even more playful than adults. As early as 2 weeks after birth, captive calves begin to swim around the heads of other dolphins while their mothers rest (McBride & Kritzler, 1951). At 5 weeks they imitate the adults as they chase rays and sharks about the tank (McBride & Hebb, 1948). At about the same age they mimic the maneuvers of adults as they tease and scare fish from their hiding places.

Imitation is probably an important factor in the precocious development of sexual play, which is a major part of the repertory of all captive dolphins. From 6 weeks of age on, males display penile erection and copulatory movements. Since they will not be fertile for several years, this behavior must be considered playful. Erection is not good evidence for true sexual motivation because dolphins appear to have considerable central control over the position of the penis. McBride and Hebb described a young male that swam upside down, towing a feather with his erect penis. In captivity young males display copulatory behavior not only with other dolphins of any age or either sex but with turtles, sharks, and even eels. Captive subadult males sometimes display sexual role differentiation, one assuming the typical male inverted posture beneath the other, which remains passive. Males have not been observed to ejaculate during sexual play. A female calf was also observed to indulge in sexual play, presenting her genital region by swimming in front of another dolphin, then turning on her side (Saayman et al., 1973). Although there are many detailed descriptions of elaborate sexual play in captive dolphins, there are no such reports from the field. Thus, the sexual play of dolphins, like that of some primates, may be an artifact of captivity.

In their play captive dolphins seem to prefer sex to violence. High-speed chases with jumping and splashing are common, but should the pursuer catch up with the pursued, their roles are simply reversed. Saayman and his colleagues noted aggressive play in the form of "open-jaw sparring" and mock "threats" in a captive group, but these behaviors were consistently low in frequency compared to other playful behaviors, such as leaping and chasing. Only mild agonism is evident in such games as keep-away, in which dolphins compete for feathers or other toys, sometimes chasing each other for more than an hour at a time (McBride & Hebb, 1948). Threats and fighting are conspicuously absent from descriptions of such games. Moreover, several ritualized forms of play depend heavily on cooperation rather than competition. In a kind of catch game described by M. C. Tavolga (1966), one infant placed a pelican feather in a water jet so that the current carried it to a second dolphin, which caught it and returned it to the jet, so that the first dolphin could catch it. Another team sport involved a fish that two dolphins chased back and forth between them, often for 15 minutes at a time. A third game was fetch, played by repeatedly retrieving an inner tube or other floating object for a human to throw. This game is of particular interest because, in at least one aquarium, it was invented by a dolphin, not a human.

In one study whenever a human interrupted the game of fetch before the dolphin was ready to quit, it would raise its head from the water and emit a series of high-pitched, grating quacks. McBride and Kritzler (1951) interpreted this display as an expression of impatience, presumably directed at the human. This phonation may also serve as a solicitation, for it usually ceases as soon as the game is resumed. Another call often used to attract the attention of a human is a whistle rising in pitch with a plateau followed by a second rise. A similar call, but without the plateau, was produced by a captive dolphin when "spoor" of other dolphins was introduced into its tank. Lilly (1963) interpreted this phonation as an attention-getting signal. Yet another auditory solicitation is an underwater exhalation that signifies "begging or inquisitiveness" (D. K. Caldwell & Caldwell, 1972, p. 470). Physical contact may also serve as a solicitation. Bateson and Gilbert (1966) found that dolphins sometimes solicit attention from those of higher rank with "beak propulsion," swimming with their beaks inserted into the receiver's genital aperture. This signal is probably derived from the nursing posture.

Because they travel and hunt in groups, dolphins must have effective means of staying in touch with one another. In the sea, which is always somewhat opaque and noisy, maintaining contact is not a trivial problem. When swimming in open water, dolphins surface to breathe more or less simultaneously. Their exhalations produce an explosive burst of white noise, which is probably audible to others at the surface. That this blowing may serve to maintain contact is suggested by the Caldwells' observation that blowing by one captive dolphin is often echoed by others in nearby tanks. Other tonic noises that may inform dolphins of one another's presence are echolocation clicks, splashes, and, when the sea is calm, rumblings of intestinal gas. The Caldwells (1972) suggest that a fluke motion sometimes acts as an "intention movement for maintaining synchronous swimming" (p. 802).

Since both captive and free-ranging dolphins respond to whistles with whistles of their own, Lilly and Miller (1961b) inferred that whistles could serve as contact signals. The Caldwells also postulated that such "call–answer sequences" (Evans, 1967) inform members of a herd of one another's locations.

T. G. Lang and Smith (1965) allowed two dolphins to communicate via an electronic audio link that was connected and disconnected at 2-minute intervals. They found that whistles were much more frequent during "connected" intervals than during "disconnected" ones. During the connected intervals, whistles formed "exchanges," defined as rapid antiphonies of three or more whistles from each subject. Three of the exchanges consisted entirely of Type A whistles, and the investigators suggested future experiments to determine whether this whistle, with its relatively simple waveform, is used primarily to maintain acoustic contact. (See Table 7-1.)

Nine of the exchanges consisted of Type B whistles from one subject and Type D whistles from the other. In seven other sequences the dolphins began with such exchanges, then exchanged whistles of other types. Lang and Smith hypothesized that Types B and D were used as "call signals" to identify the transmitter. A similar suggestion was made by Essapian in 1953. In an investigation of this possibility, the Caldwells (1965) found that 90% of the whistles produced by each of their five captive subjects were specific to that individual and had a distinctive pitch contour regardless of the loudness, duration, or rate of repetition of the whistle. Moreover, each dolphin emitted its whistle in a wide variety of situations. These situations included introduction to an unfamiliar tank, entrance of divers into the tank, feeding, and the lowering of a dead dolphin from the same school into the tank. In a later (1972) publication, they note that some dolphins have a "secondary" whistle, also peculiar to the individual and independent of concurrent behavior. These results are critical to the interpretation of Dreher's (1966) playback experiment described previously. The differences in response patterns in this experiment may have been due to differences in the frequencies of production of "call signs." These differences in turn may have been due to differences in the identity of the dolphin that was most highly stimulated by the recording.

The dolphin's great ability to mimic sounds (Lilly, 1967) may be related to the maintenance and recognition of groups. In a pattern of phonation termed a duet by Lilly and Miller (1961b) and chorusing by Dreher and Evans (1964), two dolphins whistle in unison, reproducing each other's often complex changes in pitch and amplitude exactly. E. O. Wilson (1975), elaborating on a suggestion made by Andrew (1962), has suggested that mimicry might result in convergence in the phonation patterns of the members of a group. This convergence would permit a separated individual to recognize and rejoin the others. As Wilson points out, this ability would be particularly valuable to a creature that swims in the ocean at high speeds, frequently leaving and rejoining its fellows. Such separations would be especially common during the excited milling that occurs during feeding. Saayman and his associates (1973) observed many cases in which a school of several hundred dolphins (presumably containing many groups) entered a bay, formed a single line abreast, and repeatedly swept through the bay, reassembling rapidly after each sweep. Rejoining a group after such an exercise would require the ability to recognize associates amid a cacophony of whistles, clicks, and splashes.

Maintenance of a group depends not only on the ability to locate and recognize other members but also on bonds of affiliation. Dolphins create and express these bonds by several forms of pleasurable body contact. One of the most common of these forms is allogrooming. It may be surprising to find allogrooming in a mammal without fur, but by giving up fur, the dolphin has not escaped the louse and has even acquired a new nuisance, the barnacle (Pilleri & Knuckey, 1969). Evidently, the incidence of such parasites is low, for Lilly found none, and Pilleri and Knuckey report only a single case. Lilly attributed this cleanliness to the constant shedding of the outer layer of skin. The rate of shedding is accelerated not only by a variety of caresses but also by rubbing on inanimate objects (McBride & Hebb, 1948). Social rubs are sometimes performed as two dolphins swim together, one gently rubbing its torso across the flippers of the other. In other cases one brushes another with its flukes, lower jaw, or flanks. Small particles of skin flake off the animal so treated (D. K. Caldwell & Caldwell,

1972). Pepper and Beach (1972) showed that a mild tactile stimulation from a human could act as a reward.

Besides allogrooming, several other forms of intimate contact are common among like-sexed or different-sexed dolphins, in or outside the breeding season. Saayman and his colleagues (1973) frequently saw two dolphins swim slowly together, one with the tip of its dorsal fin in the genital slit of the other. The Caldwells often observed one dolphin nosing the genitals of another or, in a more restrained display of affection, swimming side by side with their flippers touching. Often several dolphins would swim together in this way, each in contact with the next. Evidently, dolphins find such contact highly rewarding, for they actively solicit it from humans with whom they are familiar. Given the overtly sexual nature of some of these contacts, distinctions between affiliation and sex probably mean little to these affectionate and nearly pansexual creatures.

Physical contact can be more than gratifying—it can be a matter of life and death for a dolphin in distress. When a dolphin is sick or injured, it often has difficulty rising to the surface to breathe. In such cases, or if it is an infant separated from its mother, a dolphin produces a "distress whistle" (Lilly, 1963). The distress whistle has two parts. Sonagrams show that the fundamental of the first part usually rises from about 4000 hertz to 8000–20,000 hertz in approximately 1.4 seconds. There is then a pause lasting a few tenths of a second. The fundamental then drops, mirroring the initial rise. A similar pattern is seen in the harmonics, which range from about 8000 to 16,000 hertz or even higher. The whistle is loudest during the initial rise, becomes softer at higher frequencies, then again becomes louder as the pitch drops. The call is usually repeated in groups of up to 50. According to Lilly, humans can learn to distinguish the distress calls of individual dolphins.

Responses by other dolphins to a distress call are immediate. Lilly distinguishes two phases in these responses. "Primary tactics" include instant cessation of phonation, approach, and attempts to lift the victim to the surface. An exchange of whistles between victim and rescuers ensues and is followed by one of several "secondary tactics," adapted to the victim's problem. When the caller was an infant that had evidently become confused and was unable to back out of a corner, the mother steered it free, upon which it stopped producing distress whistles. In cases in which injury resulted in a "list" that allowed water to enter the blowhole, another dolphin positioned itself so as to hold the injured dolphin upright. When a sick dolphin was unable to rise to the surface even in shallow water, a rescuer pressed its flukes to the bottom, straightening the dolphin so that its blowhole was out of the water. Rescues often involved two dolphins, which took turns assisting while the other surfaced to breathe. In one such case they provided continuous support for 4 days and intermittent support for another 2 weeks.

There have been several observed cases of assistance to dolphins injured in the open sea. Siebenaler and Caldwell (1956) described a case in which a charge of dynamite stunned a member of a group of 25. The injured dolphin could swim but listed at 45°. Two adults immediately placed their heads beneath its flippers and held it at the surface. In this position the supporting dolphins could not breathe themselves, and they periodically rose to the surface some distance away. An apparently different pair instantly took their place. After several minutes the injured dolphin recovered, and the whole school swam away at high speed, jumping in unison. In a similar case described by the Caldwells (1966), a bottlenose was accidentally foulhooked by a sport fisherman. It gave a "loud squeal," upon which dolphins converged from every direction and milled around the injured animal. Perhaps as a result of this milling, the line broke, and the dolphin was freed. In their review of such incidents, the Caldwells note one case in which the rescuer was known to be male. In this case, as in several others, the attempt was directed toward a dead delphinid. Supporting behavior directed toward dead dolphins and members of other species is evidently quite common (Hubbs, 1953; Norris & Prescott, 1961).

Although distress whistles are stimulated by pain or other pathology, alarm whistles are stimulated by fear. The fearful stimulus may be a killer whale, a human, a vessel, or, in captivity, any novel object introduced into the tank. Dreher and Evans (1964) presented eight captives with a variety of presumably frightening stimuli, including a dead infant and a model dolphin towed with a string. They compared the frequencies of occurrence of various whistle contours before and during these presentations. They concluded that a sharply falling whistle (similar to the second half of the distress whistle), a fall/rise/fall pattern (Table 7-1), and a single-humped whistle (Contour 8) were associated with "concern and alarm" (p. 381). Their findings support Lilly's (1963) contention that distress and alarm whistles are easily distinguishable.

Another phonation, produced under circumstances similar to those contrived by Dreher and Evans, is a single, loud "crack" (Caldwell, Haugen, & Caldwell, 1962). Preliminary analysis shows that the crack has a rapid onset, is less than .1 seconds long, and carries most of its energy below 8000 hertz. As Caldwell and her colleagues point out, the combination of high volume, abruptness, and low frequency is well adapted to communication of alarm over long distances. They were unable to determine the mechanism of production but concluded that clicks are probably produced inside the head. It is possible that cracks are produced by the same (unknown) mechanism that produces clicks.

Since a group of free-swimming dolphins ordinarily produces a nearly constant chorus of whistles and clicks, silence can be a signal. D. K. Caldwell and Caldwell (1972) found that at the appearance of danger, a group of captive dolphins fell silent and noted that the sudden hush was as salient as a loud noise.

The first response to an alarm signal is bunching up into a tightly knit school, usually with infants and mothers at the center (McBride & Hebb, 1948; M. C. Tavolga & Essapian, 1957). What happens next depends on the nature of the danger. In open water, dolphins usually flee, either leaping at the surface or diving and changing direction under water (Pilleri & Knuckey, 1969). McBride and Hebb noted that when a tiger shark was introduced to a group of seven bottlenoses, they first swam excitedly in a tight school, then rammed the shark repeatedly, killing it within a few hours.

AGONISTIC MESSAGES

On occasion, dolphins attack each other with the same ferocity they show to sharks. Ordinarily, however, such deadly attacks are prevented by displays of dominance or threat or by less serious forms of fighting. Next to size of body and frequency of aggression, the clearest signs of dominance are seen in patterns of movement— dominant dolphins swim freely, while subordinates submit by quickly moving out of their way. Should they fail to do so, they are liable to be attacked. D. K. Caldwell and Caldwell (1967) noted that dominants often stared at subordinates and that this display generally elicited withdrawal of the inferior. They also noted (1977) a submissive posture in which the mouth is held closed and the flank presented, and pointed out that this posture is antithetical to threat.

Swimming formations also provide clues to the rank of a dolphin. Pilleri and Knuckey (1969) found that animals with large, powerful flukes (presumably males) led most groups and that calves, flanked by several females, followed. Using underwater observation vehicles, Evans and Bastian (1969) observed a stacked swimming formation, with larger animals nearest the surface and smaller and presumably younger ones lower down. They suggested that this formation was literally a dominance hierarchy, based on rank-related access to the surface. They noted that within the stack a group of what appeared to be young males all swam at the same depth, suggesting equality of rank. This observation is suggestive of rankless bachelor herds in some ungulates (see Chapter 5) and may be relevant to the question of dolphin ancestry. The Caldwells observed homosexual copulations among captive males and noted that the dolphin that intromitted became the dominant.

"Ownership" of food items is an exception to the deference generally shown an animal of higher rank. According to Norris (1967), once a dolphin has caught a fish, the dolphin can play with it or even let it go briefly without fear that another dolphin will take it. M. C. Tavolga (1966), however, noted several exceptions to this respect for possession. Other dolphins sometimes attempted to take food fish directly in front of the dominant male, which immediately asserted himself by threat or attack.

Dolphins sometimes use threatlike displays to express frustration. Dreher (1966) recorded a double-humped whistle that was almost always associated with "impatience, annoyance, or irritation" (p. 536). In similar circumstances the dolphins studied by the Caldwells (1972) smashed their flukes onto the surface, producing a loud report and often drenching the observer.

The line between dominance and threat is not easily drawn because all kinds of aggressive acts are more commonly performed by high-ranking rather than low-ranking animals. Nonetheless, there are several aggressive signals that may be used, however rarely, by animals of any rank. A common occasion for threat by the dominant male observed by Tavolga was an attempt by a younger male to copulate with the female with which the dominant was consorting. Sometimes, however, his threats seemed to occur spontaneously. But he was not the only one that used threats. Mothers in the same group often threatened males that got too close to their infants.

The mildest and most common form of threat is the jaw clap. This display, first described by McBride and Hebb (1948), has both visual and acoustic properties. The jaws open slowly, showing the teeth, then snap shut, producing a loud, percussive clap. The clap is often repeated several times in rapid succession. Jaw clapping is often associated with a posture in which the dolphin arches its back while facing its opponent, appearing to be poised for a charge. If the offender does not retreat, this charge may occur. This usually results in a short chase or a bout of open-jawed sparring (D. K. Caldwell & Caldwell, 1977; Saayman et al., 1973).

If one or the other of the dolphins is highly aggressive, the chase or sparring may culminate in a serious attack, one biting and raking the flukes, peduncle (the narrow "waist" attaching the flukes to the body), and dorsal fin of the other. More commonly, it will ram the other with its snout or slam it with a powerful blow of its flukes. Both in the wild and in captivity, such attacks often result in bruises or scars. Norris (1967) examined scars on several wild dolphins and concluded that most were produced by the teeth of other dolphins. Similarly, Saayman and his colleagues attributed the tattered dorsal fins of older dolphins to bites inflicted by conspecifics. McBride and Hebb noted that most of the scars on their captives were on areas most commonly bitten: snout and face, genital region, or peduncle and flukes. They inferred that there was considerable restraint even in serious attacks, for a male that repeatedly attacked an infant left only shallow teeth marks. Nevertheless, in captivity at least, fighting does sometimes result in serious injury or even death. When two young dolphins were introduced to a group, the only adult, a female, attacked them again and again. They eventually died of their injuries. But not all injuries result from fights, for, as we shall see, the courtship of dolphins is often quite violent.

SEXUAL MESSAGES

Dolphins do not obey most ordinary constraints on sexuality. They do not reach maturity until they are about 4 years old, but they display adult-like sexual behavior much earlier. For example, M. C. Tavolga (1966) observed a 3-month-old male that persistently attempted to copulate with its mother. Although the breeding season is limited to a 3-month period in late winter, males are sexually active all year long, often with members of their own sex. Courtship is usually initiated by males, but females use the same postures, with roughly the same frequency. The most fundamental principles of sexual etiquette are flagrantly disregarded during the breeding season, when as many as four bulls may simultaneously attempt to mate with a single cow.

The most complete description of courtship behaviors in both captive and free-ranging dolphins is that of Saayman and his associates (1973). Unless otherwise stated, this account is based on their article. As the breeding season approaches, bulls spend more and more time with the cows, often displaying erections. Because the bull's penis is rigid and, like that of some artiodactyls, erected and retracted by ligaments, it can be deployed and withdrawn quite rapidly, and erections can be sustained throughout several copulations. Males display readiness by erection and possibly by a pink ventral coloration seen in the breeding season (D. K. Caldwell & Caldwell, 1977).

No one knows whether cows advertise their estrus chemically. Although dolphins cannot smell, they can taste and may thus be able to detect substances released by receptive females. McBride and Hebb (1948) noted that when a cow was placed in a tank connected to another tank containing a bull, he almost always displayed an erection. Since the response was immediate and the gate connecting the tanks was opaque, McBride and Hebb concluded that the arousing stimulus was probably auditory. One of Dreher's (1966) subjects was similarly stimulated by recordings of Contour 3 (see Table 7-1). It is possible that this call identifies females and was responsible for the erections reported by McBride and Hebb.

Sexual advertisement by chemicals may be problematical, but advertisement by postures is not. Cows issue unequivocal invitations by swimming in front of bulls and turning on their sides, presenting their genitals. Another common female sexual solicitation is resting the head on a bull's back, then rubbing slowly across his back so that his dorsal fin caresses her belly. Like other sexual displays, these invitations are also performed by bulls.

Courtship usually begins with display swimming, one partner racing around the other, usually just below the surface, and often corkscrewing or swimming upside down. Display swimming sometimes becomes a chase, both partners corkscrewing and leaping, alternately or simultaneously. Leaps often end in a loud splash as the dolphin falls back into the water on its belly or flank.

Excited chases alternate with languid interludes. In captivity a bull may hang motionless in the water, head down and tail up, displaying an erection, while cows gather and assume the same attitude. Consorts caress each other with flippers, flukes, or the whole length of their bodies. These caresses are delivered to almost any part of the body but are concentrated on flippers, flukes, and genitals. Sometimes the pair swim slowly together, one with its dorsal fin in the other's genital slit.

Such tenderness may, however, give way at any time to frantic chasing and vicious biting, usually by the bull. The bites are delivered to the cow's head, tail, dorsal fin, and genitals. They are powerful enough to tatter the fins and scar the skin. Should the female escape, the male emits a characteristic series of "yelps" (probably a distress quack) until she returns (Wood, 1973).

Copulation begins when the male approaches the female from

below and inspects her belly while emitting clicks. He then approaches from the rear quarter with an erection and assumes a sigmoid posture (M. C. Tavolga & Essapian, 1957). According to the Caldwells (1977) and Puente and Dewsbury (1976), however, this posture is *not* associated with copulation. Rather, a mutual head-on ram at high speed is almost always followed by copulation. The male clasps the female with his flippers and attempts to intromit. The female may resist by rolling away or striking out with her flukes, or she may cooperate by remaining still. A "whimpering" may be heard at this time (Slijper, 1977), but it is not known whether it is produced by the male, by the female, or by both.

Coitus is quite brief in comparison to the lengthy and elaborate preliminaries. Tavolga and Essapian observed two types of intromission: partial, which lasted less than 10 seconds and definitely resulted in ejaculation; and complete, which lasted 30 seconds and may or may not have involved ejaculation. Pelvic thrusting occurred before, during, and after intromission. Copulation recurred at intervals of 1 to 8 minutes for more than half an hour.

The spirit of the dolphin is most clearly expressed through sex, and no narrative can do justice to the power and grace with which they glide through their erotic dance.

SUMMARY

Though alien in form and ecology, dolphins display a typically mammalian repertory of neonatal interactions. This repertory includes nursing, an infantile distress signal, contact signals, and maternal guarding and guiding. Alloparental behavior is highly developed and includes protection from other dolphins, especially the highly aggressive bulls.

The entire message system is permeated with sexual gestures adapted to other uses. In adults, play and sex are mixed with aggression, but it is usually possible to tell which of these motivational systems is dominant in any given interaction. Dolphins show tremendous variety, plasticity, and inventiveness in their play, probably more than any terrestrial mammal. Play involves considerable sexual behavior, very little aggression, and a degree of cooperation greater than that of any mammal except humans. Dolphins solicit attention with quacks, whistles, and nudges applied to the genitals. They maintain contact with one another through tonic interactions such as simultaneous surfacing and by signals, which may include a characteristic group whistle acquired by imitation. Each dolphin has a characteristic identifying whistle of its own, which forms a large fraction of its sonic output. Strong attachments between group members are maintained by a variety of forms of physical contact, many with sexual overtones. Dolphins have a distinctive distress whistle that brings other dolphins to their aid. An alarm whistle, easily distinguishable from the distress whistle, elicits an immediate tightening of the swimming formation, sometimes followed by coordinated flight.

High social status is signaled by freedom of movement, position near the surface in the stacked swimming array, and high frequency of threat and attack. Threats include jaw claps, a posture suggesting readiness to charge, charges, and a single-humped whistle. Forms of attack include ramming, slamming with flukes, biting, and jaw wrestling.

Females may advertise sexual readiness by chemical or auditory means, and they definitely use gestures and postures, including caresses and genital presentations. Most courtship behavior is displayed with equal frequency by members of both sexes. This behavior includes display swimming, leaping, a variety of caresses, and biting. When his consort withdraws, the male emits a characteristic yelp until she returns. Courtship is long, but copulation is brief. Erection is rapid, and intromission always lasts less than half a minute. Coitus is accompanied by whimpering of unknown origin and may be repeated many times over a period of about half an hour. In all, dolphins use at least 16 different types of messages.

There are several areas of nearly total ignorance within the field of dolphin communication. Dolphins show a variety of facial expressions, but their behavioral correlates have not been investigated. There is an interesting discrepancy between the results obtained by the Caldwells (1972), who found that 90% of the whistles produced by an individual were its "call sign," and the much lower rates of production for such whistles obtained by other investigators, notably T. G. Lang and Smith (1965). The situational or individual differences responsible for this discrepancy demand further study. Instrumentation that will allow identification of the source of each call in a group of captive but freely interacting dolphins would facilitate interpretation of data from playback studies like Dreher's (1966).

Finally, we return to the question with which we began our discussion of dolphin communication. Is there a remainder of unexplained signaling left after subtraction of messages common to other mammals? Yes, and it is very large. The meaning, if any, of most dolphin whistles remains a mystery. This does not mean that dolphins have a language, however. Bateson's (1966) comment is apposite: "I do not think that any animal without hands would be stupid enough to arrive at so outlandish a mode of communication" (p. 371). We should not be dismayed by Bateson's skepticism. That the evidence fails to support the belief that dolphins use language makes them more interesting, not less. Their mythic status in the modern imagination is an informative projection of our own concern with our place in the world. And a myth need not be false.

8
WOLVES
AND CATS

A dog's a dog — a cat's a cat.

T. S. ELIOT

Wolves (*Canis lupus*) and cats (*Felis catus*) represent, respectively, the *canid* (dog) and *felid* (cat) families of the order Carnivora. The form and content of their messages have been heavily influenced by their carnivorous lifestyle. Because they can kill creatures of their own size, they have evolved special inhibitory mechanisms to prevent individuals from destroying their kin. As we shall see, these mechanisms do not prevent lethal fights among wolves and cats that are not closely related. Although wolves and cats are carnivores, they have different social organizations, communications, and personalities. Wolves are highly social. Cats, although considered solitary, do often interact with others of their kind. It may be that as we learn more about other supposedly solitary species, we will find that they, like cats, are not as aloof as we have been led to believe.

Humans have established different relationships with wolves and cats. We have driven the wolf back into the most remote and inaccessible corners of its original range, which included most of the Northern Hemisphere. Elsewhere, the wolf prevails only in its domestic incarnation, the dog. Cats, on the other hand, have extended their range with the aid of humans and their vermin without much domestication. They seem to retain their original wildness even when resting in the parlor.

WOLVES, CANIS LUPUS

The social behavior of wolves is so suggestive of our own that their society has been proposed as a model for human social evolution (Hall & Sharp, 1978; Kortlandt, 1965; Read, 1923). Like humans, wolves have distinct personalities, are attached to members of their group, are often hostile to strangers, and can express subtle variations of a great variety of moods. They share so many behavioral and physical characteristics with their domesticated cousin, the dog, that visitors to zoos are often disappointed at the wolves' resemblance to the family pet. Nevertheless, the dog is no more or less than a domesticated wolf.

214

Differences between the two are so few that at least one taxonomic authority (Bohlken, 1961, cited in Mech, 1970) has suggested that the dog be considered a subspecies of the wolf.

The most significant physical differences between wolves and dogs of comparable size are in the skull, the wolf's being longer and broader with a shallower orbital angle and rounded rather than compressed tympanic bullae. Furthermore, wolves, unlike most dogs, swing their legs on the same side in the same plane (see Figure 8-1). Finally, wolves' tails are not curly, as are the tails of most dogs (Mech, 1970).

The behavioral similarities between wolves and dogs have been well documented by Zimen (1971). He compared the development of a wolf with that of a poodle and found a remarkable correspondence between both the forms and the timing of emerging behavior patterns in the two animals. In general, wolves are wilder in disposition than dogs—they are more curious, expressive, and excitable, particularly when their movements are restricted. Because of these wild characteristics, attempts to make a wolf into a pet are usually as distressing to the owner as they are cruel to the wolf.

FIGURE 8-1   A wolf showing its characteristic stride. Photograph by Roger Peters.

Part of the temptation to keep wolves as pets is the ease with which even an untrained observer can read their expressions. A number of physical features make these expressions easy to see. The generally light color of the lower part of the face sets off the black lips, which can be drawn forward, pulled back, or retracted vertically. Wide tufts of fur stick out below the ears, increasing the apparent size of the face and revealing, even at a distance, which way the wolf is looking. The fur on the brow is often dark-tipped, so that contractions of the underlying muscles produce striking vertical "frown" or horizontal "worry" lines. (See Figure 8-1.) My observations at the Brookfield Zoo, near Chicago, suggest that these facial expressions are sometimes perceived by other wolves at distances of up to 20 meters.

At greater distances the positions of ears, mane, and tail can inform wolves (or even an airborne ethologist) of a wolf's disposition. In general, elevated positions of these organs signify dominance or threat; lowered positions signify submission or friendliness. Besides serving as an organ for visual signaling, the tail probably also functions at short range as an odor wafter.

The wolf's voice is as expressive as its body. The voice is capable of a tremendous variety of pitches and volumes, from growls and squeaks inaudible only a few meters away to a howl that can carry for kilometers. Howls may sound mournful or terrifying to humans, but as we shall see, this is one area in which our sensitivity to signals from the wolf may lead us astray.

The wolf's expressiveness, like its other social characteristics, is an adaptation to the social hunting of large, fleet-footed animals. These adaptations are so successful that, in the words of Goldman (1944), "It seems doubtful whether any other species of (wild) land mammal has exceeded this geographic range, and this wolf may, therefore, be regarded as the most highly developed living representative of an extraordinarily successful mammalian family" (p. 1). Wolves were once common throughout the Northern Hemisphere except in deserts and tropical rain forests. Largely as a result of human interference with wolves or their habitat, they are now restricted to relatively inaccessible areas, including Alaska, Canada, northeastern Minnesota, northern Asia, and the mountainous regions of the Middle East and South Central Asia (Goldman, 1944). In most of these places, humans remain the wolf's only major vertebrate enemy, but rabies can sometimes wipe out most members of a pack (Chapman, 1978).

The principal prey of the wolf are large ungulates, especially deer, moose, mountain sheep, and wapiti. Wolves also take some livestock and sometimes eat smaller mammals, including beavers, hares, rabbits, and even mice (Mech, 1970). The last are consumed only as a last resort or in play. They are sometimes swallowed whole and pass through the wolf nearly intact, presumably providing only a minimal amount of nutrition. In summertime, scats reveal that wolves often feast on berries.

To hunt the speedy, agile creatures that are the staple of their diet, wolves travel on routes that encompass a large area, probably moving from one region of prey concentration to another as each is alerted to their presence. Using radiotelemetry, Mech (1973) discovered that groups of wolves in northeastern Minnesota have ranges varying in size from 125 to 310 square kilometers when prey is available. When prey is extremely scarce, the wolves may make excursions of 20 kilometers or more and then return to their original range.

Wolves are among the most social of social carnivores—the lone wolf is the exception, not the rule. Of almost 5000 Alaskan wolves seen by Rausch (1967), 91% were accompanied by other wolves. Similar figures have been obtained in Minnesota by Stenlund (1955) and Mech (1977b). As Mech (1970) points out, several factors determine the size of wolf groups. First, the advantage of hunting large prey in a group must outweigh the repulsion caused by competition among them. These two pairs of countervailing factors interact to produce groups called packs, which generally contain four to eight members. The range of pack sizes is two to 36, with an average of about five or six. Packs sometimes split up for days at a time. The subgroups remain in the territory using the same trails, then reassemble (Mech, 1970). In 1938, on the basis of his observations of wolves and their sign in northern Minnesota, Olson concluded that wolf packs are families composed of a pair of adults and several cohorts of offspring. Since then, a number of field studies have validated this conclusion (Mech, 1973; Murie, 1944; Young, 1944).

A number of studies on captive wolves have further elucidated relations within the pack. The nucleus of a pack is a pair of adults, the alpha male and the alpha female. They are usually the only members of the pack to mate. There is some evidence, however, suggesting that the alpha male does not always breed but that a male of lower rank may do so (Rabb, Woolpy, & Ginsburg, 1967). There is substantial evidence that the alpha pair actively suppress attempts by other pack members to mate (Schenkel, 1947).

In addition to the alpha pair, three other classes of pack members can be distinguished: subordinate adults; pups and yearlings; and pariahs, or omega wolves, which follow the pack, maintaining only a loose association with the other members. Schenkel described separate linear-dominance hierarchies for males and females. Recall that *linear* means that dominance relations are transitive; if A dominates B and B dominates C, then A dominates C. Rabb and his colleagues and Woolpy (1968) have described several exceptions to the separation of male and female dominance systems. The alpha female dominates most of the males, and during the breeding season and while nursing pups, she may even dominate her mate. Usually, however, the alpha male is the "lord and master" of the pack (Murie, 1944). Rabb and his colleagues have documented several situations, including the removal, death, or introduction of a wolf, that can disrupt prevailing dominance relations, but ordinarily they are fairly stable. The idiosyncrasies of wolf personality often make determination of dominance relations quite difficult, but Mech (1970) has shown that dominance relationships within a litter may emerge as early as the fourth week of life.

Mech (1973, 1977b) has shown that wolf ranges in Minnesota are territories—that is, areas that each pack uses exclusively. These territories are large, averaging about 230 square kilometers, and only a narrow buffer zone 1 or 2 kilometers wide is shared by neighboring packs. The territories are so large that encounters between packs are extremely rare, even on those few occasions when one pack does trespass deep into another pack's territory. When such encounters do occur, serious fights often result (Mech, 1977c). The contrast between the predominantly friendly relations within packs and the violent interactions between them is striking.

Lone wolves are nomadic (Mech, 1972, 1973, 1974; Mech & Frenzel, 1971). Not only do they not have territories of their own, but they actively avoid territorial centers of other wolves, traveling mainly along the peripheral buffer zones (Mech, 1972, 1973). Except when pairing up with another wolf, they restrict their use of scent marking and howling, the two major forms of long-distance communication between wolves. Doing so probably helps them to avoid detection and attack by packs (Rothman & Mech, 1979).

NEONATAL COMMUNICATION

In early spring, 3 to 5 weeks before parturition, the mother-to-be begins visiting a den. Her major activity during these visits is excavation, but as parturition approaches, she and the other adults of the pack may cache meat by burying it near the den. Sometimes the female may dig a new den, literally starting from scratch, but often an existing burrow of a fox, badger, or other mammal is used. In

Minnesota some wolves take advantage of crevices beneath boulders, thus minimizing the amount of digging necessary to provide shelter. The same den is often used year after year, but even in such cases some advance preparation is required. Dens are almost always located near water, usually on a hillside or ridge providing a view (Mech, 1970). Murie (1944) described a typical Alaskan den as follows:

> I wriggled into the burrow which was 16 inches high and 25 inches wide. Six feet from the entrance of the burrow there was a right angle turn. At the turn there was a hollow, rounded and worn, which obviously was a bed much used by an adult. . . . From the turn the burrow slanted slightly upward for six feet to the chamber in which the pups were huddled and squirming [p. 44].

At least a day before giving birth, the dam goes into the den and remains there until after the pups are born. Food is brought to her by other pack members, which carry it either in their mouths or in their stomachs, regurgitating it when they reach the den (Murie, 1944). In spite of this solicitude, not even the father is allowed to enter the den until the pups are several weeks old (Fox, 1971).

Bill Brind's excellent film *Wolf Pack*, produced in 1974 by the Canadian National Film Board, contains a remarkable sequence documenting the birth of wolf pups. The pups are born at intervals of 20 to 40 minutes. As each emerges, it is licked dry and its umbilicus nipped through. The placenta usually follows. As soon as the first pup is born, the den is filled with squeaks, moans, and rough, medium-pitched vocalizations that might be represented "mm, mm, mmmu." Soon the pups begin to nurse, pressing the area around the nipple with repeated, simultaneous thrusts of both forepaws.

J. P. Scott and Fuller (1965) have described the behavioral development of dogs, and Scott (1967) has asserted that this development is quite similar in wolves. Unless otherwise stated, the following material is drawn from Scott and Fuller.

The pups are born deaf, blind, and incapable of self-regulation of body temperature. They can, however, crawl toward a source of heat. This ability allows them to huddle with one another and their mother. They can produce a variety of vocalizations that help the mother care for them. Whines are produced when they are hungry or cold, yelps when in pain, and grunts when in contact with a nipple.

Occasionally the mother moves her pups to a new den, carrying them one by one in her mouth. Each is held gently about its middle, by the nape of the neck, or by the belly and a hip (Schönberner,

1965, cited in Mech, 1970). Besides nursing and carrying, there are two other phasic interactions, licking and regurgitation, which provide a basis for later social behavior. Newborn pups do not eliminate unless their anogenital area is licked by their mother or father or is rubbed with a warm, moist object. This inhibition probably functions to keep the den dry and relatively odor-free. According to Schenkel (1967), the posture assumed by the pup as it is licked reappears in the adult behavior that he calls passive submission.

During their fourth week of life, the pups begin to eat pre-digested food regurgitated for them by all the adults in the pack. When an adult approaches the den, it is mobbed by the pups, which lick, poke, nip, and sniff at its mouth. The adult then regurgitates food, sometimes in several loads so that all the pups have a chance to get some. Schenkel claims that the begging postures of the pups form the basis for the adult behavior that he calls active submission. Similarly, the tendency to mob approaching adults provides a foundation for the "group ceremony" performed in later life at the appearance of a dominant wolf.

Besides feeding by regurgitation, communal care of pups includes baby sitting when the mother leaves the den to hunt. In such cases another adult, male or female, lies just outside the den's entrance. Murie's (1944) description of the indignities suffered by one baby sitter leaves little doubt that maternal fondness for pups is shared by other adults.

The father participates in communal care not only by providing food but by assisting his mate in the difficult process of weaning. He helps her avoid the pups' attempts to nurse by pushing them away from her or even by pinning them to the ground by grasping their necks in his jaws. Fox (1971) interprets this behavior as "instilling discipline" and noted that by the end of weaning pups were "respectfully submissive" toward their parents (p. 163).

The den is abandoned when the pups are 8 to 10 weeks old (Mech, 1970). At this age the pups are still unable to keep up with the pack and are left at rendezvous sites, to which the adults regularly return with food. Rendezvous sites are used throughout late summer (Joslin, 1967) and are a focus for summer movements and social life.

INTEGRATIVE MESSAGES

The period of development just described corresponds to J. P. Scott and Fuller's (1965) "neonatal," "transitional," and "socialization" stages. The interactions characteristic of these stages are conducive to the formation of strong emotional bonds among siblings, parents, and other members of the pack. Many of these interactions are playful, especially during the transitional, socialization, and "juvenile" (3 months to maturity) stages. Play is one of the most common forms of social involvement among pups. Adults play too, especially during courtship, but the frequency and duration of playful episodes seem to decrease with age. Play establishes and maintains dominance relations and affiliative alliances. For this reason, no matter how capricious it may seem, wolf play is serious business. Investigative play between pups and adults is light-hearted, but most other play has elements of agonism, however ritualized and gamelike it may be.

The agonistic character of even the most frivolous forms of play is evident in the behavior of two wild pups that Mech and I observed at a rendezvous site in Minnesota. They chased each other, with the roles of pursuer and pursued reversing several times, and competed for position atop large boulders in a kind of king of the mountain. Another form of frolic described by Murie (1944) and quite common in a group of three adults that we kept in an enclosure for research purposes also involves competition. When I threw a stick, ball, deer leg, or other toy into the pen, one of the wolves, usually the female, would grab it and parade back and forth in front of the others with her head held high and a bounce in her gait. Occasionally this behavior appeared spontaneously, one of the wolves grabbing a bone and prancing around the others as though challenging them to try to take it away. All such attempts were greeted with slight snarls and other low-intensity threats on the part of the challenger.

At slightly higher levels of agonistic intensity, active and passive submission and genital licking by pups may alternate with mock fighting. This latter behavior often takes the form of jaw wrestling or scruff wrestling, each pup attempting to grasp the muzzle or nape of the other. Jaw wrestling is often accompanied by whines and growls performed with the ears flattened and extended to the side (Fox, 1971). It can easily escalate into serious fighting.

Although the line between rough play and mild fighting is somewhat arbitrary, it is possible to make a meaningful distinction. For one thing, play involves frequent role reversals, with the pursuer, for example, becoming the pursued. For another, play sequences, unlike fights, are frequently interrupted by nonagonistic behaviors such as eating or looking around. During these interruptions both wolves frequently assume the play face, panting with mouth open and lips horizontally retracted (see Figure 8-2).

This expression often precedes and solicits play. In such cases the ritualization of the panting is evident because the panting precedes exertion. Bowing by lowering the front legs and head, often while wagging the tail, is another common invitation to play and is fre-

FIGURE 8-2   This wolf shows the play face used to
express a friendly mood. Photograph by Roger
Peters. Courtesy of the Chicago Zoological Park.

quently combined with the play face or barking (see Figure 8-3). Yet
another indicator of playful intent is an exaggerated looking away,
with ears flattened and lips horizontally retracted. In this display a
wolf moves its head to one side and coyly peers back over its shoulder,
flashing the white of the eye (Fox, 1971). Finally, wolves often solicit
play by bouncing on their forelegs while pivoting on their rear ones
and simultaneously tossing their heads. The sinuous movements of
this invitation are highly effective—other wolves, especially younger
ones, seem to find them irresistible. They have a similar effect on
humans, causing even erstwhile students of wolf behavior to break
into grins.

   One of these invitations seems to work in the opposite direction; the play face can be used to solicit play from a wolf. When it
was necessary for me to handle the wolves kept for Mech's wolf project,
I would solicit play by imitating the play face, causing at least one of
them to bound over to me. During the wrestling match that usually
ensued, they treated my hands as they would another wolf's muzzle,

grasping them firmly but gently. Their control was so great that when I was bare-handed their sharp teeth never left a mark, even when I jaw-wrestled by grabbing their muzzles from above, below, or inside their mouths. In winter, when I wore thick buckskin gloves, they would bite through them, but they inhibited the bite just as their teeth began to contact my skin. This remarkable control of jaws and teeth enables wolves to play roughly with each other without causing aggression-eliciting pain.

Besides the special invitations to play, wolves use gestures and vocalizations as solicitations for attention or care. These *et-epimeletic* signals frequently evoke approach, grooming, feeding, or affection from other wolves. One of the most common of these solicitations is the raised paw, which dogs readily adapt to scratching on doors. Paw raising is frequently accompanied by whimpering or whining, "a high, though soft and plaintive sound . . . [whose] pitch varies so much, it seems ventriloquial" (Young, 1944, p. 77). The effectiveness of whimpering as a solicitation was proved by Joslin (1966, cited in Mech, 1970), who kept a wild wolf close to him, sometimes within 30 meters, for more than an hour by whimpering. Tembrock (1963)

FIGURE 8-3   The wolf on the right solicits the attention of the one on the left by lowering its head and forequarters. Photograph by Roger Peters. Courtesy of the Chicago Zoological Park.

found that whimpers are fairly pure tones with a fundamental at about 760 hertz. Fox (1971) mentions a variant of the whimper that he calls the grunt-whine. Apparently, this vocalization is commonly used to solicit grooming, and Fox states that it is frequently used by adults to halt fights between their pups. Sometimes whining is combined with the mouth movements of barking to produce what Crisler (1958) calls talking, in which a wolf "seeks your eyes and utters a long, fervent string of mingled crying and wowing, hovering around one pitch" (p. 150).

Another form of solicitation resembles bunting in deer. This sharp, upward movement of the snout is usually applied to a conspecific's chin, but dogs often direct it toward their master's hand, especially when it holds a cup of hot coffee.

Playful and et-epimeletic solicitations elicit attention and approach over distances of only a few meters. Over longer distances the howl performs a similar function, often causing separated members of the pack to assemble. Howls are tremendously variable in quality, ranging from low, monotonic moans lasting less than 1 second to wailing cacophonies going on for more than 2 minutes. Shoemaker (1917) described the latter type as follows:

> . . . the best comparison I can give would be to take a dozen railroad whistles, braid them together and then let one strand after another drop off, the last peal so frightfully piercing as to go through your heart and soul; you would feel as though your hair stood straight on end if it was ever so long . . . [p. 29].

There is plenty of evidence from the field that howling often assembles separated members of a pack. On eight occasions Murie (1944) heard wild Alaskan wolves howl while separated. In five of these cases, a reunion occurred within a short period. While observing wolves on Isle Royale, in Lake Superior, from an airplane, Mech (1966) saw a wolf on a ridge in the characteristic head-up howling posture (see Figure 8-4). Within a few minutes other members of the pack joined the wolf on the ridge. Rutter and Pimlott (1968) described three cases in which howling functioned similarly and supported Young's contention that a particular type of howl is used for assembly. Young describes this howl as "loud, deep, guttural, though not harsh" (p. 77). While radio-tracking wolves from the air in Minnesota, Mech and I observed a wolf in what appeared to be a howling posture. Within a few minutes other wolves ran to the kill from several different directions and began to feed. These cases, recorded by different ob-

FIGURE 8-4   Wolves howling. Photograph by
Roger Peters.

servers in different geographical areas, demonstrate that howls sometimes summon pack members. The probability of reunion without
howling or some other long-distance signal is very low.

The tendency to respond to a howl with approach may develop
from a pup's reaction to its mother's "ululating long whines" (Fox,
1971, p. 166). Fox heard this vocalization when a captive mother
approached her den box. Her pups tumbled out to greet her as soon
as she made the call.

Howling, however, does not always lead to assembly. Sometimes, especially when the howl is far away, the response is another
howl. In such cases, howling acts as a contact signal, merely informing
other wolves of the presence of an individual or a pack (Joslin, 1966,
cited in Mech, 1970). This process of informing is facilitated by a
phenomenon known as the refractory period. Pimlott (1966) found
that once a group of captive wolves has been stimulated to howl, they
will generally not respond again for 15 to 20 minutes. As Mech (1970)
has pointed out, the refractory period allows a pack to listen for replies.

Howling has been shown to be an effective long-distance auditory contact signal. Circling can be conjectured to be a short-distance visual contact signal. Since a wolf often disappears from view
when it lies down, it would be useful to signal that the wolf is about
to rest. The circling that often precedes lying down might be such a
resting intention, perhaps ritualized from the motions of crushing
vegetation to make a bed.

Another means of staying in touch with other wolves is provided by scent marking, which is stimulated by and oriented toward specific, salient objects (Kleiman, 1966). The four forms of scent marking by wolves are described in detail by Kleiman and by Peters (1974). First in order of frequency and importance is raised-leg urination (RLU). RLU increases the effectiveness of urine as a signal in several ways. First, it increases the "active space" of the mark by placing it above the boundary layer of still air along the ground. This can double the volume permeated by an odor (Bossert & Wilson, 1963). Second, the evaporative area increases as the urine trickles down. Third, urine on a vertical surface, like the trunk of a tree, is protected from weather. Finally, urine from an RLU is closer to the nose of the receiver than urine on the ground (Peters, 1974). Defecation can be a form of scent marking when the feces are placed on salient objects such as bushes or other scent marks. Scratching with the feet also applies an odor to the environment, and scratching is often associated with RLU and defecation (Peters, 1974). Rubbing the shoulders and back on novel odors is also a form of scent marking, but its role in maintaining contact has not been investigated. As far as the contact-maintaining function of urination and defecation is concerned, little can be added to the following account, written by Seton in 1909:

> The whole of a region inhabited by wolves is laid out in signal stations or intelligence posts. Usually there is one at each mile or less, varying much with the nature of the ground. The marks of these depots, or odor-posts, are various; a stone, a tree, a bush, a buffalo skull, a post, a mound, or anything serves providing only that it is conspicuous on account of its color or location; usually it is more or less isolated, or else prominent by being at the crossing of two trails. . . . There can be no doubt that a newly arrived wolf is quickly aware of the visit that has recently been paid to the signal post—by a personal friend or foe, by a female in search of a mate, a young or old, sick or well, hungry, hunter, hunted or gorged beast. From the trail he learns further the directions where it came and whither it went. Thus, the main items of news essential to his life are obtained by the system of signal posts . . . [p. 772].

In the 70-odd years since these words were written, scientists testing Seton's assertions have validated most of them. Seton's signal posts are often visited, investigated, and remarked by adult members of a pack. Scent posts often consist of concentrations of RLU, feces, and scratch marks, but sometimes these marks are spread out. Along commonly used trails, the average distance between marks is only about 250 meters, so a wolf moving along a trail is rarely more than 125 meters from a mark produced on the pack's last trip down the trail. As a wolf moves at its average speed of 8 kilometers per hour, it encounters and makes about one scent mark every 2 minutes. Each of the marks it encounters provides information about previous use of the trail, including how long ago the previous user passed (Peters & Mech, 1975). This information may then be used to rejoin—or, in the case of a trespasser, to avoid—the wolf that made the mark.

Besides acting as contact signals, fecal scent marks probably provide information about the identity of the wolf that produced them. Feces are coated with secretions from the anal sacs as they pass through the rectum. Analysis by gas chromatography reveals that these secretions contain at least nine major components. By coinjection I identified five of these components as short-chain fatty acids. The relative concentrations of all components varied from wolf to wolf. Furthermore, three samples from the same individual, collected over a period of about 1 year, showed a consistent pattern. Thus, the distinct anal odor of a wolf may remain relatively constant over a long period, providing an olfactory signature that might allow other wolves to recognize it or its scats. There may be a sex difference in relative concentrations, females producing lower proportions of the shorter-chain acids than males. Conceivably, this or some other difference could provide a wolf with information about the sex of an individual, whether by direct sniffing or by sniffing its feces (Eisenberg & Kleiman, 1972). Anal sniffing is commonly observed when wolves meet one another (Schenkel, 1947). Perception of anal odors is probably facilitated by tail wagging, which wafts them about. Because the composition of urine varies with hormonal and metabolic conditions, it is quite likely that odors in urine are equally distinctive. Von Uexkühl and Sarris (1931) and J. P. Scott and Fuller (1965) found that dogs respond differently to their own and others' urine marks. If dogs can make this discrimination, it is likely that wolves can do so.

Wolves' voices are also highly distinctive (Harrington & Mech, 1978; Theberge & Falls, 1967), and thus may, like odors, convey information about identity. Theberge and Falls showed that wolves could discriminate between live and recorded versions of the same howl. Since the differences between the live and the recorded howls were quite subtle, there is little doubt that wolves have the capacity to discriminate the howls of different individuals. Individual variability is preserved even in group howls because wolves tend to use different pitches. This tendency also provides other wolves with information about the size of the pack (Mech, 1970). Thus, differences in howling may allow recognition of packs as well as of individuals.

Appearance provides a third kind of cue to individual identity. Wolves have highly individual markings on their faces and tails, and in a pack each has a characteristic demeanor. Using these cues, a human observer can learn to distinguish individuals even in a large pack. It is possible that wolves also use these markings to recognize one another, but no one has as yet attempted to test this hypothesis. At night or in heavy cover, of course, these visual cues would be useless, so it is likely that the major medium for recognition is olfaction.

The same odors that wolves use to recognize one another are probably also used to recognize locations. Since wolves travel purposefully over a large area, visiting old kills, dens, rendezvous sites, and concentrations of prey, they must have some means of knowing where they are. One suggestion that scent marks act as olfactory landmarks comes from the close correspondence between the features of the environment marked by wolves and those drawn by human map makers (Peters, 1973). Wolves traveling through an area leave urine and feces in the same places a human would leave a cairn, blaze, or piece of marking tape—at trail junctions, road crossings, or direction changes. It seems likely that one function of such scent marks is facilitation of later travel decisions.

Familiarity has an emotional aspect as well as a cognitive one. A wolf in familiar surroundings is confident, at home. In contrast, when placed in new surroundings, a wolf may become restless and fearful (Fentress, 1967). Kleiman (1966) noted that novelty stimulates scent marking and has argued persuasively that scent marking renders the environment less threatening by permeating it with familiar smells.

The puzzling behavior known as body rubbing may represent the reverse of the process described by Kleiman. When rubbing, a wolf writhes sinuously, applying its mane and shoulders to novel odors. Fox (1971) suggests that rubbing renders odors more familiar and less threatening by transferring them to the wolf. The only case of rubbing that I recorded while following approximately 330 kilometers of wolf tracks in snow was evidently stimulated by the carcass of a long-dead vole (Peters, 1974). The wolves that rubbed had recently crossed the fresh trail of another pack and reversed their direction of travel almost immediately after rubbing. Thus, they may well have been in need of reassurance, and this field observation tends to support Fox's hypothesis. However, the most common stimulus for rubbing by free-ranging wolves is probably the short-chain fatty acids in urine, feces, or carrion. Rubbing in such substances would supplement the fatty acids produced in the wolf's own skin glands and would help them keep the integument pliable and waterproof. According to this conjecture, rubbing is primarily a form of grooming (Peters, 1974).

While rubbing in various odors, including perfumes, fatty acids, and urine of various species, that I placed on the floor of their enclosure, the wolf project's captive wolves always assumed what Fox calls the consummatory face. Their eyes were closed or unfocused, their lips were neutral or horizontally retracted, and their ears were extended laterally. Fox notes that this behavior is characteristic of consummatory behavior, including intromission, eating, and elimination. I often noted it when I scratched wolves or when they scratched themselves. This face seems, therefore, to express satisfaction, but its significance to other wolves, if any, is unknown.

At the opposite end of the emotional spectrum from satisfaction is distress. Low-intensity distress is expressed with the whimper described as a solicitation. Sudden or high-intensity distress is expressed by the yelp (Schenkel, 1947). This vocalization is a short, high-pitched, pure tone with a slight drop in pitch at the end. According to Fox, the yelp is a common response to pain in pups up to about 5 weeks of age, after which it is markedly inhibited. The threshold for this response must be considerably higher in adult wolves than it is in dogs, for, even when stuck with a hypodermic needle, wolves do not usually yelp.

When extremely distressed, as when placed on an operating table prior to anesthesia, dogs often express their anal sacs (Donovan, 1969). Although there is no evidence that wolves do so, the close relationship between wolves and dogs suggests that anal-sac expression may, like yelping, occur in wolves but merely have a higher stimulation threshold.

Apart from humans and bears, wolves have no animal enemies. It is therefore not surprising that they have only one alarm signal, the bark. Alarm barks are short, usually less than .1 seconds in duration (Harrington & Mech, 1978). Their energy is distributed over a wide range of frequencies, from about 320 hertz to about 900 hertz (Tembrock, 1963). These characteristics, short duration and wide frequency range, make sounds easy to localize (Marler, 1955). According to Joslin (1966, cited in Harrington & Mech, 1978) alarm barks occur singly and often terminate howling sessions. Mech (1970) interprets this pattern of occurrence as evidence for an alarm function, presumably because the bark warns the howling wolves to remain silent. Barking is sometimes elicited by a human's approach to a den or rendezvous site, but in such cases it may act as a threat rather than as an alarm. Further evidence for an alarm function comes from Fox (1971), who noted that when an unfamiliar human approached, a mother's barks, combined with growls, sent pups back into their nest box.

Wolves are friendly because they have to be. The pack that preys together must stay together, and without strong emotional bonds a pack would soon dissolve under internecine strife or the sheer magnitude of the territory. Murie (1944) described the affiliations that bind pack members together as follows:

> The strongest impression remaining with me after watching the wolves on numerous occasions was their friendliness. The adults were friendly toward each other and were amiable towards the pups, at least as late as October. This innate good feeling has been strongly marked in the three captive wolves which I have known [p. 23].

The strength of the affiliate attraction between pack members is illustrated by the appearance of a traveling pack. When seen from the air, pack members seem to be joined by invisible elastic bands. In accord with a kind of psychological Hooke's law, the further back one falls, the faster it catches up. The leaders, too, seem to feel the tension, and if they get too far ahead, they pause while the others catch up.

The ability to form such bonds was probably a major factor in the domestication of the wolf. Dogs seem to have retained this ability and treat their masters much as wolves treat the members of their pack (Mech, 1970). Mech points out that there are bonds within a pack between mates, between parents and offspring, and between litter mates. The first two types will be discussed as sexual and submissive messages, respectively. Bonds of the third type are formed during the "period of socialization" (J. P. Scott, 1967, p. 101), when pups are between 3 and 11 weeks old. Two complementary processes are involved in the development of these attachments. First, friendly contact is rewarding. During their first few months the pups are exposed to one another whenever they huddle, nurse, or play. At the same time they are ordinarily isolated from contact with members of other packs. It is possible that this bonding is in part an effect of mere exposure similar to that demonstrated in other species (Zajonc, 1971), but wolves and dogs seem to have a special capacity to form attachments during this period. According to J. P. Scott and Fuller (1965), tail wagging by dogs and wolves is an expression of the pleasure stimulated by social contact. They note that tail wagging rewards humans for initiating an interaction with a dog but caution that it is difficult to tell whether tail wagging has a similar rewarding effect on other dogs. Wolves wag their tails in many of the same situations in which dogs do, but its social significance has not been thoroughly investigated.

The second basic process in the formation of litter-mate attachments involves negative rather than positive reinforcement. J. P. Scott (1967) describes the distress caused by isolating a young wolf. This distress is relieved by reunion with litter mates, thus reinforcing tendencies to approach them.

Members of a pack frequently separate and, a short time later, reunite. Their intolerance of strangers makes each reunion a critical moment climaxing in special greetings in which wolves discharge tension and reaffirm their affiliation. These rituals occur even in captive packs as long as there is enough room or cover for one or more members to absent themselves at least symbolically. At the Brookfield Zoo these greeting ceremonies, also called group ceremonies, are especially common when wolves gather to be fed. These celebrations of affection are frantic dances to an accompaniment of whines, barks, and squeaks. The choreographic theme of these rituals seems to be the attempt of each wolf to sniff and lick the face of the alpha male while preventing others from doing so. The resultant milling and jockeying for position may facilitate recognition and, by exposing each wolf to the odors of most of the others, may promote mutual attraction (Zajonc, 1971).

The squeaks and whines that accompany group ceremonies are low in volume and high in pitch, with frequencies between 2500 and 2800 hertz in adults and around 3800 hertz in pups. Squeaks last only .1 to .2 seconds (Field, 1979), but whines may last several seconds (Harrington & Mech, 1978). Both Crisler (1958) and Fentress (1967) describe the squeak as occurring when a tame wolf met a familiar human or dog. Mech (1970) and Harrington and Mech regard the squeak as a special kind of whine that Joslin (1966, cited in Harrington & Mech, 1978) described as a component of friendly greeting. Occasionally squeaks, whines, and barks culminate in a howl, which, at least in the context of a group ceremony, seems to be an ultimate expression of social, if not musical, harmony.

Licking and nibbling directed to the face and neck are expressions of friendliness that sometimes occur during group ceremonies and sometimes occur in calmer greetings involving only two wolves (Schenkel, 1947). A ritualized form of licking called licking intention, with rapid and repeated extrusions of the tongue, often occurs at the beginning of such greetings as the wolves approach each other (Fox, 1971). In very peaceful situations a wolf may "snuffle" another with which it has a particularly intimate friendship. Schenkel described snuffling in this way:

> In snuffling the sides of the neck, the tip of the nose reaches the skin through the fur so that the nose is lost in the hairs.

The snuffler usually moves the head very slightly at the same time, moving the tip of the nose searchingly in the partner's fur [p. 91].

When it was necessary to determine whether wolves trapped in the same area were members of the same pack, I introduced them to each other in pairs. One pair of females immediately lay down together and began to snuffle, convincing me that they indeed knew each other well.

An especially tender signal of close friendship is the nose touch. During 150 hours of observation of several groups of captive wolves, I observed 75 of these "kisses." They always occurred during friendly or sexual interactions and were exchanged by wolves that tended to stay closer to each other than to other wolves in the group.

AGONISTIC MESSAGES

Affiliation and submission can be expressed independently, but they are closely related. Schenkel (1967) defined submission as the effort of the inferior to attain friendly or harmonic social integration with a combination of inferiority and a positive social tendency ("love"). Submission enables low-ranking wolves to express their attachment to high-ranking ones while reducing the danger of attack.

One form of nonassertive friendliness, active submission, has already been mentioned in connection with juvenile begging for regurgitation. As when begging, a wolf showing active submission crouches, puts its ears back, and nuzzles, licks, or gently bites the muzzle of the superior. Sometimes a forepaw is raised and strokes the superior's face or neck. The tail is held low and is often wagged laterally, sometimes so forcefully that the hindquarters swing with it (Schenkel, 1967).

Various degrees of active submission are displayed by pack members as they mob the alpha male in a group ceremony. Mech (1970) notes that group ceremonies often occur when a pack has just located prey and points out that in such cases the relationship of active submission to food begging is direct—the pack members seem to be asking the alpha to lead them to a kill.

Active submission is often used as a greeting when meeting other wolves or humans (Mech, 1970). In the latter case the wolf must rear up on its hind legs in order to reach the human's face. This greeting is often quite overwhelming and is thus liable to be misinterpreted as an attack.

Lorenz (1952) interpreted active submission as offering the superior the "bend of the neck, the most vulnerable part of his whole body" (p. 207). He claimed that this action automatically inhibits the superior's bite. There are several problems with this assertion. For one thing, the "bend" of the neck is not the most vulnerable part of a wolf's body. Compared to the throat or belly, it is well protected (by the mane) and for this very reason is the target for ritualized biting in mock fights. Second, as Mech points out, wolves using active submission sometimes *are* attacked, so its inhibitory effect is by no means automatic. Finally, Schenkel's (1967) description shows that active submission puts the teeth of the inferior at the neck of the superior, not the other way around. In Figure 8-5 the dominant wolf,

FIGURE 8-5   The subordinate wolf, on the right, displays active submission and defensive threat. The elevated tail of the wolf on the left expresses dominance. Redrawn by Heidi Reynolds from "Submission: Its Features and Function in the Wolf," by R. Schenkel, *American Zoologist*, 1967, 7(2), 319–329. Reprinted by permission of the American Society of Zoologists.

with ears and tail up, is on the left. J. P. Scott (1967) noted that the dominant is likely to be confronted not with the neck but rather with "a mouthful of snapping teeth" (p. 379).

The display that Schenkel calls passive submission is even less assertive than active submission. Ear and tail positions are the same as in active submission, but the wolf lies on its side or back, usually exposing its abdomen (see Figure 8-6). Schenkel hypothesizes that passive submission is derived from the posture assumed by pups when adults lick them to stimulate elimination. He notes that superiors often respond to passive submission by sniffing or licking the inferior's inguinal area, thus taking the role of the parent.

Schenkel describes two forms of behavior that contain elements of both active and passive submission. In these displays a wolf crouches with its hindquarters either to the side or flat on the ground and pushes its muzzle toward the superior.

The tucked-down position of the tail associated with both active and passive submission is characteristic of low social status and is part of a behavior pattern known as anal withdrawal (Schenkel, 1947). In anal withdrawal a low-ranking wolf restricts the flow of odor from its anogenital region by covering it with its tail or by sitting.

FIGURE 8-6   The wolf on its back displays passive submission. Photograph by Roger Peters, courtesy of the Chicago Zoological Park.

Tail wagging in active submission nicely illustrates the intimate re-lationship between submission and affiliation: the tail is held low to restrict emission of odor, but it nevertheless wags in social excitement.

Wolves show milder degrees of submission by facial expres-sions. One of the most common of these expressions is the submissive grin with licking intention, described earlier as a sign of affiliation. Looking away is often combined with the grin and is also used as a low-intensity appeasement. Another quick, common, and casual sub-missive signal is flattening of the ears to the side or rear (Fox, 1971). All these signs of submission reduce apparent size, negate assertion, and can thus be understood as antitheses of dominance displays or threats. Looking away, for example, is the antithesis of the threatening direct stare.

Because wolves' personalities have so many dimensions other than dominance, I sometimes feel that the better one knows a group of wolves, the less certain one is about their ranks. Schenkel (1947) underlined the complexity of dominance when he said that status disputes typically involve all the members of the pack, not just those participating directly. Thus, to predict the outcome of an agonistic interaction, one must know not only the wolves' ranks but their moods and alliances as well. For example, although the higher-ranking wolf will ordinarily win in competitions for food, attention, or toys, the most dominant of our wolf-project captives never did so. The reason was that he was shy and wary, and the two subordinates often acted in concert against him. When pieces of meat or bones were thrown into the enclosure, he would run away, and the others would pounce on the new items. Once they had grabbed the items, a zone of "own-ership" around the subordinates' mouths (Mech, 1970) prevented him from attempting to take them away.

Although leadership is a prerogative of dominance and the alpha male usually takes the initiative in waking the pack, attacking prey, and choosing travel routes, his leadership is by no means au-tocratic. Subordinates can influence his behavior by refusing to follow (Mech, 1970). Thus, both privilege and leadership, the two major components of dominance (Mech, 1970; Schenkel, 1947), have im-portant restrictions.

Dominance is usually treated as an agonistic phenomenon, but in wolves it also has a strong integrative component. At the Brookfield Zoo, for example, a low-ranking wolf would repeatedly go out of his way to approach the alpha female, invariably evoking a dominance display. It was as though he were soliciting the display to reaffirm his subordinate relationship. The friendly reciprocity that characterizes many dominance displays is illustrated by standing across, which occurs only between individuals that are intimate with each other. The dominant stands over the subordinate's forequarters as it lies down. If the dominant is a male, the subordinate may roll over and lick the dominant's penis.

Many dominance displays are, however, far less friendly than standing across. Fox (1971) describes how captive parents discipline their pups, pinning them by grasping their muzzles in their jaws and pushing them to the ground. Dominant wolves often use this technique on other adults but usually grab them by the neck. The subordinate invariably freezes.

Another forceful demonstration of dominance is riding up, in which the higher-ranking wolf places its forepaws on the shoulder or back of the subordinate. Riding up may resemble sexual mounting but does not usually include the clasping or pelvic thrusting characteristic of attempts to copulate. While riding up, the dominant may emphasize its superiority by biting the other's neck (Schenkel, 1947).

Usually a dominant does not have to go to so much trouble to control the behavior of its subordinates. At the Brookfield Zoo the alpha male often halted chases and attacks merely by raising his head and staring. This display has an even greater impact when the dominant stares while standing with elevated tail. I have seen this display freeze a young wolf in its tracks.

The elevated tail used in this display indicates an assertive mood and is thus an excellent indicator of dominance (see Figure 8-5). Tail positions are very important in displays that Schenkel (1947) calls anal presentation and anal control. A dominant walks slowly and stiffly over to a subordinate and draws alongside so that its anal region is near the subordinate's nose. At the same time the dominant puts its nose near the subordinate's anal region, inducing it to display anal withdrawal.

Woolpy (1968) and Rabb (1968, cited in Mech, 1970) have noted that RLU, in which urine is squirted on a vertical target, is performed only by dominant wolves, primarily the alpha pair. Peters and Mech (1975) supported this contention with the observation that 22 of 27 RLUs performed by wolves at the Brookfield Zoo were performed by the alpha pair, 20 of them by the male. They further showed that RLU is generally associated with assertive behaviors, including threats. On the basis of these and similar data, RLU has been described as a dominance display (Lockwood, 1979; Peters, 1974). Since RLUs are sniffed by other members of the pack, they may operate through olfaction.

Berg (1944) and Martins and Valle (1948) showed that high levels of testosterone, which increase the frequency of assertive behavior in many species (Guhl, 1961), also induce RLU in dogs. Roth-man and Mech (1979) speculate that metabolites in wolf urine may enable conspecifics to detect differences in endocrine state. Thus, RLU may provide olfactory cues signifying dominance.

Bekoff (in press) suggests that the RLU posture, but without excretion of urine (raised-leg display, or RLD), acts as a visual signal in dogs. Lockwood's (1979) observation that captive dominants urinate without raising their legs when their pack mates are asleep lends credence to the notion that this display may have visual significance in wolves. Similar remarks may apply to ground scratching with stiff-ened legs and raised mane, a form of marking and display that has both visual and olfactory consequences and is similarly associated with assertive behavior (Peters, 1974). Ewer (1968) suggests that such dis-plays may have evolved from displacement digging, which often ac-companies marking and fighting.

Mech (1970) has alluded to the self-legitimizing effect of high rank, stating that mere "enjoyment" of rank helps to maintain it. Since displays of dominance often elicit reciprocal submission, which in turn reinforce the dominance display, there does indeed seem to be a positive-feedback loop that stabilizes dominance relationships.

Because of this relatively stable dominance hierarchy, threats, which can be performed by wolves of any rank, can be distinguished from dominance displays, which are the prerogative of the higher-ranking animal. The security of an alpha's rank enables it to control the behavior of its pack mates without recourse to high-intensity threats. Conversely, wolves of low rank rarely have the confidence necessary to back them up. Thus, high-intensity threats are most frequently performed by wolves of middle rank. In one such display, called the bite threat, the head is high and the neck arched. The lips are vertically retracted, baring the teeth and wrinkling the muzzle. The mouth is open, and its corners are pulled forward. The forehead is swollen and wrinkled with muscular tension, and the ears are tilted forward. The tongue is rhythmically extruded and withdrawn. The impression of explosive readiness is heightened by stiffened legs, bris-tling mane, and trembling tail. Another form of threat, the ambush, is not as dramatic, but like the bite threat, it may precede a genuine attack. When ambushing, a wolf, usually dominant over the recipient, crouches as though ready to spring. This threat, like the bite threat, is often performed at some distance from the receiver. Both displays, however, often cause the receiver to freeze or shy away (Fox, 1971; Mech, 1970; Schenkel, 1947).

Milder threats can be expressed with any of the components of the bite or ambush displays, sometimes with minor variations. For example, the tail may be bristled and whipped from side to side as the mane is erected. A bristling tail or mane may have olfactory as well as visual effects. There are concentrations of apocrine and sebaceous glands under the mane and near the base of the tail (Schaffer, 1940). Mech (1977a) has detected a distinct odor associated with these concentrations. Peters (1974) proposed that dorsal and caudal piloerection serves to release these odors by increasing the effective area of the hairs from which they are released.

Growls or barks may accompany these and other threat displays, but sometimes these vocalizations are used alone. Growls are low and coarse, with an energy peak around 800 hertz and the rest of the energy falling between 250 and 1500 hertz (Tembrock, 1963). Short growls, sometimes called snarls, last only a few tenths of a second, while longer ones last for several seconds. Neither snarls nor longer growls are very loud, and they seldom carry beyond about 200 meters (Joslin, 1966, cited in Harrington & Mech, 1978). Schenkel (1947) and Fox (1971) found that growls were directed mainly at wolves of lower rank, but I have heard them made by the lowest-ranking member of a group, particularly when defending food.

Barks used as threats are similar to those used in alarm but are longer in duration and occur in sequence rather than singly. On several occasions Joslin howled within 200 meters of a pack. In such cases he was often approached by wolves that barked in threat. He described one such sequence of barks as a series composed of one or two short barks followed by a longer bark and several lower, softer barks. This series was repeated 37 times in less than half an hour. Pauses of about 1 minute between each series were punctuated with growls. Harrington and Mech often heard barks combined with howls ("bark-howls") under similar circumstances.

Subordinate wolves often resort to defensive threats when submission fails to appease a wolf of higher rank. These defensive threats include snapping with the teeth bared, shifting the direction of gaze, laying the ears back, and "skulking" with the neck and back arched and the tail drawn between the hind legs. The lower the wolf's rank, the farther from its opponent it performs this display. Animals of high or intermediate rank may bark and snap less than 1 meter from the dominant, but animals of low rank retreat as far away as 10 to 15 meters to do so. The many combinations of offensive and defensive components that occur in threat are shown in Figures 8-7 and 8-8.

Defensive threats are often used by wolves of low or middle rank when other members of the pack gang up on them. Such attacks are often directed at wolves that are fighting and usually put an end

FIGURE 8-7   Facial expressions of the wolf: (a)
and (b) normal expressions of a high-ranking
animal; (c) and (d) anxiety; (e) and (f) threat; and
(g) and (h) suspicion. Redrawn by Heidi Reynolds
from "Ausdrucks-Studien an Wölfen:
Gefangenschaft-Beobachten," by R. Schenkel,
Behaviour, 1947, *1*(2), 319–329. Reprinted by
permission of N. V. Boekhandel & Drukkerij
Voorheen E. J. Brill, Leiden, Netherlands.

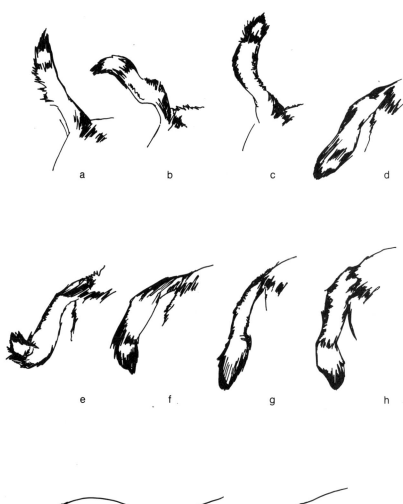

to the fight (Mech, 1966; Schenkel, 1947; Woolpy, 1968). Schenkel and others have interpreted ganging up as an example of "energy displacement." This interpretation treats "aggressive energy" as a physical or chemical force that automatically increases in strength until it finds an outlet. By emphasizing the apparent spontaneity of aggression, this conception draws attention away from its behavioral and environmental contingencies. In the absence of any physiological correlate for aggressive energy, this "hydraulic" or "flush-toilet" model has met with considerable criticism (Berkowitz, 1962).

Mech (1970) suggests that ganging up ensures that fights, whatever their motivational basis, end quickly in submission rather than injury. This safety factor may sometimes be needed, but serious fights within a pack are probably not frequent. During 150 hours of observation of captive wolves, I saw only two fights. Both these fights were highly ritualized, consisting mainly of seemingly inhibited bites to the head, slams with the hips, and rearing on the hind legs.

Serious fighting seems to be limited to special circumstances: encounters with outcasts and strangers (Mech, 1966, 1970) and captivity, which prevents losers from escaping. Rabb and his colleagues (1967) reported attacks on wolves while they were immobilized by a copulatory tie but did not say how serious these assaults were. Since wolves' major weapons are their teeth (see Figure 8-9), they fight by biting. In a fight that I observed between captive wolves, I noted that most bites were directed toward the head, neck, and chest.

FIGURE 8-8   Expressive positions of the wolf's tail: (a) self-confidence in social intercourse; (b) certain threat; (c) imposing attitude (with sideways brushing); (d) normal attitude (situation entirely without social pressure); (e) a not-entirely-certain threat; (f) normal attitude (similar to "d"), particularly common during eating and observing; (g) depressed mood; (h) between threat and defense; (i) actively casting oneself down (with sideways brushing); (j and k) strong restraint. Redrawn by Heidi Reynolds from "Ausdrucks-Studien an Wölfen: Gefangenschaft-Beobachten," by R. Schenkel, Behaviour, 1947, 1(2), 319–329. Reprinted by permission of N. V. Boekhandel & Drukkerij Voorheen E. J. Brill, Leiden, Netherlands.

FIGURE 8-9   The teeth of a wolf are deadly
weapons. Social rituals usually prevent them from
inflicting harm in conflicts within the pack.
Photograph by Roger Peters.

Because fights between members of different packs are so often
serious, wolves may associate the scent marks and howls of other packs
with danger. The resulting aversion may contribute to the territorial
spacing of packs. Scent marks are well adapted to the advertisement
of territory. As mentioned in the section on integrative messages, they
are distinctive, so wolves can tell when a scent mark they encounter
is not their own. Moreover, the rate of RLU is twice as high along
the edges of territories as in the centers, in part because wolves scent-
mark at an extremely high rate after encountering sign of another
pack (Peters, 1974). These phenomena, combined with the previously
mentioned tendency to mark at strategic locations throughout the
territory, ensure that trespassers are quickly informed of the presence
of the residents. Direct evidence for aversion to scent marks of neighbors is difficult to obtain, but in three of four cases I observed in which
one pack encountered sign of another, the encountering pack soon
turned back into its territory. In one of these cases, a pack turned back
while in hot pursuit of a deer that they had already wounded. Thus,

although scent marks do not form an impenetrable barrier, the motivation to remain on familiar turf is sometimes stronger than hunger.

The other main means of advertising a territory is howling. Rutter and Pimlott (1968) describe a case in which two packs howled at each other, then moved directly away from each other. Harrington and Mech (1978) found that this response is not unusual. Sometimes, however, the packs remain within earshot of each other, and interpack howling sessions continue for hours (Joslin, 1966, cited in Harrington & Mech, 1978). Such sessions may merely prevent approach rather than induce retreat.

Harrington and Mech cite several studies showing that wolves can hear howls at distances up to 15 kilometers. Thus, a howl could cover an area of 130 square kilometers. This area is larger than the smallest Minnesota wolf territory measured by Mech (1973) and more than a third of the area of the largest. Clearly, howls are loud enough to be effective territorial advertisements.

Mech (1970) mentions the tendency of wolves to howl on different notes and hypothesizes that this tendency may enable them to "gain at least a relative idea about the number of wolves in a chorus" (p. 323). If this is so, the greater the apparent number of wolves, the more intimidating a group howl should be. It may be no accident, then, that humans tend to exaggerate the apparent number of participants in a howl. This tendency is shown by the following excerpt from the *Memoirs* (1885) of General Ulysses S. Grant:

> On the evening of the first day out from Goliad we heard the most unearthly howling of wolves, directly in our front. . . . We could not see the beasts, but the sound indicated they were near. . . . It appeared that there must have been enough of them to devour our party, horses and all, at a single meal. . . . My companion asked, "Grant, how many wolves do you think there are in that pack?" . . . Suspecting that he thought I would overestimate the number, I determined to show my acquaintance with the animal by putting the estimate below what possibly could be correct, and answered "Oh, about twenty," very indifferently. He smiled and rode on. In a minute we were close upon them, and before they saw us. There were just two of them [p. 77].

Howling and scent marking play complementary roles in the maintenance of wolf territories. Scent marks are long lasting, but howls have greater range. The scent mark shows where a pack is likely to show up, while the howl shows where it is. Both advertisements are exciting to wolves, but neither operates in a stereotypical fashion. Rather, both signals provide information that other wolves can, and usually do, use to avoid encounter.

SEXUAL MESSAGES

It is often said that wolves mate for life. This is one of the few popular beliefs about wolves for which there is substantial evidence. Data gathered by Rabb and his associates (1967) show that in captivity both males and females prefer the same mates over periods as long as 6 years. Since free-ranging wolves do not usually mature until they are at least 2 and rarely live beyond the age of 10, this 6-year period of relative monogamy accounts for much of adult life.

Since the breeding pair in a pack travel together throughout the year, they need not advertise to find each other. Their sexual messages serve mainly to announce the approach of estrus, to synchronize their reproductive physiologies, to strengthen their bond, and, according to a recent study by Mech, Rothman, Colin, and Seal (1978), to suppress reproduction by other pack members. Lone wolves, of course, must find a mate but do not advertise to do so. Rather, they maintain a low profile by refraining from howling and scent marking and by eliminating away from major trails. They seem to locate mates by reading scent marks of packs and moving into unoccupied areas where the chances of meeting another wolf of the opposite sex are enhanced (Rothman & Mech, 1979).

The farther north wolves are, the later their breeding season (Mech, 1970). At the Brookfield Zoo the breeding season begins in late January and lasts through mid-February (Rabb, 1968, cited in Mech, 1970). In northern Minnesota it starts about the same time but lasts until almost the end of February (Mech & Knick, 1978). In both locations, copulations are observed over a period of about 3 weeks.

As early as mid-October, wolves at the Brookfield Zoo begin to display affectionate behavior. This precourtship, like the affiliative messages exchanged throughout the year, is anteriorly directed but includes some new forms: bunting snout with snout, gentle jaw wrestling, and mutual rubbing of head and neck (Schenkel, 1947). Some affiliative forms, such as snuffling, increase in frequency, especially between males and females (Peters, 1974).

Precourtship becomes true courtship when females begin to advertise the approach of estrus. One major medium for this advertisement is vaginal odor. Males at the Brookfield Zoo begin to increase the rate at which they sniff females' genitals in December. A high-ranking female responds to such investigations by standing and raising her tail, but low-ranking ones move away or display anal withdrawal. Some of the attractiveness of the female's rear is attributable to the odor of anal sacs. Donovan (1967) found that male dogs are sexually aroused by anal-sac secretions of estrous females but not by secretions from females that are not in heat.

Other odors that attract males come from blood, vulval secretions, or urine, all of which can be held by the pouchlike invaginations of the external genitals or by the surrounding fur. Females may show vaginal blood as early as 6 weeks before mating (Young, 1944), but at the Brookfield Zoo they do not do so until, at most, 2 weeks before the first copulation of the season (Rabb et al., 1967). Rothman and Mech (1979) found blood in urine marks presumably made by females throughout the Minnesota breeding season. Since vaginal blood comes from the lining of the uterus, its odor is likely to convey information about reproductive readiness. Odors from apocrine or sebaceous glands in the vestibule or on the vulva may supplement the odor of blood. Whether or not blood and glandular secretions contain sexual attractants, urine certainly does. Urine from estrous dogs is highly attractive to males even when it is obtained by catheterization (Beach & Gilmore, 1949; Doty & Dunbar, 1974) and is thus uncontaminated with uterine or vulval material. Blood and glandular secretions are thus not essential to the appeal of female anogenital odor—traces of urine caught in the folds of the vulva may suffice to attract males. Beach and Merari (1968, 1970) showed that vaginal secretions of estrous dogs elicit sniffing and licking by males. but Beach and Merari's stimuli may have contained traces of urine. M. Goodwin, Gooding, and Regnier (1979) have shown that methyl-p-hydroxybenzoate is a sexual stimulant in estrous-dog urine. It would be interesting to see whether vaginal odor contains attractants other than those found in bladder urine.

Male wolves seem to be even more attracted to female urine than they are to the females themselves. The reason is probably that the relatively large surface area of voided urine creates a powerful odor. During the breeding season, females further increase the stimulus value of their urine by RLU (Peters, 1974). Rothman and Mech found blood in about one-quarter of all RLUs examined during the Minnesota breeding season. They point out that since males made many of the RLUs they found, the frequency of occurrence of blood in female RLUs must have been considerably higher than 1 in 4. This high frequency of appearance would allow blood to play a major role in the advertisement of reproductive condition.

Rubbing by males may complement RLU by females by increasing the stimulus value of urine that is deposited on the ground. Captive males commonly rub in female urine during the breeding season (Peters, 1974). This response increases the stimulus value of the urine by distributing it over the large surface area of the male and by exposing the male to the odor as he "wears" it. Rotten meat, another natural stimulus for rolling, contains many of the short-chain fatty acids that have been shown to be sexual attractants in other mammals (Michael, Keverrene, & Bonsall, 1971). Thus, body rubbing in novel substances by wolves, like rubbing by cats in catnip (Palen & Goddard, 1966), may be a sexual response that can be triggered accidentally. This conjecture, however, has not been tested, so the motivation and function of rubbing remain a mystery.

Besides rubbing, common male responses to female urine are sniffing, licking, and RLU. Males and females often make pairs of RLUs when traveling together during the breeding season (Peters, 1974; Rothman & Mech, 1979). Female RLUs, together with male RLUs as responses to female urine, are partially responsible for the increased rate of RLU production characteristic of the breeding season (Peters, 1974). Rothman and Mech discovered several interesting facts about RLU in newly formed couples. When pairs of RLUs are produced, the female usually urinates first. The male then sniffs her urine and marks it with an RLU of his own. The female often sniffs the resulting double mark. Rates of production of both single and double RLUs are at their peak when a pair begins to form their bond. Furthermore, newly formed pairs perform RLUs only when they are together. Finally, Rothman and Mech noted that double RLUs were not performed by pairs that did not mate. For these reasons they concluded that RLU, especially double RLU, plays a critical role in the formation, maintenance, and advertisement of the pair bond and in the synchronization of reproductive processes.

These investigators also found that high rates of RLU are often associated with mutual chases during courtship. In captivity, because movements are limited in scope, these chases often take the form of a dance. Mech (1970) quotes a description of one such dance observed at the Brookfield Zoo in a personal communication by Rabb (1968): "The male starts dancing around the female, lowering his front quarters like a playful dog, and wagging his tail. He may also nip the female's face, ears, and back, and mount her side, after which he tries to mount her from the rear." Females may also initiate courtship dances, but, according to Rabb and his colleagues (1967), they do so only one-third as often as males. Female invitations often consist of placing forepaws, head, or neck across the male's shoulders. Sometimes the message is less subtle, as in this description by Schenkel (1947): "With raised tail, the rutting alpha-bitch moves in a feathery dance step, whimpering or 'singing' tenderly. . . . Meanwhile, she moves her genitals in slow, minute, pendulum-like movements in a vertical direction" (p. 106).

Such courtship dances occur throughout the breeding season, but fewer than 3% of them result in copulation (Rabb et al., 1967). Usually, the female resists the male's attempts to mount by running away, tucking her tail between her legs, or sitting. Only after lengthy courtship and at the height of estrus does the female show receptivity by standing with her tail to the side. The male approaches from the rear, clasps her about the hips with his forelegs, and places his penis in the vagina. He then treads with his hind legs, probably to facilitate deeper penetration (Rabb, 1968, cited in Mech, 1970). If the mount is performed correctly, a copulatory tie is formed by the swelling of the bulbus glandis at the base of the penis and the contraction of the vaginal sphincter (Fuller & Dubuis, 1962). The male dismounts by stepping over the female's back, leaving them tied rear to rear. The tie is quite strong. From the air Mech saw two wolves try "desperately" to break apart, but they could not. Rabb has observed ties that lasted up to 36 minutes, and Mech observed a 15-minute tie on Isle Royale. Semen flows most of the time a pair is tied (Harrop, 1955).

Both the length of courtship and the copulatory tie have been proposed as devices for bonding (Mech, 1970). However, struggles to separate are not consonant with this hypothesis, and the findings of Zajonc (1971) on the attractive effects of mere exposure strongly suggest that prolonged and intimate contact ought to create a strong attachment, even without a tie. Finally, Mech and Knick (1978) found that during the day a pair sleeps closer together during the breeding season than at other times.

Outside the breeding season, pair bonds are still there, though, and are maintained through postnuptial courtship. Haber (in personal correspondence to Fox, 1971) has observed mates frolicking together well after the breeding season and has seen one present the other with such small food items as squirrels, pushing them toward the other with its nose. As parturition approaches, male and female work together, catching food near the den. Once the pups are born, the business of rearing ensures continued close contact between the mates. By fall, when the pups begin to fend for themselves, the cycle of courtship recommences.

SUMMARY

Wolves have one of the most elaborate message systems of any nonprimate, with a total of 25 different message types in addition to neonatal interactions. These neonatal interactions include cleaning and nursing as well as special signals for distress and satisfaction. During the neonatal period, socially stimulated elimination provides a basis for later passive submission. Similarly, begging for food lays a foundation for active submission and the group ceremony. Discipline is instilled by pinning, which is also used as a dominance display between adults.

Play is predominantly agonistic and includes several ritualized forms of competition. A variety of visual and vocal signals serve as invitations to play, while others act as solicitations for care or attention. Howls are used to assemble the pack or simply to announce presence. A similar function is performed by RLU, defecation, and scratching, all of which also carry information about the identity of the transmitter. These scent marks further function in promoting familiarity with the environment. Wolves can express extremes of hedonic tone with a facial expression for satisfaction and yelps for pain and distress. A variety of gestures and rituals foster affiliations within the pack by exposing pack members to one another. These affiliative messages include tail wagging, group ceremonies, whines, squeaks, licks, snuffles, and nose touches.

Affiliative signals are often mingled with active and passive submission and their intermediaries. These signs of submission are antitheses of dominance displays, which include standing across, staring, elevating the tail, anal presentation, and riding up. Olfactory displays of dominance include RLU and scratching. Various levels of threat usually succeed in intimidating without physical contact. The bite threat is particularly effective, but sometimes piloerection and growling suffice. A pattern of defensive threat is sometimes used to ward off attack by a wolf of higher rank. A system of territorial advertisement based mainly on scent marking and howling minimizes contact between wolves of different packs, which are liable to fight to the death. Fights within a pack, on the other hand, usually end quickly in ganging up, which leads to submission.

Courtship is prolonged and serves to advertise and maintain the pair bond while synchronizing reproductive processes in the two mates. Friendly courtship-like activities take place before and after the breeding season. RLU, anogenital sniffing, and rubbing in urine expose both partners to urine and vaginal blood, which may contain primer pheromones in the form of hormones or their metabolites.

As a result of this enormous social complexity, anthropologists and ethologists continue to be intrigued with the problems and adaptations of this wide-ranging, social predator.

THE DOMESTIC CAT, *FELIS CATUS*

Cat owners often feel that their animal is not so much a pet as a wild creature that accepts their care but does not depend on it. They are right—unlike fully domesticated livestock, cats have only relatively recently begun to associate intimately with humans, and they have never been subjected to rigorous breeding for characteristics that render them unable to survive on their own. The *feral* populations of *catus* in North American and European town and country are testimony to the independence of the species. According to Zeuner (1963), the major changes in morphology as a result of domestication are slightly reduced body and tooth size and a broader, shorter face. The major behavioral change is an increase in tolerance of familiar conspecifics (Leyhausen, 1965a, 1965b). Even after generations of domestication, cats have not lost their lethal precision, which, like the lock work of a target pistol, is the more fascinating for being rendered in miniature. Even such famous breeds as the Siamese, the Persian, and the Angora differ from their wild cousins mainly in coloration and length of coat.

The most probable wild ancestor of *catus* is F. *libyca*, the tabby-coated "yellow cat" of Africa and the Middle East. Another pretender is the European wild cat, F. *silvestris*. Dental similarities favor *libyca*, but the domestic cat's baculum (penile bone) resembles that of *silvestris*. Since cats of all three persuasions interbreed readily, the problem of ancestry is complex (Zeuner, 1963).

The best available evidence suggests that the cat was domesticated by the Egyptians toward the end of the Old Kingdom (17th century B.C.). Cats were kept as objects of veneration as much as controllers of pests. Paying cats the respect that their independence demanded, Egyptians shaved their eyebrows in mourning whenever one died (Zeuner, 1963). Once domesticated by the Egyptians, cats spread quickly and are now found, feral and as pets, on every continent.

Cats' short history as domestics is only a footnote to a long evolution as highly specialized carnivores. Cats can detect their prey at light levels one-sixth those required by people. Although their acuity is not as great as ours, they share with us at least a rudimentary form of color vision (Mello & Peterson, 1964). The marathon training necessary to elicit a color discrimination suggests, however, that cats do not ordinarily use this type of information.

A cat's hearing, too, is more sensitive than people's, especially at high frequencies (Neff & Hind, 1955). These high frequencies are especially important in finding prey. Leyhausen (1956b) has shown that cats are especially responsive to high-pitched rustles, such as those made by mice. Little is known about the cat's use of smell in detecting prey, but cats seem to hunt by orienting their eyes and ears rather than by sniffing around like dogs.

Once the cat has detected its prey, a sequence of predatory mechanisms unwinds with deadly grace. Taking advantage of every shred of cover, the cat closes the distance between it and the prey. It does this with great speed in what Leyhausen calls the slink-run. Crouching behind the last bit of cover, it treads with its hind legs as though testing for traction, then rushes, at first in a slink-run, then, if necessary, in full gallop. The gait is digitigrade, which means that the cat runs on tiptoe, increasing the effective length of its legs. Finally, it pounces, keeping the hind legs on the ground for stability. The claws, ordinarily retracted to keep them sharp, are extended and fasten like hooks into the victim's back. The kill is quick as the sharp canines are driven between the vertebrae of the neck, popping them apart and severing the spinal cord (Leyhausen, 1965b). Ewer (1973) relates the ability to perform this killing bite to an intricate system of dental mechanoreceptors, high-speed afferent neurons, and precisely controlled jaw muscles.

The cat's finesse does not falter even as it devours its prey. Cats can slice meat neatly with their carnassial teeth, specially modified to work like pruning shears (Ewer, 1973). They can leave even the delicate bones of a pheasant perfectly clean but in position, like an articulated skeleton for an ornithology lab.

Cats' food requirements are as specialized as their technique for fulfilling them. They are as close to being pure carnivores as any mammal. Most of the approximately 30 million domestic cats in the United States are fed fish and meat scraps and commercial cat food. Feral cats sometimes eat small quantities of grass or other vegetable matter, but the staple of their diet is small rodents (Errington, 1936). The cat's major enemies are automobiles, dogs, and human practitioners of "predator control."

Cats are considered solitary carnivores, and many people perceive them as being independent or even aloof. Nevertheless, their daily travels do bring them into contact with one another from time to time, and on such occasions they do engage in a variety of social interactions. The paradoxical character of cat society is captured in the title of an article by Leyhausen (1965a), "The Communal Organization of Solitary Mammals." Unless otherwise stated, the following description of cat social structure is drawn from this source.

A cat's range consists of all the places and paths that it uses. Within this range there is a "first-order home" and sometimes one or more "second-order homes" where it may find (or be provided with) shelter and food. Surrounding the first-order home is a core area that also contains favorite places for resting, grooming, and basking. This area is defended and is consequently only rarely visited by other cats. A complex network of commonly used trails links the core area with other places where the cat hunts, rests, and meets other cats. One country cat studied by Leyhausen and Wolff (1959) used an area of about 1 square kilometer.

When two cats meet in the outer zone where their ranges overlap, they avoid each other, one waiting until the other is well out of sight before proceeding. If they do not see each other in time and come upon each other suddenly, a fight results. Such clashes eventually result in a dominance order, demonstrated on subsequent close encounters by the hasty retreat of the loser. Usually one such fight serves to establish ranks between cats. But even among females, which are less tolerant of each other than males, this dominance is not absolute. For example, priority, not dominance, determines access to a path or place. Even a high-ranking cat will not interfere with an inferior that has arrived there first. When a new male arrives in the neighborhood, he must fight for a position in the male hierarchy. Since a new male arrives quite frequently, the hierarchy rarely solidifies.

Although cats do not form year-round pairs, some females mate only with a particular male, even a low-ranking one, season after season. Thus, the dominance system does not allow monopoly by high-ranking males, and any healthy male has a chance to reproduce.

Cats are remarkable in that they communicate with humans with a greater variety of signals than they use with one another. Of 16 phonetically distinguishable vocalizations recorded by Moelk

(1944), six were used only with people. The remaining ten were used with both humans and other cats. Moelk explains the greater responsiveness of kittens to humans by pointing out that humans can respond to them as individuals and apply rewards and punishments at crucial moments. This remark applies to adult cats as well. In any case, the following discussion will be confined to vocalizations between cats.

NEONATAL COMMUNICATION

Like several of their wild relatives, domestic cats tend to have their litters either in early spring or in midsummer (P. P. Scott & Lloyd-Jacob, 1955). During the last 2 weeks of pregnancy, the mother-to-be spends more and more time visiting and resting in dry, enclosed, and undisturbed nooks and crannies. She generally becomes intolerant of strangers and even her own young of previous litters. Ewer (1968) observed an abrupt reversal of this intolerance just before a cat gave birth. Within half an hour, she stopped attacking her offspring and began to lick and caress them. Evidently, hormonal processes associated with the approach of parturition caused this dramatic switch in motivation.

Schneirla, Rosenblatt, and Tobach (1963) conducted extensive studies on the period of reciprocal stimulation during and after the delivery of 38 litters. The following description of neonatal interaction is drawn from their account. In a case that they describe as typical, four kittens were born at half-hour intervals. All the kittens were licked dry within about half an hour after birth. The mother interrupted her licking to devour each placenta as it was passed, and she usually severed the umbilicus without special effort.

During the rest period following delivery of the last kitten, a number of processes begin to form a bond between mother and offspring. The mother protects the litter by lying with her abdomen toward them and her legs extended, forming a "U." This provides a warm "bay" that helps the kittens orient themselves toward her as they paddle forward with their forelegs. They are further assisted by continued licking directed into the bay and usually find their way to the nipples within an hour.

Nursing occupies up to 70% of the first week after parturition. The mother leaves her litter only briefly and on her return wakes the huddling kittens by licking them and emitting a call that Moelk (1944) transcribed as [mhrn]. The mother then presents herself for nursing. The kittens stimulate the flow of milk with a "milk tread," alternate pushes of the forepaws on the skin on either side of the teat (Ewer, 1968). By the end of their second day, the kittens have developed a "teat order," each one preferring a particular nipple. A kitten will often attach itself to the first nipple it encounters, but if the nipple is not the one it "owns," it will allow itself to be displaced, emitting only a small growl or [my] of protest (Moelk, 1944; Rosenblatt, 1972). Ewer (1959) interprets teat ownership as an adaptation that reduces squabbling and injury—kittens' claws are already sharp at this age. Both olfactory and tactile clues are involved in teat preference, but the former, according to Ewer (1961) and Rosenblatt, are more important. Washing the mother's abdomen temporarily prevents kittens from finding their preferred nipples. A similar effect was obtained by destroying the kittens' sense of smell. When nursed by artificially scented and textured nipples, newborn kittens preferred nipples with odors and surfaces similar to the ones they had been fed with.

A kitten's attachment to its nipple is physical as well as psychological. If the mother gets up and moves off too suddenly, the kittens may be dragged a short distance before they let go. In such cases they give a sharp, high, distress cry, which Moelk transcribes as [mi-i] or [mi-ou]. If they do not give this cry, mothers often ignore them, but when they do cry, the mother returns them to the nest (Leyhausen, 1956a). Young mothers carry their young any which way, but older mothers, even those giving birth for the first time (primiparous), grasp them "correctly," by the nape of the neck (Ewer, 1973).

Mothers seem to recognize infancy with the aid of anogenital odor. Ewer (1968) observed a mother that sniffed and licked the anogenital region of a strange kitten whenever she approached it from the rear. When the kitten turned and presented its face, however, she attacked it. The kitten turned again, presenting its anogenital region, and the female began to lick it again. This reversal of motivation was repeated every time the kitten turned around.

Throughout their first few weeks of life, kittens remain dependent on their mother for all the necessities of life, including nutrition, warmth, and even elimination, which must be stimulated by licking (Rosenblatt & Schneirla, 1962). The mother ingests the kittens' feces during this period (Wemmer & Scow, 1977). When they are 1 week old, their eyes begin to open. At 3 weeks, kittens begin to move around outside the nest, and their dependence on their mother begins to wane. At this age they may even be nursed by another mother if her litter was born on roughly the same date they were.

Kittens in the stage of beginning independence start to rely on one another for warmth, huddling together during the mother's

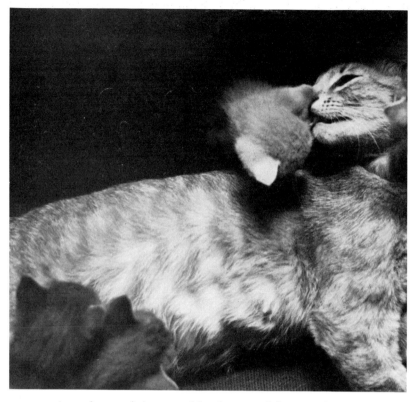

A mother cat licks one of her kittens while two others nurse. Photograph by Brian McMahan.

lengthening absences. They can also depend on one another, rather than the mother, for a feeding summons. Kittens purr as they nurse, and the purrs of the first kitten to get to the mother upon her return alert its litter mates and bring them running (Ewer, 1968, 1973). They can distinguish their mother from other females probably by smell (Schneirla et al., 1963). Odor is also important in the development of their ability to find their way back to the nest; when Rosenblatt (1972) washed the floor of their home area, kittens could no longer find their way back. Later on, however, they began to rely on visual cues, and washing had no such effect.

Vocal and visual signals become increasingly important as kittens grow older. A short, low, alarm growl from their mother sends kittens into hiding, where they remain until the danger has passed. When meeting her 9-week-old kittens, a mother greets them with the [*mhrn*] call, to which the kittens reply in kind (Moelk, 1944). Another signal used by young kittens is the hissing defensive threat with bent back and bristling fur. This threat is often used by tiny kittens to intimidate dogs. It is a tribute to the effectiveness of this display that it is often successful even when the dog is ten times the kitten's size.

Many people believe that mother cats teach their kittens to hunt. This is true only in the sense that she provides them with the opportunity to learn. This "progressive" education begins when, during the fifth week after birth, the mother, another female, or in some breeds a male brings dead prey to the nest. Since the mother has by this time begun to resist their attempts to nurse, the kittens are motivated to pay attention to the food. At first the adult simply eats the food in the presence of the kittens, but a few days later it may leave some for them. When they are about 10 weeks old, the mother brings them prey that is still alive but usually injured. She assists them only by retrieving it if it escapes (Ewer, 1973). Although she does not train her kittens to kill, the mother's role in their learning is considerable. Ewer notes that kittens that do not learn to kill during this period show no interest in prey as adults. She also found that older kittens are allowed to follow and assist their mother as she hunts. That imitation may play an important role in such learning is shown by Chesler (1969), who demonstrated that kittens can learn to press a bar for food more quickly if they are allowed to observe their mother do so than if they are deprived of such an opportunity. Mutual stimulation among litter mates is also important in learning to kill. Leyhausen (1965b) has shown that competition is often necessary to provide the excitement that allows a kitten to overcome its initial reluctance to attack.

INTEGRATIVE MESSAGES

Given the importance of learning and competition in the development of predatory skill, it is not surprising that much play among young cats takes the form of mock hunting. Many of these forms of play are also good practice for combat with other cats. For example, kittens' first form of play is chasing and pouncing on their mother's tail. She often responds to this pestering with a threatening wail, [wa-ou]. Later on, kittens play with one another, ambushing, stalking, and chasing. The chases often culminate in pawing, hugging, rolling, and tumbling about. These are all behaviors that will serve the kittens well when they deal with prey. Sometimes, however, the chases end when both kittens throw themselves on their backs, as adults do when defending themselves from another cat. Even when a mother plays with her young, the threats, pounces, and other playful actions are nearly identical to those used in earnest. The information that these actions are playful is conveyed by a special play-solicitation posture. Cats use a bowing posture, with front legs extended and rear end elevated, to solicit play from each other (E. O. Wilson, 1975). Otherwise, playful actions can be recognized by their brevity and mixture of motivations. Ewer (1968) noted that kittens ambush litter mates as though they were prey, then fight them as conspecifics, and finally flee as from a predator. One form of behavior seen only in play is jaw wrestling, which Leyhausen (1965a) vividly describes as two kittens grabbing each other's mouths "like the forks of a universal joint" (p. 26). They remain still for an instant, then let go.

Should jaw wrestling or some other form of play get too rough, one of the kittens gives a short shriek of distress that Moelk (1944) represents [mi-y]. This cry usually causes the other to desist or even to make up for its boorishness by briefly grooming the injured party. This sequence is particularly common in older cats, whose play often escalates into fighting. Older cats sometimes avoid misunderstandings of playful intent by bowing in solicitation at the beginning of their interaction. The forelegs are stretched out forward, lowering the forequarters, while the hind legs elevate the rear. The eyes are round and the ears erect (E. O. Wilson, 1975). At extremely high levels of distress, cats express their anal sacs, producing a "pungent, unpleasant odor" (Wemmer & Scow, 1977, p. 764).

When they are about 9 weeks old, cats begin to use the [mi-a:ou] call to solicit attention from nearby conspecifics (or humans). They also give this call when sitting or walking alone. In such cases it probably functions as a contact signal, helping a cat avoid surprise encounters that might cause a fight. Vocalization is, however, ancillary to cats' keen vision, which even at night provides their major means of detecting one another. The importance of vision in regulating contacts is shown by the fact that a cat's first response to another is to sit and watch it, often at distances of up to 100 meters. Only after the other cat has passed out of sight will the watcher proceed. When two cats approaching a trail junction from different directions spot each other simultaneously, they both sit down. Eventually one of them will proceed, or they both head back whence they came (Leyhausen, 1965a).

Scent marking provides two other forms of contact messages. Both males and females scent-mark by spraying urine, but males do it more commonly. When a male sprays an object, he sometimes sniffs it first, then backs up to it, extends his hind legs, and raises his tail.

He bends the tip of his penis back so that the thin jet of urine is directed backward at an angle of about 30° above the horizontal. Sometimes he wiggles his tail tip as he sprays.

Although the evidence is indirect, it appears that odor from the anal glands supplements the odor of the urine itself. Schaller (1972) found that when lions spray, the urine passes over the anal glands and picks up odorous fluids from them. The course of urine is similar in the cat and the lion, and sprayed cat urine has a musky odor similar to the odor of anal glands, so it is likely that in cats, too, anal-gland secretions are mixed with urine as it is sprayed.

Urine is usually sprayed against a salient vertical target such as a tree, fence post, bush, or wall. After spraying, the cat often rubs his lips, chin, or cheek on the target and then on other, nearby objects, thus further spreading odors in the urine. By such rubbing, the cat may also disseminate substances from the hypertrophied sebaceous glands that are concentrated on the parts of the face he rubs (Ewer, 1968).

The spatial distribution of such marks and the typical responses of other cats to them suggest that they are not territorial advertisements, but that they simply inform the receiver that the marker has passed. Many scent marks are located along trails used by several different cats. The marks are not confined to a cat's territory. Furthermore, there is no evidence that cats find one another's scent marks aversive. Usually the receiver sniffs the mark, often moving its nose over the target for several seconds. Then it may spray the target itself or simply proceed on its way.

Leyhausen and Wolff (1959) suggested that cats may be able to tell how fresh a mark is and that marks may function like railway signals: "Fresh mark = section closed, going on may lead to a hostile encounter. Less fresh mark = proceed with caution. Old mark = go ahead. The individual, before passing a mark, usually covers it with his own, thus 'closing the section' " (p. 670). Leyhausen (1965b) further suggests that scent marks may identify the marker. Ewer and Wemmer (1974) hypothesized that the height of a civet cat's scent mark might convey information about the sex of the marker, male marks presumably being higher than females'; Wemmer and Scow (1977) suggest that urine conveys sexual and individual identity and that head and neck rubbing, scraping, and scratching may convey individual identity.

So far, however, there do not seem to be any published accounts of individual differences in urine or glandular secretions that could encode such information. There is, however, behavior suggesting that body odors, including those from the anal region, are used to recognize conspecifics. Leyhausen (1956a) describes a typical greeting between strangers as follows:

> . . . they slowly approach each other and stretch, the foreparts somewhat lowered, nose to nose, but not touching. The ears meanwhile are friendly and inquisitive and are held slightly forward. The tail hangs peacefully. Both try to sniff the other from head and neck to the rear and to feel the other with their vibrissae. These sniffs are often linked with flehmen. Usually, however, one withdraws from the other while simultaneously trying to sniff it under the tail. Thus, they circle each other [p. 18].

Anal sniffing is confined to meetings with strangers; in encounters between members of the same family or other cats that know each other well, a fleeting nose-to-nose greeting suffices.

Both strangers and familiars often greet one another with various vocalizations, including [mhrn] and [grr] (Moelk, 1944). Since the voices of different cats can be distinguished by humans, it is likely that cats themselves can recognize one another by the same means.

When greeting results in recognition, cats may exhibit friendly behavior in the form of *Köpfchengeben*, in which one cat pushes its forehead and then its neck, shoulders, back, and flanks against the other as it passes alongside. *Köpfchengeben* is often reciprocal, and sometimes a thud is audible as the cats bring their heads together. In an extremely friendly greeting both cats purr while performing reciprocal *Köpfchengeben*. The purring continues as they begin to groom each other, each licking the other's face, ears, and neck. Sometimes they give the neck extra attention in the form of nibbling, but they generally ignore parts of the body to the rear of the shoulders (Leyhausen, 1956a). Because it occurs while a cat is being groomed and because of its early association with nursing, the purr used in friendly greeting can be considered a sign of satisfaction. Sometimes, however, it precedes physical contact and is thus also a sign of peaceful friendliness. It is a graded signal and becomes louder and rougher with more energetic grooming.

The use of three different behaviors to express affiliation may seem surprising in a mammal considered a solitary carnivore. Although they hunt alone and do not form close-knit groups, cats do seem to have a considerable capacity for forming attachments. They can, for example, become quite fond of a particular human. Apart from the sexual bond that sometimes brings the same pair together season after season, this capacity is largely untapped in cats' "natural" lives. It is not a product of domestication, for Smithers (1968) found that *F. libyca*, the cat's wild ancestor, can also form strong personal bonds.

Another major exception to the cat's supposedly solitary existence is the "social gathering" described by Leyhausen (1965a). Even outside the breeding season, groups of half a dozen or more cats of both sexes often gather at dusk. These meetings always occur on common ground or on the border between two or more core areas. Some cats sit quietly, looking around, while others exchange greetings or groom each other. These gatherings sometimes last for hours, and then the cats go home one by one. Leyhausen noted that these gatherings are generally unmarred by agonism apart from a few mild displays of dominance. The peaceful character of the gatherings is especially remarkable because these same cats will, under other circumstances, ordinarily avoid, threaten, or fight one another. The peace is maintained by friendly behavior and by mild displays of dominance and submission.

AGONISTIC MESSAGES

According to Leyhausen (1956a), adult cats never have "a true submissive attitude" (p. 25). Nevertheless, there are a number of behaviors typically performed by the social inferior that convey lack of aggressive intent. One of these is the head twist, possibly an intention movement derived from the defensive back roll described later. Another is anal presentation. The subordinate presents its rear by raising its tail, walking past the dominant, and stopping with its rear under the dominant's nose. The lower-ranking cat may then display an infantile form of behavior, kneading the ground with its forepaws as in the milk tread. If the superior shows signs of hostility at this approach, the inferior sits down and pointedly looks away.

Displays of dominance are complements and antitheses of these signs of submission. A dominant cat exercises anal control by holding its head high and stiffly approaching the subordinate's rear, inducing anal presentation. If the inferior does not cooperate in this manner, the superior sits and stares at the subordinate until it presents or withdraws. Ewer and Wemmer (1974) suggest that the twitching tail of the spraying posture may be a visual sign of dominance. The twitching can be considered antithetical to the immobility or slow, sinuous waving of the submissive cat's tail.

If withdrawal is prevented by a physical or psychological obstacle—for instance, the presence of offspring—approach may be rejected with a low-intensity defensive threat. The defender crouches and swings the head back and forth, with the lips slightly retracted but not baring the teeth, and the ears back but not laid flat. While in this position, a cat may produce the explosive hiss commonly referred to as a spit or may strike out with a forepaw. At slightly higher levels of fear, the cat crouches, draws its head in, and dilates its pupils.

When both fear and aggression are very high, the cat displays the well-known bent-back posture, with bristling fur, open mouth, dilated eyes, flattened ears, and elevated tail. Leyhausen (1956a) interprets the form of this threat as the outcome of opposing motivations to flee and attack; the anterior part of the body tends to retreat while the posterior advances. As a result, the back flexes and the hind legs move around so that the cat presents its flank. This impressive display is emphasized by long, low hisses. Unlike spits, hisses are graded in intensity (Wemmer & Scow, 1977).

Leyhausen has analyzed the postural and facial displays of defensive threat according to their proportions of aggression and fear. Figures 8-10 and 8-11 show examples of these threats, with aggression increasing from left to right and fear increasing from top to bottom. The postures just described correspond to the three lower frames on the main diagonal of Figure 8-11.

Another common consequence of mixed motivations is displacement scratching by standing on the hind legs, reaching up the trunk of a tree or some other vertical surface, and raking downward with alternate flexions of the forelegs. Although this behavior, sometimes called claw sharpening, does in fact sharpen the claws (Wynne-Edwards, 1962), it is often used as a display to end an encounter or to deter an attack.

When aggression is unmixed with fear, the cat displays the postures, gestures, and vocalizations of pure, or offensive, threat. This is especially likely when the opponent is a stranger, a trespasser, or a sexual rival. In offensive threat both opponents often show the same behaviors, appearing at times to be mirror images of each other. They stiffen all four legs, and since the rear ones are longer, their rear ends are elevated. Their heads, ears, and tails are also held high, increasing their apparent size. Keeping their legs stiff and pausing after every step, they stride slowly toward each other. As they close in, their steps become shorter and the pauses between them longer. They whip the tips of their tails laterally, meanwhile producing a graded sequence of vocal threats of increasing intensity: snarling, growling, wailing, and screaming. The snarl is a short growl, which is a low, prolonged, rough vocalization produced deep in the throat. The wail can be represented [wa-ow] and is a vowel-like sound produced by opening and closing

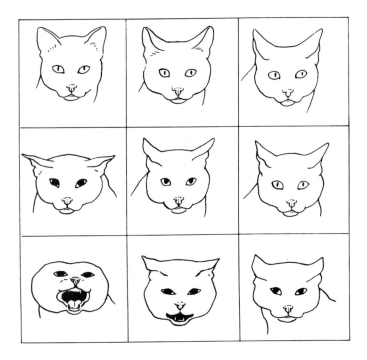

FIGURE 8-10 Facial expressions of the cat. From left to right, increasing aggressiveness. From top to bottom, increasing defensiveness. From "Das Verhalten der Katzen," by P. Leyhausen, *Handbuch Zoologie*, 1956, 8(10), 17–34. Reprinted by permission of Walter de Gruyter & Co., Berlin, Germany.

the mouth while growling. At times its ordinarily low pitch is gradually raised into a high-pitched scream (Moelk, 1944; Wemmer & Scow, 1977).

A variant of the wail, with a low, almost purring quality, is used as a challenge. Alone or in groups of two or three, established residents confront newcomers or adolescents with this call. According to Leyhausen (1956a), acceptance of this challenge is the most common occasion for serious fighting.

Competition for estrous females is another common cause for combat. Females attract several males 1 or 2 days before full receptivity, and the rivals fight intermittently until she accepts one of them. Since the spoils of such a tournament often include a chance to reproduce, these fights are in earnest, and serious injuries are common. The noise

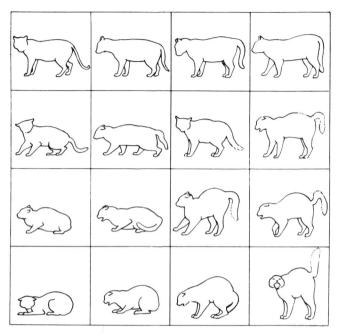

FIGURE 8-11   Postures of the cat. From left to right, increasing aggressiveness. From top to bottom, increasing defensiveness. From "Das Verhalten der Katzen," by P. Leyhausen, *Handbuch Zoologie*, 1956, 8(10), 17–34. Reprinted by permission of Walter de Gruyter & Co., Berlin, Germany.

and vigor of the fray are, however, out of proportion to the injuries inflicted, which are usually confined to the relatively well-protected area around the shoulders.

In a typical fight the opponents circle each other, yowling in threat. Then one springs at the other, raking it with the forepaws. Only in rare, extremely serious attacks, as when a mother is defending her kittens, are the cat's most lethal weapons, the teeth, brought into play. In such cases the attack is without preliminaries—she simply springs at the intruder's neck and attempts to deliver a killing bite. In response to either sort of attack, the besieged cat throws itself on its back, a position from which it can employ all four paws as well as its teeth. The attacker then follows suit, and the two lie on their sides, flailing and screaming. They may then rise and reenact the sequence of offensive and defensive threats. Alternately, the less aggressive may take flight, the victor in hot pursuit and raking the loser with its claws whenever it gets close enough (Leyhausen, 1956a).

Even when a fight is not serious enough to cause injury, it is probably frightening and painful enough to motivate the loser to avoid similar encounters. One way to do so is to stay out of other cats' core areas, where they spend most of their time. Sign left by the other cats may facilitiate such avoidance. Two types of marks are said to be concentrated at the periphery of core areas and may thus serve as territory advertisements. According to Leyhausen, the claw-sharpening display shreds the bark of trees, leaving salient visual marks, which are concentrated at territorial borders. Since encounters between neighbors are most likely to occur at such borders, displacement scratching should be most likely to occur there. Thus, it is unclear whether the distribution of these marks is the cause or the effect of defense of core areas. It is not known whether feces act as territory markers, but the behavior of F. silvestris, a close wild relative of catus, is suggestive. Silvestris is known to bury its feces within its territory but to leave them exposed at its edges. Domestic cats do not always bury their feces, but it is my impression that they are least likely to do so when near or beyond the edges of their core areas. Even when they do cover feces, secretions from their pedal glands (Schaffer, 1940) may transmit an olfactory signal.

SEXUAL MESSAGES

Female cats generally come into estrus twice a year, once in the winter and once in the spring. If a female loses her litter, she immediately comes into heat again. She remains receptive for 3 to 4 days but becomes attractive to males a day or two before she will accept them. Both vocalizations and odors are responsible for this attraction. According to Moelk (1944), the female's mating cry comes in two forms—a mild [mhrn-a, ou] and an intense [o-o':ə]. Both are produced repeatedly and alternately, often for hours at a time.

Odor, however, may be more important than sound, for, according to Ewer (1968), scent alone attracts the males of urban populations. Some of the odor that interests males comes from concentrations of large sebaceous glands on the female's face. She rubs these glands on objects that are then sniffed, often at great length, by males. Michael (1961) noted that males watch females as they rub and then attempt to mount. He concluded that facial rubbing is in part a visual display. Males are also attracted by her urine, sniffing it and displaying flehmen. On the basis of such observations, Verberne and DeBoer (1976) concluded that skin-gland secretions and urine convey information about the female's hormonal condition. Michael found that the anogenital odor of an estrous female repels males.

Males also have sebaceous complexes on their faces, which are sniffed by females, perhaps to acquire information about the male. The response of some cats to catnip or its active ingredient, nepeta-lactone, may be related to sexual stimulation of the female. Cats of both sexes respond to catnip by rubbing and rolling on the ground, behavior otherwise displayed only by females at the height of estrus. Palen and Goddard (1966), who investigated this phenomenon, found that estrous females were more highly stimulated by catnip than non-estrous females were but pointed out that the catnip might not act directly; that is, both catnip and estrus might increase the irritability of the skin, making the cat more likely to rub and roll. So far, no one has isolated a chemical similar to nepeta-lactone in the skin glands of male cats.

When advertising by smell, song, or both, the female is seldom alone for long. Males come from all over the neighborhood, announcing their arrivals with two-syllabled "mild coaxing forms of the [mhrn] pattern" (Moelk, 1944, p. 202). These calls soon form a polyphonic chorus with accompanying growls, wails, and snarls as fighting breaks out among the rivals. The result is the caterwauling of mixed male threat and sexual advertisement that so often enlivens warm spring evenings.

If the female is not ready to mate, she rejects each of her suitors by spitting, growling, striking out with a forepaw, or simply retreating. Eventually the retreats from one of the toms become shorter and shorter. This coquettish flight is the first sign that she has agreed to mate with one of the contenders. This fortunate animal is not always the victor—Leyhausen (1956a) has observed cases in which, while the winner confidently sniffed around, the female sneaked off with her favorite.

Once the female has made her choice, coquetry gives way to invitation as she rubs her head on the ground, then rolls in front of the tom. He approaches, emitting purrs, attempting *Köpfchengeben,* and sniffing and licking her head and vulva. At this stage he may again be repulsed by bats with the forepaws. Her claws, however, are at this stage usually retracted. Finally she presents herself by lying prone, sliding her rear legs back, raising her rump slightly, and moving the root of her tail up and to the side. The male mounts quickly, biting the fur on her nape. He treads with his hind feet, positioning his penis. When, often after several attempts, he succeeds in orienting it properly, he thrusts once, growling as he ejaculates. The female responds with a distress shriek, then whirls and attacks. The male leaps backward just in time to avoid her paws. The female's shriek is probably a response to the pain of intromission, caused in part by short, backward-pointing spines on the penis. These spines probably assist in providing the stimulation necessary to induce ovulation. If the partners know each other well, intromission is probably no less painful, but the contretemps is avoided, and each partner peacefully licks its genitals after withdrawal. In any case, the female soon begins to rekindle her partner's ardor, or at least to express her own, by purring, *Köpfchengeben*, or rolling. A second copulation soon ensues, but after two or three of them, the male's interest begins to wane. This decline is met by increasingly brazen advances; she displays *Köpfchengeben*, rubs against him, throws herself in front of him in the mating position, and even insinuates herself beneath him. If these attempts fail, as eventually they must, she may carry on with another male. Although individual mounts are quite brief, a female's sexual activity may go on for days (Leyhausen, 1956a).

SUMMARY

In spite of their supposed solitude, cats interact through a variety of olfactory, auditory, visual, and tactile signals. The foundations for several of these messages lie in neonatal milk treading, purring, recognition of familiar odors, and threats. The mother uses a [mhrn] as a greeting, which wakes the kittens on her return.

In slightly older cats social play provides practice for catching prey and combating conspecifics. Such play is often solicited with a bow and terminated with a cry of distress when it becomes too painful. Older kittens and adults solicit attention and announce their presence with the familiar [mi-a:ow] call. This call is aided in establishing contact between cats by scent marking with urine and facial glands. These vocal and olfactory signals may also convey sexual or individual identity. Most incongruous with conventional notions of cat personality are *Köpfchengeben*, reciprocal grooming, and purring, all signs of friendliness, and their frequent occurrence at peaceful social gatherings.

Even when their mood is not particularly peaceful, cats avoid violence by displays of submission, including the head twist, anal presentation, and the milk tread. Dominants display complementary behavior: staring, elevated head, and anal control. When these displays fail to keep the peace, a variety of defensive threats may still ward off attack. These threats include crouching, baring the teeth, spitting, and the characteristic bent-back posture. These fearful expressions may be mixed with varying degrees of aggression. At extremely high levels of these opposing motivations, there may be "sparking over" to displacement scratching. Ordinarily, however, a display contains a mixture of elements expressive of tendencies toward both fight and flight. Pure, or offensive, threat involves stiffened limbs, a whipping tail, snarling, growling, and wailing. A form of this threat appears as a challenge to adolescent or newly arriving males and in sexual rivalry, in which it also serves as a male sexual advertisement. Fighting is common only in these three situations. Severe injury or death may result from such fights, but attack is usually restricted in target and weapons. Cats can reduce the risk of such fights by avoiding trespassing in one another's core areas. They may be able to do this by reading claw marks and feces, which are, according to some observers, more commonly seen at territorial edges.

Females advertise ensuing estrus with odors and distinctive mating cries, [mhrn-a, ou] and [o-ó:ə]. Odors from sebaceous glands and urine are also effective in attracting males, which fight one another for access to the female. The female chooses her mate, and lengthy courtship follows. Female courtship displays include coquettish flight, rubbing, and rolling, culminating in presentation. The male uses *Köpfchengeben* and [mhrn] cries. A distress cry from the female announces intromission and almost immediate ejaculation. Tactile stimulation from continued copulation, often with more than one male over a period of days, results in ovulation.

Cats' messages are encoded in eight vocal, 25 visual, and three olfactory forms (Leyhausen, 1960, cited in Eisenberg, 1973). Wemmer and Scow (1977) add seven forms of tactile communication, giving a total of 43 forms. As my analysis shows, these 43 forms encode 22 different types of messages.

9
NONVERBAL COMMUNICATION AND LANGUAGE

There's language in her eye, her cheek, her lip.
Nay, her foot speaks; her wanton spirits look out
At every joint and motive of her body.

WILLIAM SHAKESPEARE

267

This chapter deals with two related topics. The first, nonverbal communication, has a rich empirical and phenomenological literature. In contrast, the second, the evolution of language, is rich only in speculation. Perhaps an analysis of human nonverbal communication, combined with the preceding analyses of communication in other mammals, can shed some light on the matrix within which language evolved. This approach is likely to be fruitful if, as I hope to show, human nonverbal communication conveys the same kinds of meanings as the communications of other mammals. Such a demonstration would suggest that language evolved in the context of a nonverbal system that is still open to observation.

NONVERBAL COMMUNICATION IN HUMANS

To what extent is human nonverbal communication compa-rable to the communications of other mammals? Superficially, it is far more complex. Human nonvocal signals, including facial expressions, postures, gestures, and physical contacts, number about 136 (Bran-nigan & Humphries, 1972). Nonverbal sounds, including grunts, moans, gasps, whistles, shrieks, and at least four kinds of cries, bring the number of nonverbal signals to 150, almost twice the number used by the vervet. When we categorize these human signals by meaning, we find that most, if not all, fit into 22 of the 30 message types already described. In the following account I will not attempt to deal with all 150 signals but will show that a few of the more important ones send the same information as the messages of our mammalian cousins. For a more complete description of human nonverbal communication, I recommend books by Birdwhistell (1970), Mehrabian (1971), and Morris (1977).

NEONATAL COMMUNICATION

Wolff (1969) showed that babies have three types of cries, each signifying a different kind or intensity of distress: hunger, anger, and pain. Evidently, the quality of the cry also encodes identity, for Lind (1971) found that mothers respond differently to the pain cries of their own, versus others', babies. Lind also showed that mothers respond to their babies' hunger cries with an increased flow of milk in the breasts. Maternal identity is communicated by odor, for MacFarlane (1977) has shown that nursing infants can discriminate the odor of their mothers' breasts from the odors of other nursing mothers. Another example of human elaboration of the neonatal mammalian theme is the baby's smile, which Eibl-Eibesfeldt (1970) has described as a signal that enhances the formation of an affiliative bond between mother and child. Morris (1971) suggests that human infants become "imprinted" on the sound of the maternal heartbeat. Infant cooing is generally considered a sign of pleasure or satisfaction.

INTEGRATIVE MESSAGES

Human integrative messages are culturally controlled but emerge from the mammalian matrix. Children's play includes many familiar games such as king of the mountain, tug of war, follow the leader, tag, and hide and seek, which are also played by deer, wapiti, rats, wolves, and vervets.

Many nonverbal signals are paralinguistic, which means that they are used with and modify the meaning of language. Mehrabian (1971) calculates that 93% of the information exchanged in a typical conversation is coded not in words but in paralinguistic and other nonverbal signals. One such signal is the identity of the speaker. The voices of adults communicate not only their maturity but their individual identities. With the aid of the sound spectrograph, forensic technicians can develop voice prints, which are admissible as evidence in some courts. Humans are supposedly uninterested in one another's smells, but Russell (1976) has shown that humans can identify the sex of the donor of armpit odor at frequencies well above chance.

Contact calls vary from culture to culture and often involve words or phrases such as *over here*. The meaning of the message is encoded not only in the words but in the location of the sender. Linguists refer to the phatic function of speech, by which they mean that language is often used to maintain contact, not to convey other information. Mates often use idiosyncratic whistles, chirps, and clicks as greetings to establish contact or to maintain contact even when they are in the same room. Underlying the banality of meteorological comments and sweet nothings is an ancient mode of interaction in which each person acknowledges the presence of the other. Humans also use visual signals to stay in touch with one another. Over short distances, direction of gaze is an important contact signal. Over longer distances, humans use semaphore-like movements of their arms to apprise one another of their location.

Humans often familiarize themselves with their environment with simple marks such as cairns or blazes, but sometimes they adopt a paralinguistic strategy and use graffiti. It is tempting to speculate about the motivation and function of such marks. Such speculation might begin with the fact that, like the scent marks of other mammals, graffiti are distinctive, often displaying the name and calligraphy of the author, thereby identifying that person to others. In forests almost all graffiti are of this "Kilroy was here" type. In cities, where the aerosol can is the tool of choice, graffiti, like scent marks, express not only identity but threat, dominance, and sexual readiness, often in the same phrase. Presumably, the authors obtain a sense of power, identity, or reassurance when encountering their own graffiti. Some of the millions of dollars spent to erase such marks might be better used to understand, control, or redirect their underlying motivations.

Since humans are much less hairy than other primates, they do not depend on allogrooming to form and express bonds of affiliation. Instead, they embrace in ways that are prescribed by their culture. Whatever their form, whether an *abrazo* or a ritualized kiss on the cheek, embraces expose the participants to each other's odors and touches. On the basis of comparative evidence, smiling and laughing have been described as agonistic signals. In some cultures, however, they seem to be social reinforcers that create bonds of attraction (Darwin, 1872).

When receiving an embrace or other consummatory stimulus, humans express satisfaction with moans, grunts, or a consummatory face. Humans, like their hairier cousins, have not been shown to respond to these signs of satisfaction in any particular way, but it is likely that they reinforce conspecifics who provide consummatory stimuli.

Human alarm signals show considerable variation on the basis of culture and class. Alarm signals are usually words or phrases, but the meaning is encoded as much in such paralinguistic features as abruptness of onset, loudness, and brevity as it is in the meaning of the words. Tomkins (1963), following Darwin (1872), describes facial expressions that appear when the sender is startled or afraid: the eyes and mouth open, the eyebrows rise, the pupils dilate, and the skin lightens.

Low-intensity distress signals, such as weeping, may or may not involve words. Weeping by adults is generally lower in pitch than that of infants but otherwise resembles the infant's hunger cry. Darwin describes contraction of muscles around the eyes, spasmodic respiration, and heaving of the shoulders in both infants and adults. Humans may send low-intensity or medium-intensity olfactory distress signals. A pilot study performed in our laboratory suggests that human subjects can discriminate the armpit odors of fearful donors from those of excited but not fearful donors. High-intensity distress screams generally do not involve words. Like the distress cries of other mammals, they are high-pitched, prolonged, and often repeated.

Human solicitations vary from culture to culture. There is, however, some cross-cultural consistency in such signals as raised pitch in speech, tugs at garments or arms, pouting or "O"-shaped mouth, and widened eyes (Birdwhistell, 1970).

Human assembly messages are usually verbal and are almost completely determined by culture. Apart from loudness, they seem to have few features in common.

AGONISTIC MESSAGES

In most parts of the human agonistic-message system, the mammalian theme is stated clearly. In territorial advertisement and fighting, however, the theme is lost in cultural counterpoint. Human territorial advertisements are governed by legal prescriptions. Like the scent marks of wolves and pikas, human "No Trespassing" signs are concentrated at the edges of lots, but they are visual, not olfactory. Graffiti may occasionally act as territory advertisements, as when they proclaim the dominant position of a gang in a particular urban sector ("Dukes rule"). A gang's turf, however, differs from the typical mammalian territory in several important ways, especially in that only some conspecifics (members of other gangs) are excluded. Human fighting is more highly ritualized than that of other mammals. Boxing has analogues in other species, but the Oriental martial arts, for example, are far more regulated by tradition.

Human threats, on the other hand, are quite similar to those used by other mammals. Fist shaking is widely distributed across cultures and is a direct demonstration of readiness to fight. Morris's (1971) description of the human "threat face" applies to several of the mammals previously discussed, especially the wolf. The eyebrows are brought together, the forehead muscles are contracted to produce wrinkles, and the lips are pursed. In defensive threat, humans raise the eyebrows, wrinkle the forehead, and pull back the corners of the mouth. Higher-intensity offensive threats are expressed with snarls and growling vocalizations, sometimes, but not always, in the form of words.

Human dominance displays also show our mammalian ancestry. They generally increase apparent size without preparation for fighting. Broadening of the chest, squaring of the shoulders, and adoption of an elevated position are all common human expressions of dominance. The stare used as a dominance display by several other species is also seen in many different human cultures.

As might be expected, human signs of submission are often antithetical to displays of dominance. Withdrawal, cringing, kneeling, prostration, and lowering of the head all decrease apparent size. Looking away, lowering the eyelids, and lowering the gaze are antitheses of staring. Most of these gestures are seen in humans of different cultures (Eibl-Eibesfeldt, 1970).

SEXUAL MESSAGES

Human sexuality is governed by culture, but the government is not totalitarian. Our ancestry shows in every phase of sexual behavior, from advertisement through copulation. There is evidence suggesting that in our sexual advertisements we humans may not be totally dependent on visual and auditory signals, as has often been supposed. Female vaginal odors may provide clues to human sexual readiness, for these odors are most attractive around the time of ovulation, when conception is most likely (Doty, Ford, Preti, & Huggins, 1975). Some of the attractiveness of these odors is due to short-chain fatty acids (Comfort, 1971). These substances are probably formed by bacterial action on sebaceous and apocrine secretions that collect in the complex folds of the labia. Some of the individual differences in these odors that Comfort describes may be attributable to the great variability in the configurations of these folds. The same series of short-chain fatty acids has been claimed to act as a sexual attractant in rhesus monkeys (Michael, Keverrevene, & Bonsall, 1971; Michael, Bonsall, & Warner, 1974). Many of them also occur in the anogenital regions of wolves (Peters, 1974). The same bacterial processes may operate on male secretions that can collect under the foreskin. Evidence that such chemicals may be able to influence females is provided by the fact that females' sensitivity to their odors varies with their menstrual cycles. Humans are also susceptible to sexual arousal by purely visual signals, as is demonstrated by the commercial success of numerous illustrated periodicals without other value.

During the courtship phase some humans, with tremendous variability on the basis of culture and status, display nasonasal, orooral, and orogenital contacts, which may facilitate male ejaculation or female orgasm, increasing chances of conception (Masters & Johnson, 1966). A similar function has been proposed for the vocalizations sometimes produced by either or both partners during copulation. Hamilton and Arrowood (1978) compared such human vocalizations to those of baboons and gibbons. In all three primates the vocalizations are elaborations of the heavy breathing that accompanies vigorous copulatory movements. In all three primates female vocalizations were more complex than male vocalizations and began later in the copulatory episode.

This brief discussion of human nonverbal communication has only scratched the surface of an enormous and rapidly growing body of research. It should, however, suffice to make the point that, especially in our neonatal, integrative, agonistic, and sexual behavior, we are still mammals. As a final note, I should mention Bateson's (1966), comment that, even at our most cerebral stage, we are still communicating about relationships of dominance, attraction, and dependency. He noted that when he gives a lecture about cetacean communication, he is not only trying to convey abstract information but also asking for attention and respect. In Bateson's case, both attempts met with success.

THE EVOLUTION OF LANGUAGE

Humans are language-using mammals, and even their nonverbal communication shows the pervasive effects of this complex adaptation. In this section I shall first describe some of the more important features of language, then describe some theories on how it might have evolved. Whenever possible, I will relate these theories to the preceding analyses of communication in other mammals.

FEATURES OF LANGUAGE

In the early 1900s the Linguistic Society of Paris banned all papers on the origin of language from its journals. At that time there were insufficient data to warrant further speculation on this topic. In the years since this prohibition took effect, there has been valuable work on this problem, and it now seems possible that someday we may have a real understanding of how language evolved.

One of the most important areas in which strides have been made is the nature of language itself. Languages are codes for the transmission and processing of information. They are highly flexible codes compared to the signal systems of nonhumans. Their flexibility is the result of the use of four different levels of patterning. As you listen to a conversation in a language with which you are unfamiliar, what you hear is a continuous stream of sounds, corresponding to the physical, or phonetic, level of patterning. Eventually, some sounds seem to recur. You may hear the second speaker repeat a pattern spoken by the first. Although their voices are physically different, you have learned the distinguishing features of some patterns. These features occur at the *phoneme* level. You still have no idea what is being discussed. The phonemes you have learned are in fact meaningless, but they can be combined to form meaningful patterns called *morphemes*, which include not only words but patterns like "ing," denoting an action *going* on. The fourth level of patterning is grammatical. The same morphemes can occur in different orders, producing different meanings. For example, "Og killed the gnu" does not mean the same as "The gnu killed Og." Moreover, the use of word order to convey meaning implies that some orders are meaningless or not allowed: "Killed Og gnu the." Multiple levels of patterning contribute to the flexibility of language by shortening codes. For example, 20 phonemes uttered in sequences of five could provide unique morphemes for 20^5 = 3.2 million different referents.

The systems of grammatical rules that restrict word sequences in various languages have been analyzed according to the psychological processes that underlie their production and comprehension. There are two major schools of thought about these processes. Stimulus-response psychologists, notably B. F. Skinner (1957), assert that in learning a language one learns what words most often follow other words, progressing through approximations in which the next word chosen depends only on the preceding word, then on the preceding two words, and so forth. There are several major objections to this "finite-state" or "stochastic" model. First, the learner would have to be exposed to too many sequences to learn even the simplest grammatical rules. Second, children do not utter successive approximations to a language. Their first word sequences are grammatical (Braine, 1963; Lenneberg, 1967). Third, the stochastic model cannot explain our dual comprehension of ambiguous sentences like "They are flying planes," which has two definite and distinct meanings. Fourth, the stochastic model cannot explain comprehension of nested clauses in which sequential dependencies give no clue about meaning—for example, "The cat you said I fed died" (Chomsky, 1957).

For these reasons most linguists accept the necessity of what is called a phrase-structure grammar, in particular, Chomsky's transformational grammar. Chomsky states that sentences are organized hierarchically into phrase structures similar to the diagrams of sentences done in grammar school. These structures are generated in speaker and listener alike by a series of transformations on key, or "kernel," sentences. For example, the two-phrase structures underlying "They are flying planes" are "they/are flying/planes" and "they/are/flying planes." These two kernels provide the basis for our understanding of numerous transformations: passive, "The planes are being flown"; negative, "They are not flying planes" or "The planes are not flying"; interrogative, "Are they flying planes?" and so on.

Chomsky's theory is attractive to psychologists because it imputes similar processes to speaker and listener. Neisser (1966), for example, presents an excellent case for the view that we understand speech in the same way we produce it, by constructing sounds, words, and sentences out of the raw material of memory or sensory data, rather than by rote, passively matching input to stored representations.

Grammar is one of four "design features" (Hockett, 1960) that allow human languages to transmit enormous amounts of information. A second design feature, displacement, refers to the fact that language often refers to objects or events remote in space and time. A third design feature is duality of patterning, referring to the fact that in speech meaningless elements (phonemes) combine in different ways to produce different meanings. Thus, the "b" in *bat* and the "r" in *rat*, meaningless by themselves, distinguish the flying mammal from the crawling one. Because of grammar, displacement, and duality, language also displays open-ness, referring to the fact that we typically produce and understand sentences that have never occurred before, like this one.

Lenneberg (1967) argues that these characteristics are part of a species-specific adaptation that allows children to learn language merely by being immersed in it. Recently, however, it has been claimed that experiments by Fouts (1973), the Gardners (1969), Premack (1971), Rumbaugh and Gill (1976), and others have undermined the uniqueness of human language ability. These investigators have taught chimpanzees and a gorilla to use symbols to communicate with humans and with one another. The symbolic productions of these apes exhibit grammar, displacement, duality, and open-ness. The chimpanzees trained by Rumbaugh and his colleagues demonstrate grammar because they produce symbols (by means of a computer terminal) in different orders to express different meanings. Chimps in each of the laboratories commonly refer to past events, exhibiting displacement. Stokoe (1972) notes that American Sign Language, used by the chimpanzees studied by Fouts and the Gardners, consists of signs that can be broken down into combinations of motions and positions called cheremes. American Sign Language, therefore, demonstrates duality. Finally, the productions of these chimpanzees are often original, or "open." Washoe, the chimpanzee trained by the Gardners and now studied by Fouts, learned the sign "open" for doors and spontaneously applied it in an original "sentence" to a faucet. This is open-ness in a double sense. Four of the most critical design features of language—grammar, displacement, duality, and open-ness—do not discriminate between human language and the symbolic productions of other apes. Nevertheless, humans have exploited the information-bearing capacity of language far beyond the levels thus far attained by any known nonhuman. If the difference is one of degree, the degree is very large. Moreover, recent publications by Terrace, Petitto, Sanders, and Bever (1979) and the Sebeoks (1980) challenge the claim that apes use syntactic rules to create novel "sentences." These authors ascribe the performance of Washoe, Sarah, and other "talking" apes to cues emitted unconsciously by the human interlocutor.

THEORIES ON THE EVOLUTION OF LANGUAGE

Even if one regards symbolic behavior by chimpanzees and gorillas as examples of the class of behavioral curiosities that includes bowling pigeons and motorcycling bears, we are forced to admit the potential for complex symboling in creatures whose cranial capacities, dentition, and ecology are no more human than those of chimpanzees and gorillas. The task of a theory of language evolution, therefore, is to show how contingencies in the environment of some apelike creature might have selected for the ability to learn complex sequences of vocalizations referring to objects, locations, actions, or events. Theories of language origin have focused on naming, on more complex reference, or on grammar. There are six theories on the origin of naming.

The gestural theory of Wundt (1900) and Paget (1930) attempts to explain the origin of naming on the basis of physical similarities between articulatory movements and the motions and shapes of external objects. For example, *spike, spit, spark, and speck,* and 81%

of all English monosyllables beginning with "sp," refer to a point or something coming to a point. Were it possible to demonstrate such relationships in several unrelated languages, as Paget tried to do, the theory might have some merit, but as R. Brown (1958) has shown, many of Paget's examples are arbitrarily chosen. Whether or not the presumed similarities exist, the preceding review of mammalian communication demonstrates that mammals do not use articulatory mimicry of motions and shapes in their environment. This theory is, therefore, without foundation from a comparative point of view.

The onomatopoeic theory suggests that the first words were imitations of natural sounds that came through repetition to symbolize the object imitated. Evidence presented in favor of this notion is usually in the form of subtle similarities between words and the sounds emitted by their referent, but this does not show that these similarities were instrumental in the original formation of the words, especially since onomatopoeia rarely translates. Furthermore, onomatopoeia does not seem to occur in the referential signals of nonhuman mammals. The snake chutter, for example, does not sound like a snake. This theory, too, is without comparative support.

The third theory of the origin of naming is based on learning theory. Thorndike (1933), for example, tried to account for the origin of naming with the principles of contiguity and reinforcement. Associations between random vocalizations and environmental events would be formed, for example, when a grunt of satisfaction was paired with the food being eaten. The grunt would then come to stand for the food. We have seen processes somewhat similar to this one in the association by fawns and kittens of signals that precede nursing with the act of nursing itself. Moreover, the principles of contiguity and reinforcement have taught chimps and gorillas to communicate with symbols.

R. Brown's (1958) speculations about the first acts of naming suggest the belief of Mel Brooks's character the "2000-year-old man" that the sounds of words should resemble their meanings. He believes that *strawberry* is too big a word for a small fruit and should be replaced by *pleep*. *Bed* is too short and hard and should be replaced by *ffrrllm*. Brown writes in *Words and Things* (1958):

> The front vowels . . . are associated with smaller magnitudes than the back vowels; . . . large black crows gave the low cry we write *caw* while smaller, quicker birds *chirped* and *twittered* their higher notes. . . . The sensible attributes of the nonlinguistic world may tend to cluster, and man could symbolize

the visible or tactile attributes of such a cluster with auditory attributes from the same cluster. . . . we could learn a principle of intersensory appropriateness in naming [p. 132].

This is a sort of nonmotor gestural theory. Unfortunately, there are too many exceptions to this general clustering to make the theory very interesting. Elk whistle and frogs boom. Moreover, it is difficult to see how visual and auditory associations could evolve into language, since, as Hockett (1960) points out, most word/object relationships are arbitrary.

Brown and Hildum (1956) ran subjects who associated the sounds "dee" and "daw" with small and large circles, respectively, at greater than chance frequency. All the subjects spoke English, however, so the associations may reflect English habits rather than clustering of environmental attributes. Bentley and Varon (1933) had subjects free-associate to sounds. They found no tendency toward the kind of clustering required by Brown's theory.

In 1964 Hockett and Ascher proposed an ingenious mechanism for the origin of naming. They describe the hypothetical "opening" of a call system similar to that of apes into the beginnings of language. A call system consists of a small number of sounds, each emitted as an involuntary emotional response to a particular kind of situation—say, alarm or aggression. The call system opens by blending:

Suppose that ABCD means "food here," while EFGH means "danger coming." Finding both food and danger, the hominid comes out with ABGH. . . . In ABCD the part AB now means "food," and the part CD means "no danger"; in EFGH, EF now means "no food" and GH means "danger," while ABGH means "food and danger" because AB and GH have acquired the meanings just mentioned. . . . Lacking duality . . . these pre-morphemes became so similar to others that keeping them apart . . . was too great a challenge . . . . pre-morphemes came to be listened to not by their acoustic gestalts, but in terms of smaller features of sound that occurred in them in varying arrangements [p. 144].

This description is reminiscent of Struhsaker's (1967d) description of vervets' combining of "woof" and "waa" calls to produce "woof-waa" calls and "aarr" and "rraugh" calls to produce "aarr-rraugh" calls.

There are many other examples of the blending of behaviors under the control of different drives. Lorenz (1966) describes the ritualization of conflicting drives in the "inciting" ceremony of several common ducks. In inciting, some of the elements of aggressive threatening (head orientation toward the enemy) are combined with elements of fearful escape (body oriented toward the mate). Lorenz's analysis is thus formally similar to what Hockett and Ascher describe. Their "spoonerism" theory therefore finds some support in the communicative acts of nonhuman animals.

Hockett and Ascher's proposal is the most thoroughly articulated theory to date. Because it is so speculative, it is difficult to evaluate. One approach would be to learn more about primate call systems, looking for other examples of opening. Another would be to try to produce in a group of primates the kinds of conditions most likely to elicit blending—for example, conflicting drives.

Much recent speculation has been based on the notion that cooperative hunting placed special demands on the communication system of our inarticulate ancestors. Such speculations invoke the complexities of hunting as the basis for words referring to prey, locations, and strategies. For example, Coon (1962) and Etkin (1963) assert that pithecanthropes had language because coordinated hunting would be impossible without it. But wolves (Mech, 1970) and chimpanzees (Goodall, 1963) carry out coordinated hunts, presumably without language. Cooperative hunting is not by itself a sufficient explanation for the origin of reference.

Another kind of theory deals not with the origins of words but with grammar. Grammar is a set of hierarchical structures expressed as sequences of actions. The same is true of many other activities, particularly tool making and use. The processes involved in making tools of the Acheulean level of complexity involves locating materials, removing materials, preparing pounding block and chipper (tools for making tools), roughing out the shape of the biface, and chipping the cutting edge. These activities must be performed in sequence (you cannot chip until you find the materials). Each of these activities involves complicated motor sequences that also must be performed in sequence. It is improbable that Acheulean man made tools by rote repetition of lower-level motor sequences. Planning was involved. Miller, Galanter, and Pribram (1960) describe the similarities between phrase-structure grammars and plans of action like those just described. It seems that the plans required to make an Acheulean biface are as complicated as those involved in writing this sentence. It is, therefore, likely that hominids of 600,000 years ago were at least capable of language.

Plans for sequences of actions are mediated by the limbic system (Miller et al., 1960), which is present in completely developed form in chimpanzees. Analysis of the relationship between plans for language and plans for tool making could proceed by studies of the limbic system, and by "grammatical" analysis of the techniques used by archeologists such as Crabtree (1967), who manufactures Paleo-lithic-type tools.

Kalmus (1969) and Vowles (1970) have compared the hier-archical organization of complex animal performances such as shell opening and courtship to the structure of sentences. On the basis of this comparison, Kalmus concludes that the mammalian brain was preadapted to language. One problem with this argument is that, as Vowles demonstrates, hierarchical "phrase structures" occur in the behavior of birds, so the avian brain is also preadapted to language. Nevertheless, such "grammatical" theories can make use of the trend toward greater redundancy, climaxing in humans, which emerges from the analysis of mammalian communication. Perhaps language-like structures became useful when any one of the many threatening or sexual signals could be used in a given position in a sequence.

Jerison (1975) and Peters (1974, 1978) have speculated that the origins of grammar might be found in the cognitive map (Kaplan, 1973), a hierarchical mental structure that represents locations and routes connecting them. Underlying these speculations is the fact that human hunters/gatherers, as perhaps our hominid ancestors did, use larger territories than any other living primate. Presumably, they had the opportunity to form more complex mental maps than primates with smaller ranges. Once hominids learned to associate sounds such as cries of distress or grunts of satisfaction with certain locations, their shared cognitive maps would provide a basis for production and com-prehension of sequences of such utterances. Hierarchical structures can provide a grammar only when they are shared and the cognitive map provides a plausible basis for such sharing. Perhaps the cognitive-map theory of language evolution will, like its predecessors, fall from grace. It may, however, pave the way for other theories that make sense from linguistic, information-processing, and comparative points of view.

10
SUMMARY
AND
CONCLUSIONS

If a lion could speak, we would not understand him.

LUDWIG WITTGENSTEIN

If, as E. O. Wilson (1975) suggests, an understanding of "what an animal is really trying to communicate" is the "grail of zoosemiotics" (p. 216), this book has been a crusade for that grail. Has the crusade been worth the effort? Are the problems faced by mammals of the seven different orders discussed here similar enough to allow us to use a small number of categories to classify their messages, or are these categories bound to be "endlessly proliferated" (Wilson, 1975, p. 216)? In the preceding chapters I have tried to show how the signals used by these mammals can be assigned to 30 message types that can be defined consistently and objectively. Although the number of different forms of signals used by each mammal varies widely, from about 18 in pikas to about 150 in humans, the range of meanings is relatively small, from 10 in pikas to 25 in wolves and humans. The scheme used in this book has also been applied to the message systems of tenrecs, mice, cottontails, hares, bighorn sheep, horses, bears, coyotes, baboons, and chimpanzees. The number of message types displayed by these mammals falls within the range defined by pikas and wolves. Mammalian communication evidently displays considerable parallelism and convergence, at least in terms of the kinds of information transmitted.

The crusade for the zoosemiotic grail is on course if and only if my functional or analogical categories have been objectively defined and consistently applied. In the following section I will briefly review the message types, describing how signals used by each of the mammals fits the type. This review will demonstrate that mammals not only say the same things to one another but that in many cases they say similar things in similar ways. This discussion is keyed to Table 10-1, which shows the types of messages used by each mammal, and to Table 1-2, which defines these types in terms of stimulus and response contingencies. Each type is defined by a particular kind of "information," defined as a reduction in uncertainty in the receiver's response array.

REVIEW OF THE MESSAGE TYPES

NEONATAL COMMUNICATION

Similarities among mammals are most evident in the messages exchanged by mother and neonate. The forms of neonatal tonic and phasic interactions are quite constant across orders. Some of these interactions, such as postpartum nursing and licking, are not merely analogous but truly homologous and thus immune to Wilson's criticism that judgments of analogy are inherently subjective.

Postpartum licking is performed by mothers in all the species discussed except the dolphin. According to Ewer (1968), this licking performs many functions. It frees the neonate's head from the fetal sac, allowing the neonate to breathe. Facial licking also facilitates breathing by stimulating the receptive field of the trigeminal nerve. Other functions include sanitation, drying, removal of odors that might attract predators, and stimulation of elimination and peripheral circulation. In *precocial* species, maternal licking contributes to the formation of a bond between infant and mother. Like most other tonic and phasic neonatal interactions, licking results in mutual stimulation—tactile for the infant and gustatory and olfactory for the mother. In some species, licking seems to play a role in the development of the mother's ability to recognize her young (J. P. Scott, 1972).

Mothers of most terrestrial species, except camelids, consume the placenta (Ewer, 1968). This behavior has been recorded in all the mammals discussed in this book except dolphins, which accompany their neonates to the surface as the placenta sinks. The placenta contains lactogenic hormones and other substances implicated in the development of maternal behavior (Rosenblatt, 1970). It is therefore interesting that both father and mother beavers eat the placenta and both care for the kits.

Nursing, like licking, involves mutual stimulation. The mother's nipples induce sucking, and sucking, treading, or bunting stimulates the flow of milk. In cervids and dolphins, bunting is often quite forceful and eventually contributes to the mother's intolerance of attempts to nurse.

In some species, hedgehogs, shrews, and deer, for example, nursing sometimes provides a literal attachment that allows the mother to lead her young. Otherwise, she must carry them. Hedgehogs,

TABLE 10-1
NUMBER OF SIGNALS IN EACH MESSAGE TYPE

SPECIES

MESSAGES	HEDGEHOG	SHREW	TREE SHREW	RAT	BEAVER
NEONATAL					
Infant distress	3	2	1	2	2
Infant identity	1		1	1	
Infant affiliation		1	1	2	3
Infant satisfaction					
Neonatal contact	1	4	2		
Maternal assembly		1			
Maternal identity				1	
Maternal alarm					
INTEGRATIVE					
Play	1	1	5	2	4
Contact	2	3	2	1	1
Affiliation			2	3	4
Alarm		1	1	2	1
Familiarization	2	1	1?	1	
Identity		1	1	2	
Distress	1	2		2	1
Solicitation			1		1
Satisfaction					1
Assembly					1
AGONISTIC					
Fighting	1	3	2	3	2
Offensive threat	2	9	4	3	6
Defensive threat		1	3	3	
Dominance		1	3	3	1
Submission			2	5	1
Territory advertisement		1		1	2
SEXUAL					
Male advertisement	2	3	1	3	1
Female advertisement		2	1	3	1
Courtship	1	2	3	4	4
Synchronization					
Suppression			1	1	
Copulatory signal	2	1	1	1	
NUMBER OF TYPES	12	19	19	22	18
NUMBER OF FORMS	17	~43	~33	~41	~30

RABBIT	PIKA	WAPITI	DEER	VERVET	DOLPHIN	WOLF	CAT	HUMAN
	1?	1	2	8	1	2		3
4		1	1	1			1	2
		1	3	4		1	2	2 = 2
				1		1		1
		3	2		1		2	
							1	
2		1	1				1	2
			1			1	1	1
1		7	5	≥3	5	60	2	large
2	1	3	10	2	3	4	6	≥2
2		1	3	3	2	9	3	≥1
1	1	5	10	6	3	1		≥1
3	5	2	4			4		1
3	1		4	≥6	2	1	3	2
1	1	1	1	1	1	1	2	1
				8	3	5		≥2
						1	1	≥2
						1		
3	1	3	3	1	3	3	2	large
4		16	15	5	2	7	7	≥3
		1	1	8		3	6	
2		7	11	7	2	9	2	≥3
2		7	3	5	1	7	3	≥2
3	5			9		4	2	large
	1	7	5	1	3	1	3	≥2
1		1	1	4	4	3	4	≥2
4	1	4	4	1	5	3	3	large
1								1
						1		
2						1	1	2
18	10	19	20	20	16	25	22	25
30	18	~60	~67	79	≥41	~122	43	~150

285

shrews, rats, beavers, wolves, and cats carry their young by mouth, but beavers sometimes carry their young with their forepaws, using a bipedal gait, or on their backs, using a quadrupedal gait. Vervets carry their young by gathering them up, allowing them to cling to their ventral surfaces. Dragging by nipples, like carrying by mouth, can be regarded as homologous among species that practice it. The alternative modes of infant transport, however, are probably merely analogous. Little subjectivity is involved in this judgment, since the major function is the moving of young to a new location. Precocial species, such as dolphins, deer, and wapiti, and fossorial (burrowing) hiders, such as rabbits and pikas, do not have to transport their young.

The first signals used by mammalian infants are cries of distress. Most species have more than one such cry, with different cries corresponding to different levels of distress. The bleat of the fawn, which becomes two-syllabled at high levels of distress, is an example. In some species the distress cry is graded in intensity. Hedgehog infants have separate hunger and separation cries. Shrews and beavers have low-intensity and high-intensity calls. Rat pups have two distress calls, one of which is ultrasonic. It is likely that the hunger and lost cries of some species are merely low-intensity and high-intensity distress cries, respectively. In other species, such as humans, the evidence suggests that different cries refer to qualitatively different states (Wolff, 1969). This also seems to be the case with young vervets, which use at least five, and possibly nine, different distress calls.

Humans find most infantile distress calls unpleasant, especially those of the insectivores, rats, deer, wapiti, cats, and wolves. In some cases this subjective response is attributable to the waveform of the cry, which often has a square-wave component. Like the distress signals of adults, infantile distress cries can be objectively defined in terms of eliciting stimuli and parental response. Since infantile distress signals differ in context, form, and response from those of adults, they convey information different from that of adult distress signals and are considered a separate type of message.

In contrast to the unpleasant character of calls that signify infantile distress, maternal vocalizations that assemble or establish contact with young are generally pleasant or neutral. Collias (1960) describes them as "relatively soft, low-pitched, brief and repetitive" (p. 282). This generalization applies, to some extent, to maternal whining in beavers, the "shrew's caravan" command, and the "mhrn" of the mother cat.

Infantile vocalizations that signify satisfaction inform the mother that no particular response is required. They simply let her know that the infants are there and are well. They thus act as contact signals and may also have some affiliative consequences. Such calls are never loud. The twittering of young shrews, like the purrs of kittens and the contact calls of wapiti calves, inform mothers of their off-springs' location while minimizing the risk of attracting predators. Contact between mother and infant is also maintained by odor, a short-range signal well adapted to the task. In species that hide their young, such as deer and rabbits, odor provides mothers with their major means of finding their young.

In some species, odor is also extremely important in recognition of young by their mothers. Tree shrews mark their young with sternal-gland secretions and devour any unmarked young they may find. Inguinal and anogenital odors in rabbits, cats, wolves, deer, and wapiti seem to play a similar role. Rat mothers, on the other hand, readily adopt pups or even entire litters, so it may be that they do not discriminate among individual infants.

Odors are probably also important in recognition of parents by young. Müller-Schwarze's (1971, 1972) experiments with young blacktail deer and Rosenblatt's (1972) experiments with kittens show that in some species this recognition is a result of experience with particular odors during nursing.

In some species, such as tree shrews, rabbits, deer, and cats, infants not only recognize but are also strongly attracted to maternal odors. Because this seems to be due primarily to the effects of early exposure to these odors, the attachment process has been termed imprinting. The similarity to imprinting in precocial birds is striking, particularly in cases in which later sexual preference has been shown to depend on early exposure. Nevertheless, there are some important differences between the attachment processes of mammals and those of birds, particularly in the greater length of the mammalian critical period.

INTEGRATIVE MESSAGES

Play provides a transition from infantile to adult interaction. Ewer (1968) has shown that play can be objectively defined by several quantifiable characteristics, including frequency of interruption, role reversal, repetition of appetitive actions without consummation, and degree of exaggeration of gestures. She has argued persuasively that play has its own motivation and is independent of other motivational systems, and she has shown that it is neither a precocious display of specific-action energy nor a vacuum activity. She argues that play performs several functions: discovery and practice of behavior patterns and fostering of social development. This social function includes the formation of affiliations and dominance relationships. All of Ewer's comments apply to the play of the mammals discussed in the preceding chapters.

Social play can be divided into sexual, agonistic, affiliative, and imitative interactions. Hedgehog play is mainly sexual, that of water shrews, tree shrews, rats, and beavers mainly agonistic. Rabbits rarely play, and their frolics are difficult to classify, but they seem to contain elements of all four types. Deer and wapiti, like wolves and cats, engage mainly in agonistic and imitative play with some sexual and affiliative elements. Vervets display agonistic, sexual, and imitative forms of play. They sometimes engage in structured play in the form of hide and seek. Dolphin play is mainly sexual and imitative. Beavers, wapiti, wolves, cats, and vervets have special postures, facial expressions, and vocalizations that announce playful motivation.

Unlike neonatal contact signals, which are always short-range, the behaviors that adults use to stay in touch with one another may

These grizzly bears are engaged in playful jaw-wrestling, with open-mouth threat, a combination of messages often seen in canids. Photograph by Roger Peters.

be long-range or short-range. Short-range signals include twittering by shrews, the characteristic leathery odor of hedgehog feces, tail flicks by tree shrews and mule deer, rump patches of cervids, and tonic noises in several species. Scent marks of rats, wapiti, deer, wolves, and cats act as relatively long-distance contact signals because, when fresh, they tell the receiver that the marker has just moved off. Other long-distance contact signals include the "ank" of the pika, the howl of the wolf, and the song of the dolphin. The distinguishing character-istics of contact signals are their emancipation from other motivational systems, frequent emission when engaged in maintenance activities, and lack of response other than reply on the part of the receiver. Contact signals inform receivers that there is a conspecific in the vicinity and that it is not distressed or otherwise aroused. The only mammal discussed that does not have a special contact signal is the rabbit. The reason may be that the costs of emitting such a signal in terms of increased vulnerability to predation outweigh the benefits in terms of social integration. Moreover, it is not obvious what use a rabbit would make of the information encoded in such a signal. Rabbits live in colonies within a restricted range and are rarely far from other colony members.

Unlike contact signals, which are often produced when trans-mitter and receiver are apart, affiliative messages typically involve intimate contact in which the participants touch, smell, or taste each other. Such contacts promote affiliation by expressing nonaggression and probably by evoking a circular Zajonc effect, in which mere ex-posure increases attractiveness, promoting further exposure. For ex-ample, allogrooming, which occurs in tree shrews, rats, beavers, rab-bits, deer, wolves, and vervets, not only promotes hygiene but allows the participants to touch, taste, and smell each other as they carefully comb and investigate the other's fur. *Köpfchengeben* by cats and flank and fluke rubbing by dolphins probably have similar effects, although the tactile mode is predominant. In many of these species, the affil-iative bond results in favorite grooming partners, as one would expect on the basis of the hypothetical positive-feedback mechanism. Affil-iative contacts occur in peaceful contexts and are much gentler than fighting. They can be distinguished from sexual messages by their lack of seasonality and by the fact that they are at least as likely to be homosexual as heterosexual.

Alarm signals are responses to potential danger or to any sudden and unusual change in the environment. Typically, receivers orient toward or scan for the transmitter or the stimulus, freeze, flee, aggregate, hide, or engage in some other defensive behavior. These defensive behaviors are seen as responses to such diverse signals as

Prairie dogs use nasonasal contact to form and express bonds of affiliation. Photograph by Roger Peters.

postures, gaits, odors, and cries. Collias (1960) notes that alarm cries are generally loud, harsh, and high-pitched. His generalization applies to most of the alarm calls described in the preceding chapters. Only two mammals, the hedgehog and the cat, lack alarm signals in adult interactions. These two more-or-less solitary mammals probably have little to gain by warning nearby conspecifics, which are not likely to be closely related.

Any creature whose resources or enemies are more likely to be in some places rather than others can potentially benefit from familiarity with its environment. In some cases these benefits of marking outweigh the costs in terms of energy and increased vulnerability to predation. These marks make the animal feel at home (Ralls, 1971) and may, in some species, facilitate route choice (Peters, 1979). In general, affective consequences are much easier to demonstrate than cognitive ones. When an animal investigates its own marks, it is reasonable to conclude that mere exposure renders these marks attractive. Thus, we find that animals as different as rats and wolves apply their odors to trails, especially to choice points, and spend considerable time and energy investigating and renewing them. Similar behavior has been noted in pikas, rabbits, and deer.

The same distinctive odors that are used as scent marks often supplement physical appearance and voice to convey information about sexual, maturational, or individual identity. Determining that such information is used by the receiver is often difficult and has not in fact been done for many of the mammals considered here. It does not suffice to show that differences on the basis of sex, status, or individuality exist. The receiver must be capable of making the relevant discrimination, and there must be evidence that this discrimination has behavioral consequences in nature. These three criteria have, however, been met in a number of cases. Rats, wolves, pikas, rabbits, deer, vervets, and dolphins have been shown to respond differently to odors, sounds, or appearances of animals of different sex and age

Friendly interactions between elephants, like those of other mammals, often involve investigation of the face. Photograph by Roger Peters.

classes under more-or-less natural conditions. Of the mammals discussed here, only beavers have not been shown to identify themselves in this way. Probably, future research will find individual, sexual, or age differences in the chemical composition of castoreum or the spectral composition of "talking."

When danger is imminent or when they are injured, some adults emit cries different from those of alarm or infantile distress. Thus, the guttural distress cry of the beaver is different from the tail slap or the whine of infants; the shriek of the pika from the alarm "ank"; and the scream of the deer from its alarm snort or the bleat of fawns. With the exception of the dolphin, conspecifics do not generally come to the transmitter's aid, so the function of distress calls is open to speculation. Olfactory distress signals in the form of anal-sac discharges by wolves and cats may frighten conspecifics that detect the odor later on (Donovan, 1969). Should other evidence of this sort become available, it may become necessary to regard distress signals as ultra-high-intensity alarms.

If, on the other hand, it turns out that the dolphin is not unique in responding to distress signals with succor, distress signals will have to be considered a form of solicitation. Solicitations seem to be requests, for they are expressions of drive states that cease when an appropriate consummatory stimulus is provided by a conspecific (or, in some cases, a human). Consummatory activities, such as grooming, feeding, or approach, then ensue. Some examples of solicitations are dancing by beavers, whining by wolves, meowing by cats, and the "war-hor-hor" of the vervet.

During or immediately after consummatory behavior, some mammals express satisfaction by nasal whining. Cats do so by purring, wolves and wapiti by displaying a consummatory face, with lowered eyelids and relaxed lips. The effects of such expressions on conspecifics have not been demonstrated.

Wolves and beavers sometimes produce calls that result in the assembly of dispersed members of the family. Such unequivocal responses are rare among mammals but sufficiently common to justify a special "assembly" message type.

AGONISTIC MESSAGES

Fighting, defined as injurious or potentially injurious physical contact, is the most common agonistic message type in the mammals I have discussed. With the exception of the highly territorial wolves, pikas, beavers, and the harem-herding wapiti, fighting is ritualized in ways that reduce the chance of fatal injury. One common form of

Antlers and horns, like those of the fallow deer (above) and the ibex (below) are often large in proportion to body size. Photographs by Roger Peters. Ibex photograph courtesy of Chicago Zoological Park.

fighting that appears in several mammalian orders is boxing, in which the combatants rear on their hind legs and flail with their forelimbs. The wide phylogenetic distribution of this form of fighting suggests that boxing by mammals of different forms may be homologous.

Next most widespread among agonistic messages is threat. Many offensive and defensive threats illustrate Darwin's (1872) principle of serviceable associated habits. These threats express a readiness to fight with postures, gestures, or facial expressions that prepare the transmitter for attack. Other threats, such as growling, illustrate Darwin's principle at least to the extent that they are not incompatible with attack or with other threatening expressions. Offensive threats are associated with approach by the transmitter, while defensive threats are associated with approach by the receiver. Both result in either withdrawal or cessation of approach by the receiver. Only the pika lacks special threat displays other than approach. Ordinarily, approach or territorial vocalizations by a resident cause trespassers to flee. Perhaps the absence of threats to substitute for fighting are partially responsible for the fact that fights among captive pikas are so often fatal.

Dominance displays are distinguished from threats because dominance displays increase the transmitter's apparent size or otherwise render it more imposing without showing readiness to fight (Walther, 1977). For example, anal control by wolves and cats demonstrates the ability of the sender to penetrate the receiver's space but is not an effective preparation for attack. Dominance displays are performed only by transmitters whose rank is approximately equal to or higher than that of the receiver. The only mammals considered here that do not have dominance displays are hedgehogs and pikas.

Submissive displays are often responses to, and the antitheses of, dominance displays or threats. Lying down, for example, expresses submission in rats, beavers, and wolves and is the antithesis of displays that increase apparent size. When submissive signals appear in peaceful situations, they allow approach or display of affiliation by reducing the probability of attack. Some submissive displays, such as anal presentation by tree shrews and vervets and active submission by wolves, seem to invoke sexual or parental responses incompatible with aggression. As Ewer (1968) points out, however, submission is not, as Lorenz (1952) suggests (see Chapter 8), a result of "some pre-existing, preadaptive inability to attack the defenseless" (p. 183).

Territorial advertisements are scent marks, displays, or vocalizations that reduce the probability of trespass. Such signals are most likely to be sent when the transmitter is in its own territory or when trespass is detected. Scent marks used as territory advertisements

Carnivores, like this bear, often threaten with an open mouth. Photograph by Roger Peters.

are often concentrated around the perimeter of the territory. I know of no species in which the typical response to territorial advertisements is immediate retreat. More commonly, the receiver displays signs of conflict and decreased confidence.

SEXUAL MESSAGES

The sexual-message system, like the neonatal system, contains many similar, and possibly homologous, signals. For example, short-chain fatty acids seem to act as female sexual advertisements in mammals in several different orders. Scent marking is a common form of advertisement in both sexes, but males tend to advertise with sound as well as with scent. Auditory sexual advertisements by males occur in hedgehogs, pikas, wapiti, cats, and possibly in dolphins. Male responses to female scent marks include scent marking in hedgehogs, rats, rabbits, deer, wolves, and cats. Wolves, rats, and cats rub in

Like most mammals, elephants engage in olfactory investigation of genitals before mating. Photograph by Roger Peters.

estrous-female urine, while deer, elk, and cats respond to female urine with flehmen.

Courtship messages contain elements of play, solicitation, affiliation, dominance, and submission. They are usually terminated by copulation. Only the pika copulates without courtship, and courtship is prolonged in hedgehogs, rabbits, deer, wolves, cats, and dolphins. Strenuous or prolonged courtship may demonstrate a male's physical fitness as well as the strength of his motivation, particularly in species in which the male invests in his offspring after conception (R. A. Fisher, 1930). According to Ewer (1968), however, the major function of courtship is to ensure that a male will be available and ready when the female is fertile. Ewer hypothesized that in species in which the response of the unreceptive female is flight, courtship will take the form of more-or-less ritualized pursuit. Ewer also suggested that, in spiny species, the female's transition to receptivity should be sudden. She further proposed that when the receptive female responds by fighting, male courtship should contain elements of agonistic displays, such as dominance or submission. The courtship behaviors of the mammals considered in previous chapters support Ewer's hypotheses. Antlerless female cervids cannot effectively fight males, so their only defense is flight. The tending and harem herding of deer and wapiti are examples of courtship by pursuit. The abrupt capitulation of the female hedgehog supports Ewer's second hypothesis. In species such as beavers, wolves, and cats, in which the female's weapons are about as effective as the male's, males court by displaying their size and strength and by using playful and submissive signals to show that they are not really aggressive.

Some mammals produce distinctive noises during copulation. These noises include spine rattling by male hedgehogs, twittering by female shrews, and yowling by male and female cats. The function of these sounds is unknown, but W. J. Hamilton and Arrowood (1978) have suggested several possibilities, including stimulation of the partner. In some species such stimulation may facilitate ejaculation, ovulation, or vaginal contractions contributing to the movement of sperm through the cervix.

Synchronization and suppression of sexual activity by chemical and visual signals have been proposed or established in tree shrews, rats, rabbits, and wolves. The extent to which such phasic interactions occur in other species is open to investigation.

RECAPITULATION

In this brief summary of message types, I have tried to show how meaning, as an objective characteristic of the message, can be defined by form, context, and response and used to compare communications by mammals of different orders. It has not been necessary to create new categories for the messages used by members of each order. I hope that these categories will prove useful to investigators working with other species in these and other mammalian orders, but it is probably not profitable to take this typology very seriously. Defining and counting types of meaning have been a little like counting the number of angels dancing on the head of some scholastic's pin. The point is that, no matter how you divide them up, there aren't that many angels. Most mammals have about 20 different things to say to one another. Communication promotes fitness in only about 30 different ways: by warning, identifying, locating, and so on.

The most widespread message types are play, contact, alarm, fighting, threat, dominance, sexual advertisement, and courtship. The least widespread are satisfaction and assembly.

The number of message types used by highly social species, such as wolves and vervets, is greater than the number used by less social species, such as pikas and hedgehogs. Otherwise, the numbers of types used by different mammals are remarkably constant. Only the number of forms varies.

The wolf, even though using a smaller number of signals, apparently exchanges more different kinds of information than a highly social primate, the vervet. This conclusion reminds me of Kortlandt's (1965) comment that the wolf might provide valuable insight into the problems faced by our ancestors because of the demands of a life based on social hunting.

SOME PROPERTIES OF MAMMALIAN COMMUNICATION

Before I turn to a discussion of relationships among the message categories, there are several general properties of mammalian communication that deserve mention. Bateson (1966) states that mammalian communication is primarily about relationships of dominance, dependency, and affiliation. He refers to this relational aspect of communication as the μ function. As an example, he describes the dominance display in which a male alpha wolf punishes a subordinate adult as a parent disciplines a pup, by pressing its neck to the ground with his jaws. Bateson translates the meaning of this display not as "Don't do that" but as "I am your senior adult male, you puppy." Bateson goes on to suggest that mammals' attention to the μ function of conspecifics' behavior was an important selective force in the evolution of mammalian intelligence. The μ function is widespread but not universal. Alarm and familiarization, in particular, do not seem ever to involve communication about the kinds of relationships involved in the μ function. Most other message types, however, do seem to involve a μ function, either centrally or peripherally.

A second general property of mammalian communication, according to Bateson, is analogue coding, in which the intensity of some other physical parameter of a signal varies with some magnitude in the μ functional relationship being expressed. Thus, louder, longer, or more frequently repeated signals for distress, solicitation, affiliation, and so on stand for stronger statements of dependency, attraction, or other relations. Exceptions to this generalization are found in the use of digital coding for different motivational states—for example, threat versus dominance or satisfaction versus solicitation. Neither member of either of these pairs is simply a more-or-less intense version of the other.

Every message system examined here shows some degree of redundancy; that is, the same meaning is generally encoded in more than one form in any given species. For example, visual threats are often effective by themselves but are sometimes supplemented or replaced by signals in other channels. It is possible that these other messages transmit subtle differences in meaning, but the response to different forms of the same message seems the same. As we develop more refined techniques for the recording and analysis of responses, this generalization may have to be modified. Highly social species seem to differ from less social species mainly in having more forms for each meaning.

The converse of redundancy is punning, in which a single form may have several meanings. When receiving such a form, a mammal responds on the basis of context or experience. Thus, the short "ank" of the pika, the most accomplished punster of the mammals considered, elicits reply if the receiver is in its territory and results in hiding if the receiver is away from the protection of the rock slide. The prevalence of punning implies that the meaning of a message is often modified by context and memory. Punning illustrates the parsimony of evolution; new behavior will not emerge as long as there are old behaviors that can be adapted to new meanings without too great a cost in terms of misunderstanding.

RELATIONSHIPS AMONG MESSAGE CATEGORIES

An examination of attempts to organize animal communication supports E. O. Wilson's (1975) statement that categories of meaning are bound to be "endlessly proliferated." Table 10-2 displays the lack of congruence among several such attempts. There is little agreement among these authors about the kinds of messages exchanged or even about what the criteria for classification should be. For example, J. P. Scott's (1972) "epimeletic," or care-giving, category includes tonic and phasic maternal care, affiliative allogrooming, and presumably alarm calls. His "et-epimeletic," or care-asking, category includes distress, solicitation, and satisfaction signals. "Allelomimesis," or imitation, includes leadership and contact and refers to a mode rather than a form or meaning of interaction. Ewer's (1968) system is based on meaning but includes "scent marking," a category based on the form, not the meaning of a message. Smith's (1969) system is also based on meaning, but one of his categories, "probability," refers not to a separate category of meaning but modifies the meanings of mes-

TABLE 10-2
SYSTEMS OF CLASSIFICATION OF MESSAGES

MESSAGES	WILSON (1975)	SMITH (1969)	COLLIAS (1960)	SCOTT (1972)
			AUTHOR	
neonatal	distress	bond-limited	parent–young	epimeletic and et-epimeletic
play	play invitation	play		allelomimetic
contact	contact	locomotion	contact	allelomimetic
affiliation	grooming	association	greeting, contact	epimeletic
alarm	alarm	escape	announce enemy	et-epimeletic
familiarization				eliminative
identity	recognition	identification	sex recognition	investigative
distress	distress	escape	announce enemy	et-epimeletic
solicitation	begging food		anticipation	et-epimeletic
satisfaction			pleasure	et-epimeletic
assembly	assembly			allelomimetic
fighting	threat, dominance	attack	fighting	agonistic
offensive threat	threat	attack, frustration	threat	agonistic
defensive threat	threat	escape	repulsion	agonistic
dominance	status signaling		repulsion	agonistic
submission	submission	nonaggressive		agonistic
territory advertisement			attract female	agonistic
male advertisement	sexual	copulation	attract female	sexual
female advertisement	sexual	copulation		sexual
courtship	sexual	copulation	precoital	sexual
synchronization	sexual		reproductive	sexual
suppression	sexual			sexual
copulatory signal	sexual	copulation		sexual

sages by conveying information about the likelihood of a given action. Collias (1960) assigns animal sounds to five functional categories, one of which, "parent–young," corresponds to my neonatal system. The other four, "food," "predators," "sex and fighting," and "aggregation or movements," do not correspond exactly to either types or systems in my analysis. Collias's approach shows how animal sounds can be related to basic survival activities, but he does not attempt to deal with modalities other than sound, or to focus only on mammals. Wilson stated that objective zoosemantic typologies are impossible but profitable, and then built one. His categories are explicitly functional and deal with animals of several phyla. I have borrowed heavily from all these writers.

Two basic problems underlie the confusion surrounding all these attempts to classify communicative acts. First, punning creates problems for any treelike hierarchy because the same form of behavior must often be attached to more than one branch. Second, categories of meaning may themselves be grouped in different ways. Submission, for example, is considered an agonistic message because it is a common response to threats or dominance displays. In wolves, vervets, and several other mammals, however, submissive displays sometimes occur in friendly contexts and facilitate amicable interactions. Therefore, it might also be grouped with affiliation and other integrative acts. Like punning, the possibility of alternative groupings of message types creates difficulties for any treelike hierarchy of forms and messages.

Both problems can be solved with the aid of another kind of hierarchical structure, the semilattice. In a tree an element or category at any level belongs to only one category at the next-higher level. In a semilattice, on the other hand, any element or category can belong to several categories at the next-higher level. In algebraic language the rule that generates a tree is as follows: a collection of sets is a tree if, and only if, for any two sets A and B in the collection, either A is contained in B, or A and B have no elements in common. In contrast, a collection of sets forms a semilattice if, and only if, for any two sets A and B in the collection, the set of elements A and B have in common is in the collection. In a semilattice not only can sets at any level belong to more than one set at the next-higher level, but multiple membership is realized by assigning existence to the set of elements common to more than one set. Thus, when we find scent marking used to establish contact, familiarity, and dominance as well as to advertise identity, territory, and sex, we can give a name, specific presence, to this emergent cluster of functions.

TABLE 10-3
A SEMILATTICE OF MESSAGE TYPES AND MESSAGE SYSTEMS

MESSAGE SYSTEMS

NEONATAL	INTEGRATIVE	SEXUAL	AGONISTIC
infant distress	distress	synchronization	
maternal alarm	solicitation		
	alarm		
neonatal contact	contact		suppression
	familiarization		fighting
			offensive threat
infant identity	identity	male advertisement	territory advertisement
maternal identity		female advertisement	dominance
infant affiliation	play		
	affiliation	courtship	
maternal assembly	assembly		defensive threat
			submission
infant satisfaction	satisfaction	copulatory signal	

Note: Some types are grouped under more than one system. Proximity of types indicates encoding in common forms and similarity in information content.

Mammalian messages form a semilattice in which sets of individual acts belong to sets called message forms. Message forms, in turn, are grouped into message types, and message types into message systems. Table 10-3 shows the relationships among message types and message systems. The large dashed rectangle represents the emergent system called specific presence. Another emergent system is affiliation, play, and courtship, enclosed in a smaller rectangle. Other emergent systems are not shown for the sake of simplicity. The solid lines with play and submission show that these message types belong under integrative, sexual, and agonistic systems. Table 10-3 demonstrates that it is possible to organize forms and meanings into a hierarchical structure that can not only accommodate but explicate multiple groupings.

The semilattice allows us to formalize multiple relationships among meanings. With its aid, we can see that our confusion about relationships among meanings is the result of applying Aristotelian tree-like hierarchies to a structure much richer in associations than any tree. A semilattice of objectively defined meanings allows us to validate our intuition that the voices of lions and humans share some meanings. This commonality does not involve speech, for lions do not speak. But if they did, we would understand them.

GLOSSARY

action-specific energy: A hypothetical force motivating a particular kind of behavior.

adaptation: Any aspect of an individual organism that promotes its success, particularly reproductive success, in its natural environment.

agonism: Behavior associated with conflict, including aggression, submission, and defense.

allogrooming: The grooming of an animal by a conspecific.

alloparental behavior: Parental behavior toward, or care of, young by individuals other than the parents.

altricial: Term used for young that are born relatively helpless and needing extensive parental care (for example, canids). Opposite of *precocial.*

analogues: Anatomical or behavioral features produced by convergent evolution—that is, by adaption to similar problems.

appetitive: Variable behavior adapted to discovery of a stimulus that elicits consummatory behavior.

artiodactyl: An even-toed, hoofed mammal of the order Artiodactyla (for example, deer).

behavior: Publicly observable action; activity of muscles and glands of external secretion.

biome: A major regional community type (for example, tundra, rain forest).

canid: A member of the dog family, including foxes, coyotes, and wolves, as well as the domestic dog.

cercopithecid: A member of the family of Old World monkeys including macaques, baboons, and langurs, as well as vervets.

cervid: A member of the deer family.

cetacean: An aquatic mammal of the order Cetacea (for example, whales, dolphins).

cloaca: A combined urogenital and rectal orifice.

conspecifics: Members of the same species.

consummatory: Relatively fixed behavior elicited by food, mates, or other stimuli that reduce drives.

coprophagy: The eating of feces.

critical period: A period relatively early in life during which an animal has the opportunity to develop an ability or attachment.

diastema: A gap between front and cheek teeth which allows rodents to gnaw without wearing out the cheek teeth or swallow the material gnawed.

digital: A form of coding using qualitatively different signs.

dimorphism: Differences between the sexes in size and body form beyond differences in reproductive organs.

displacement: Transfer of action-specific energy from one target to another.

dorsal: Pertaining to, or located on or near, the back. Opposite of *ventral.*

drive: A hypothetical force motivating a particular type of behavior.

echolocation: Perception by reflected sound.

enuration: Urination by one individual on another individual.

epimeletic behavior: Care-giving behavior.

estrus: Period of maximum sexual activity and, usually, fertility in female mammals.

et-epimeletic behavior: Care-soliciting behavior.

ethogram: A list of behaviors characteristic of a species.

ethology: The study of animal behavior.

felid: A member of the cat family, including lions, tigers, and lynx, as well as the domestic cat.

feral: Wild, especially referring to domestic animals in an untamed or wild state.

flehmen: A retraction of the lips often associated with olfactory investigation.

homoiothermy: Internal temperature regulation, or "warm-bloodedness."

homologues: Anatomical or behavioral features from the same ancestral source. See *analogues.*

imprinting: Rapid formation of an attachment to any stimulus present early in life.

induced ovulation: Release of an ovum (egg) by stimulation. Found in cats, martens, racoons, otters, and the European rabbit.

inguinal: Pertaining to the region of the groin.

insectivore: A member of order Insectivora, which includes shrews, moles, hedgehogs and tenrecs. More loosely, animals like tree shrews which subsist largely on invertebrates.

integument: The organ composed of the hair and skin.

lagomorph: A member of the order of rabbits, hares, and pikas.

leporid: A member of the family of rabbits and hares.

linear-dominance hierarchy: A system of social rank in which A dominates B, B dominates C, and so on throughout the group. A "pecking order."

lordosis: The concave curvature of the spine that, in females, elevates the genitals.

low-stretch: A horn display used by a higher-ranking bighorn sheep to threaten a lower-ranking one.

mnemonic: Related to the memory or the improvement of the memory.

morpheme: The minimal unit of meaning in a language.

morphology: The structure or anatomy of an organism; also, the study of the structure of organisms.

murid: A member of the Old World rat family.

neonatal: Pertaining to a newborn mammal.

olfaction: The sense of smell or act of smelling.

omnivore: An animal that eats both plant and animal matter.

parental investment: The investment of time, energy, risk, and resources by parents in their offspring.

pelage: The coat, or fur, of a mammal.

perianal: Around, or in the area of, the anus.

perineum: Superficial area between the anus and the urogenital passages.

pheromone: In mammals, a chemical signal detected by olfaction.

phonation: Producing a sound, not necessarily a vocalization.

phoneme: A class of sounds perceived as equivalent.

phylogeny: The development or evolutionary history of a taxon.

precocial: Term for young that are relatively independent and able to move about soon after birth (for example, cervids). Opposite of *altricial.*

prehensile: Adapted for grasping or seizing, especially by wrapping around.

refection: The reingestion of the first feces (cecal pellets) by lagomorphs and shrews.

reversal-shift discrimination: A form of concept learning in which all instances of a concept no longer belong to the concept.

ritualization: Adaptation of a behavior pattern for communication.

rut: Seasonal sexual arousal: from the Latin word *rugitus,* "to rear."

scent mark: Odor applied to the environment or a conspecific so as to render the odor conspicuous.

semiotics: The study of signs and their meanings.

sexual dimorphism: A physical difference between the males and females of a species other than reproductive organs (for example, size, color).

spronking: An alarm display with a characteristic bouncing run. See *stotting.*

stotting: An alarm gait in which an animal bounds with stiffened legs. Performed by antelopes, pronghorns, and cervids.

taxon: A group of organisms at any level of the taxonomic hierarchy.

tupaiid: A member of the tree-shrew family.

twist: A threatening horn display used by bighorn sheep, similar to the low-stretch display.

vacuum activity: Consummatory behavior in the absence of the usual stimulus.

ventral: Pertaining to, or located on or near, the belly or abdomen (the lower surface of most animals). Opposite of *dorsal.*

viviparous: Giving birth to live young (instead of laying eggs).

vocalization: A sound produced by vocal chords.

zoosemiotics: The study of animal communication as an evolved adaptation.

REFERENCES

Alberts, J. R., & Galef, B. G. Olfactory cues and movement: Stimuli mediating aggression in the wild Norway rat. *Journal of Comparative and Physiological Psychology*, 1973, 85, 233–242.

Alcock, J. *Animal behavior*. Sanderland, Mass.: Sinauer, 1975.

Aleksiuk, M. Scent-mound communication, territoriality, and population regulation in beaver. *Journal of Mammalogy*, 1968, 49, 759–762.

Allee, W. C. *The social life of animals*. New York: Norton, 1938.

Allee, W. C. Group organization among vertebrates. *Science*, 1942, 95, 289–293.

Allee, W. C. Dominance and hierarchy in vertebrate societies. *Collogue Internationale du CNRS*, 1950, 34, 157–181.

Allin, J. T., & Banks, E. M. Functional aspects of ultrasound production by infant albino rats. *Animal Behaviour*, 1972, 20, 175–185.

Altman, S. A., & Altman, J. *Baboon ecology: African field research*. Chicago: University of Chicago Press, 1970.

Altmann, M. Social behavior of elk, *Cervus canadensis nelsoni*, in the Jackson Hole area of Wyoming. *Behaviour*, 1952, 4(2), 116–143.

Altmann, M. Naturalistic studies of maternal care in moose and elk. In H. Rheingold (Ed.), *Maternal behavior in mammals*. New York: Wiley, 1963.

Andrew, R. J. Evolution of intelligence and vocal mimicking. *Science*, 1962, 133, 585–589.

Anonymous. The rabbit (Poem). In H. Wells (Ed.), *One thousand and one poems of mankind*. Atlanta: Tupper and Love, 1953.

Aristotle. *Historia animalia* (D'arcy W. Thompson, trans.). Oxford: Clarendon Press, 1910.

Barash, D. The social biology of the Olympic marmot. *Animal Behavior Monographs*, 1973, 6(3), 172–245.

Barfield, R., Busch, D., & Wallen, K. Gonadal influences on agonistic behavior in the male domestic rat. *Hormones and Behavior*, 1972, 3, 247–259.

Barfield, R. J., & Geyer, L. A. Sexual behavior: Ultrasonic post-ejaculatory song of the male rat. *Science*, 1972, 176, 1349–1350.

Barker, R. G. *The stream of behavior*. New York: Appleton-Century-Crofts, 1963.

Barnett, S. A. Experiments on "neophobia" in wild and laboratory rats. *British Journal of Psychology*, 1958, 49, 195–201.

Barnett, S. A. *The rat: A study in behavior*. Chicago: Aldine, 1963.

Bartlett, D., & Bartlett, J. Beavers. *National Geographic*, 1974, 145, 716–732.

309

Bastian, J. The transmission of arbitrary environmental information between bottlenose dolphins. In R. G. Busnell (Ed.), *Animal sonar systems, biology and bionics.* Jouy-en-Josas: Laboratoire de Physiologie Acoustique, 1967.

Bastian, J. Further investigation of the transmission of arbitrary environmental information between bottlenose dolphins. *Naval Undersea Weapons Center,* Technical Publication Number 109, San Diego, Calif., 1968.

Bateson, G. Problems in cetacean and other mammalian communication. In K. S. Norris (Ed.), *Whales, dolphins, and porpoises.* Berkeley: University of California Press, 1966.

Bateson, G., & Gilbert, B. *Whaler's Cove dolphin community: An interim report.* Makapuu Point, Oahu, Hawaii: Oceanic Institute, 1966.

Beach, F. A. Evolutionary changes in the physiological control of mating behavior in mammals. *Psychological Review,* 1947, *54,* 297–315.

Beach, F. A. The snark was a boojum. *American Psychologist,* 1950, *5,* 115–124.

Beach, F. A., & Gilmore, R. W. Response of male dogs to urine from females in heat. *Journal of Mammalogy,* 1949, *30,* 391–392.

Beach, F. A., & Jaynes, J. Studies of maternal retrieving in rats. *Journal of Mammalogy,* 1956, *37,* 177–180.

Beach, F. A., & Merari, A. Coital behavior in dogs: IV. Effects of progesterone in the bitch. *Proceedings of the National Academy of Science,* 1968, *61,* 442–446.

Beach, F. A., & Merari, A. Coital behavior in dogs. *Journal of Comparative and Physiological Psychology,* 1970, *70,* 1–22.

Beer, J. R. Movements of tagged beaver. *Journal of Wildlife Management,* 1955, *19,* 492–493.

Behse, J. H. Dissolution of the pheromonal bond: Waning of the approach response by weaning rats. *Physiology and Behavior,* 1977, *18*(3), 393-397.

Bekoff, M. Scent marking by free ranging domestic dogs: Olfactory and visual components. *Biology of Behavior,* in press.

Benedict, F. G. *Vital energetics.* Washington, D.C.: Newhouse, 1938.

Bentley, M., & Varon, E. J. An accessory study of phonetic symbolism. *American Journal of Psychology,* 1933, *45,* 76–86.

Berg, I. A. Development of behavior: The micturition pattern in the dog. *Journal of Experimental Psychology,* 1944, *34,* 343–368.

Berkowitz, L. *Aggression: A social psychological analysis.* New York: McGraw-Hill, 1962.

Bermant, G. Response latencies of female rats during sexual intercourse. *Science,* 1961, *133,* 1771–1773.

Berne, E. *Games people play.* New York: Grove Press, 1964.

Bertin, L. *Larousse encyclopedia of animal life.* London: Hamlyn, 1967.

Birdwhistell, R. L. *Kinesics and context.* Philadelphia: University of Pennsylvania Press, 1970.

Blount, W. P. *Rabbits' ailments.* Idle, Bradford, England: Fur and Feather, 1945.

Bohlken, H. Haustiere und Zoologische systematik. *Zeitschrift Tierzüchtung Züchtungs Biologie,* 1961, *76*(1), 107–113.

Bolten, T. R. The role of wet meadows as wildlife habitat in the Southwest. *Journal of Range Management*, 1970, *23*(4), 272–273.

Bolwig, N. Observations and thoughts on the evolution of facial mimicry. *Koedoe*, 1959, *2*, 60–69.

Bossert, W. H., & Wilson, E. O. The analysis of olfactory communication among animals. *Journal of Theoretical Biology*, 1963, *5*(3), 433–469.

Bourliere, F. *The natural history of mammals* (H. M. Parshloy, trans.). New York: Knopf, 1964.

Bowyer, T. *Social behavior of Roosevelt elk during rut*. Unpublished master's thesis, Humbolt State University, 1976.

Bowyer, T. Personal communication, July 1, 1978.

Bradt, G. W. Beaver colonies in Michigan. *Journal of Mammalogy*, 1938, *19*, 139–152.

Brain, C. K. Observations on the behavior of vervet monkeys. *Zoologica Africana*, 1965, *1*(1), 13–27.

Braine, M. The ontogeny of English phrase structure: The first phase. *Language*, 1963, *39*, 1–13.

Brannigan, C. R., & Humphries, D. A. Human non-verbal behavior, a means of communication. In N. Blurton-Jones (Ed.), *Ethological studies of child behaviour*. Cambridge: Cambridge University Press, 1972.

Brind, B. (Producer). *Wolf Pack*. Montreal: National Film Board of Canada, 1974. (Film)

Broadbrooks, H. E. Ecology and distribution of the pikas of Washington and Alaska. *American Midland Naturalist*, 1965, *73*(2), 299–335.

Browman, L. G., & Hudson, P. Observations on the behavior of penned mule deer. *Journal of Mammalogy*, 1957, *38*, 247–253.

Brown, A. M. Bimodal cochlear response curves in rodents. *Nature*, 1970, *228*, 576–577.

Brown, B. A. Social organization in male groups of white-tailed deer. In V. Geist & F. Walther (Eds.), *The behaviour of ungulates and its relation to management*. Morges, Switzerland, IUCN Publications, *24*(L), 436–446, 1974.

Brown, D. H., & Norris, K. S. Observations on captive and wild cetaceans. *Journal of Mammalogy*, 1956, *37*, 311–326.

Brown, R. *Words and things*. New York: Free Press, 1958.

Brown, R., & Hildum, D. C. Expectancy and the identification of syllables. *Language*, 1956, *32*, 411–419.

Browne, T. *Hydriotaphia*. London, 1658.

Brownlee, R. G., Silverstein, R. M., Müller-Schwarze, D., & Singer, A. G. Isolation, identification, and function of the chief component of the male tarsal scent in black-tailed deer. *Nature*, 1969, *221*, 284–285.

Bubenik, A. B. Beitrag zur Geburtskunde and zu den Mutter-Kind Beziehungen des Reh- und Rotwildes. *Zeitschrift Saugetierkunde*, 1965, *30*, 65–128.

Bunn, D. S. Fighting and moult in shrews. *Journal of Zoology*, 1966, *148*, 580–582.

Burckhardt, D. Kindliches Verhalten als Ausdrucksbewegung. *Revue Swisse Zoologie*, 1958, *65*, 311–316.

Burghardt, G. M. Defining "communication." In J. W. Johnston, Jr. (Ed.), *Communication by chemical signals*. New York: Appleton-Century-Crofts, 1970.

Burt, W. H. Territorial behavior and populations of small mammals in southern Michigan. *Miscellaneous Publications of the Museum of Zoology of the University of Michigan*, 1940, *45*, 1–58.

Burt, W. H., & Grossenheider, R. P. *A field guide to the mammals*. Boston: Houghton Mifflin, 1964.

Cahalane, V. H. *Mammals of North America*. New York: Macmillan, 1947.

Caldwell, D. K., & Caldwell, M. C. Epimeletic (care-giving) behavior in Cetacea. In K. S. Norris (Ed.), *Whales, dolphins, and porpoises*. Berkeley: University of California Press, 1966.

Caldwell, D. K., & Caldwell, M. C. Dolphins, porpoises, and behavior. *Underwater Nature*, 1967, *4*(2), 14–19.

Caldwell, D. K., & Caldwell, M. C. Senses and communication. In S. H. Ridgway (Ed.), *Mammals of the sea: Biology and medicine*. Springfield, Ill.: Charles C Thomas, 1972.

Caldwell, D. K., & Caldwell, M. C. Cetaceans. In T. Sebeok (Ed.), *How animals communicate*. Bloomington: Indiana University Press, 1977.

Caldwell, M. C., & Caldwell, D. K. Individualized whistle contours in bottlenose dolphins, *Tursiops truncatus*. *Nature*, 1965, *204*, 404–409.

Caldwell, M. C., Haugen, R. M., & Caldwell, D. K. High-energy sound associated with fright in the dolphin. *Science*, 1962, *134*, 1873–1876.

Calhoun, J. B. Mortality and movement of brown rats (*Rattus norvegicus*) in artificially super-saturated populations. *Journal of Wildlife Management*, 1948, *12*, 167–171.

Calhoun, J. B. Population density and social pathology. *Scientific American*, 1962, *206*(2), 139–148.

Campbell, B. G. *Human evolution*. Chicago: Aldine, 1966.

Carpenter, C. R. A field study of the behavior and social relations of howling monkeys. *Comparative Psychology Monographs*, 1934, *10*(2), 1–168.

Chalmers, N. R. Comparative aspects of early infant development in some captive cercopithecines. In F. E. Poirier (Ed.), *Primate socialization*. New York: Random House, 1972.

Chapman, R. C. Rabies: Decimation of a wolf pack in arctic Alaska. *Science*, 1978, *201*, 365–368.

Cheatum, E. L., & Morton, G. H. Breeding season of white-tailed deer in New York. *Journal of Wildlife Management*, 1946, *10*, 249–263.

Chesler, P. Maternal influence in learning by observation in kittens. *Science*, 1969, *166*, 901–903.

Chomsky, N. *Syntactic structures*. The Hague: Mouton, 1957.

Christian, J. J. Control of population growth in rodents by interplay between population density and endocrine physiology. *Wildlife Diseases*, 1959, *2*, 1–38.

Claesson, A., & Silverstein, R. M. Chemical methodology in the study of mammalian communication. In D. Müller-Schwarze & M. Mozell (Eds.), *Chemical signals in vertebrates*. New York: Plenum, 1977.

Collias, N. E. Classification of animal sounds. In W. Lathan & W. Tavolga (Eds.), *Animal sounds and communication*. Washington, D.C., American Institute of Biological Science, 1960.

Comfort, A. The likelihood of human pheromones. *Nature*, 1971, *239*, 432–449.

Coon, C. *The origin of race*. New York: Knopf, 1962.

Cousteau, J., & Diole, P. *Dolphins*. New York: A and W Publishers, 1975.

Cowan, I. M. Distribution and variation in deer of the Pacific coastal region of North America. *California Fish and Game*, 1936, *22*, 155–246.

Cowan, I. M., & Geist, V. Aggressive behavior in deer of the genus *Odocoileus*. *Journal of Mammalogy*, 1961, *42*, 522–526.

Crabtree, D. E. Notes on experiments in flint knapping. *Tebiwa, The Journal of the Idaho State University*, 1967, *10*(1), 1–32.

Crisler, L. *Arctic wild*. New York: Harper & Row, 1958.

Crowcroft, P. Notes on the behaviour of shrews. *Behaviour*, 1955, *8*, 63–80.

Crowcroft, P. *The life of the shrew*. London: Reinhardt, 1957.

Dagg, A. I., & Taub, A. Flehmen. *Mammalia*, 1970, *34*, 386–395.

Darling, F. *A herd of red deer*. Garden City, N.Y.: Doubleday, 1964. (Originally published in 1937.)

Darwin, C. *On the origin of species*. New York: Mentor, 1958. (Originally published in 1859.)

Darwin, C. *The expression of the emotions in man and animals*. Chicago: University of Chicago Press, 1965. (Originally published in 1872.)

Davis, D. E., & Golley, F. B. *Principles in mammalogy*. New York: Reinhold, 1963.

DeMore, P. P., & Steffens, F. E. The movements of vervet monkeys with their ranges as revealed by radio-tracking. *Journal of Animal Ecology*, 1973, *411*, 677–687.

Deutsch, J. A. Nest building behaviour of domestic rabbits under semi-natural conditions. *British Journal of Animal Behaviour*, 1957, *5*(2), 53–54.

De Vos, A. P., Brokx, P., & Geist, V. A review of social behavior of North American cervids during the reproductive period. *American Midland Naturalist*, 1967, *77*(2), 390–417.

Dixon, J. S. A study of the life history and food habits of mule deer in California. *California Fish and Game*, 1934, *20*(4), 181–354.

Donovan, C. A. Some clinical observations on sexual attraction and deterrence in dogs and cattle. *Veterinary Medicine*, 1967, *62*, 1047–1051.

Donovan, C. A. Canine anal glands and chemical signals. *Journal of the American Veterinary Association*, 1969, *155*, 1995–1996.

Doty, R. L., & Dunbar, I. Attraction of beagles to conspecific urine, vaginal and anal sac secretion odors. *Physiology and Behavior*, 1974, *12*, 825–833.

Doty, R. L., Ford, M., Preti, G., & Huggins, G. R. Changes in the intensity and pleasantness of human vaginal odors during the menstrual cycle. *Science*, 1975, *190*, 1316–1318.

Douglas-Hamilton, I. *Among the elephants.* New York: Viking Press, 1975.

Downing, R. L., & McGinnes, B. S. Capturing and marking white-tail fawns. *Journal of Wildlife Management,* 1969, *33,* 711–714.

Dreher, J. J. Linguistic considerations of porpoise sounds. *Journal of the Acoustic Society of America,* 1961, *33,* 1799–1800.

Dreher, J. J. Cetacean communication: Small group experiment. In K. S. Norris (Ed.), *Whales, dolphins, and porpoises.* Berkeley: University of California Press, 1966.

Dreher, J. J., & Evans, W. E. Cetacean communication. In W. N. Tavolga (Ed.), *Marine bio-acoustics.* New York: Macmillan, 1964.

Dryden, G. L., & Conaway, C. H. The origin and hormonal control of scent production in *Suncus murinus. Journal of Mammalogy,* 1967, *20,* 150–173.

Dudzinski, M. L., & Mykytowycz, R. Analysis of weights and growth rates of an experimental colony of wild rabbits, *Oryctolagus cuniculus* (L.). CSIRO Wildlife Research, 1960, *5,* 102–115.

Eadie, W. R. The dermal glands of shrews. *Journal of Mammalogy,* 1938, *19,* 171–174.

Eibl-Eibesfeldt, I. Über die Jugendentwicklung des Verhaltens eines männliches Dachses *(Meles meles* L.). *Zeitschrift für Tierpsychologie,* 1950, *7(3),* 327–355.

Eibl-Eibesfeldt, I. *Ethology.* New York: Holt, Rinehart & Winston, 1970.

Eimerl, S., & DeVore, I. *The primates.* Chicago: Time-Life, 1965.

Eiseley, L. The long loneliness. *American Scholar,* 1960, *30,* 57–64.

Eisenberg, J. F. Mammalian social systems: Are primate social systems unique? *Symposium of the Fourth International Congress of Primatologists,* 1973, *1,* 232–249.

Eisenberg, J. F., & Kleiman, D. G. Olfactory communication in mammals. *Animal Review of Ecology and Systematics,* 1972, *3,* 1–32.

Eliot, T. S. The ad-dressing of cats (Poem). In *The complete poems and plays.* New York: Harcourt Brace Jovanovich, 1934.

Errington, P. L. Notes on food habits of southern Wisconsin house cats. *Journal of Mammalogy,* 1936, *17,* 64–65.

Essapian, F. S. The birth and growth of a porpoise. *Natural History,* 1953, *62,* 392–399.

Essapian, F. S. Observations on abnormalities of parturition in captive bottle-nose dolphins, *Tursiops truncatus. Journal of Mammalogy,* 1963, *44,* 405–414.

Esser, A. H. (Ed.). *Behavior and environment: The use of space by animals and men.* New York: Plenum Press, 1971.

Estes, R. D. The role of the vomeronasal organ in mammalian reproduction. *Mammalia,* 1972, *36,* 315–341.

Etkin, W. *Social behavior and organization among vertebrates.* Chicago: University of Chicago Press, 1963.

Evans, W. E. Vocalizations among marine mammals. In W. N. Tavolga (Ed.), *Marine bio-acoustics.* New York: Pergamon Press, 1967.

Evans, W. E., & Bastian, J. Marine mammal communication: Social and ecological factors. In H. T. Andersen (Ed.), *The biology of marine mammals*. New York: Academic Press, 1969.

Evans, W. E., & Prescott, J. H. Observations of the sound production capabilities of the bottlenose dolphin. *Zoologica*, 1962, *47*, 121–128.

Ewer, R. F. Suckling behaviour in kittens. *Behaviour*, 1959, *15*, 146–162.

Ewer, R. F. Further observations on suckling behaviour in kittens. . . . *Behaviour*, 1961, *18*, 247–260.

Ewer, R. F. *Ethology of mammals*. New York: Plenum Press, 1968.

Ewer, R. F. *The carnivores*. Ithaca, N.Y.: Cornell University Press, 1973.

Ewer, R. F., & Wemmer, C. The behaviour in captivity of the African civet *Civettictis civetta*. *Zeitschrift für Tierpsychologie*, 1974, *34*, 359–394.

Fedigan, L. Social and solitary play in a colony of vervet monkeys. *Primates*, 1972, *13*(4), 347–364.

Fentress, J. Observations on the behavioral development of a hand-reared male timber wolf. *American Zoologist*, 1967, *7*, 339–351.

Field, R. A perspective on syntactics of wolf vocalizations. In E. Klinghammer (Ed.), *The behavior and ecology of wolves*. New York: Garland, 1979.

Finstad, G. *A field study on elk*. Unpublished manuscript, 1976. (Available from R. Peters, Department of Psychology, Fort Lewis College, Durango, Colo. 81301.)

Fisher, A. E. Effects of stimulus variation and sexual satiation in the male rat. *Journal of Comparative and Physiological Psychology*, 1962, *55*, 614–620.

Fisher, R. A. *The genetical theory of natural selection*. Oxford: Clarendon Press, 1930.

Fouts, R. Acquisition and testing of gestural signs in four young chimpanzees. *Science*, 1973, *180*, 978–980.

Fox, M. *Behaviour of wolves, dogs, and related canids*. New York: Harper & Row, 1971.

Frings, H., & Frings, M. *Animal communication*. New York: Blaisdell, 1964.

Fuller, J. L., & Dubuis, E. M. The behaviour of dogs. In E. S. Hafez (Ed.), *Behaviour of domestic animals*. London: Ballière, Tindall & Cox, 1962.

Gardner, R., & Gardner, B. Teaching sign language to a chimpanzee. *Science*, 1969, *165*, 664–672.

Gartlan, J. S. Sexual and maternal behavior of the vervet monkey, *Cercopithecus aethiops*. *Journal of Reproduction and Fertility*, 1969, Supplement No. 6, 137–150.

Gawienowski, A. M. Chemical attractants of the rat preputial gland. In D. Müller-Schwarze (Ed.), *Chemical signals in vertebrates*. New York: Plenum Press, 1977.

Geist, V. Ethological observations on some North American cervids. *Zoologische Beitrage*, 1966, *12*, 219–250.

Geist, V. *Mountain sheep*. Chicago: University of Chicago Press, 1971.

Goldman, E. A. *The wolves of North America* (Vol. 2). New York: American Wildlife Institute, 1944.

Golley, F. B. Gestation period, breeding and fawning behavior of Columbian black-tailed deer. *Journal of Mammalogy*, 1957, *38*, 116–120.

Goodall, J. Feeding behaviour of wild chimpanzees. *Symposia of the Zoological Society of London*, 1963, 10, 1–39.

Goodrich, B. S., & Mykytowycz, R. Individual and sex differences in the chemical composition of pheromone-like substances from the skin glands of the rabbit (*Oryctolagus cuniculus*). *Journal of Mammalogy*, 1972, 53(3), 540–548.

Goodwin, G. G. The rabbit. *Collier's encyclopedia* (Vol. 19). New York: Macmillan Educational Corporation, 1975.

Goodwin, G. G. Rabbit. *Encyclopedia Americana* (Vol. 23). Danbury, Conn.: Americana Corporation, 1976.

Goodwin, M., Gooding, K. M., & Regnier, F. Sex pheromone in the dog. *Science*, 1979, 203, 559–561.

Gould, E. Communication in three genera of shrews (Soricidae): *Suncus, Blasina,* and *Cryptotis*. *Communications in Behavioral Biology*, Part A, 1969, 3, 263–313.

Graf, W. The Roosevelt elk. *Port Angeles Evening News*, August 7, 1955, p. 105.

Graf, W. Territorialism in deer. *Journal of Mammalogy*, 1956, 37(2), 165–171.

Grant, E. C., & Chance, M. R. A. Rank order in caged rats. *Animal Behaviour*, 1958, 6, 183–194.

Grant, E. C., & Mackintosh, J. H. A comparison of the social postures of some common laboratory rodents. *Behaviour*, 1963, 21, 246–259.

Grant, U. S. *Personal memoirs*. New York: private, 1885.

Grundlach, H. *Beobachten bei Versuchen zur Wildschudenverhütung am Rehwild*. Kronberg, Germany: Von Opel-Freigehege, 1961.

Grzimek, B. *Grzimek's animal life encyclopedia*. New York: Van Nostrand, 1955.

Guhl, A. M. Gonadal hormones and social behavior in infrahuman vertebrates. In W. C. Young (Ed.), *Sex and internal secretion*. Baltimore: Williams and Wilkins, 1961.

Guthrie, R. D. A new theory of mammalian rump patch evolution. *Behaviour*, 1971, 38, 132–146.

Hafez, E. S. E. (Ed.). *The behaviour of domestic animals*. Baltimore: Williams and Wilkins, 1969.

Haga, R. Observations on the ecology of the Japanese pika. *Journal of Mammalogy*, 1960, 41(2), 200–219.

Hall, R. L., & Sharp, H. S. (Eds.) *Wolf and man*. New York: Academic Press, 1978.

Hamilton, W. D. Geometry for the selfish herd. *Journal of Theoretical Biology*, 1971, 31(2), 295–311.

Hamilton, W. J., & Arrowood, P. C. Copulatory vocalizations of chacma baboons (*Papio ursinus*), gibbons (*Hylobates hoolock*), and humans. *Science*, 1978, 200, 1405–1408.

Harder, W. Zur Morphologie und Physiologie des Blindarmes der Nagetiere. *Verhalten Deutsche Zoologische Geschichte*, 1949, 54, 95–109.

Harper, J. A., Harn, J. H., Bentley, W. W., & Yocum, C. F. The status and ecology of the Roosevelt elk in California. *Wildlife Monographs*, 1967, 16, 1–47.

Harrington, F., & Mech, L. D. Wolf vocalization. In R. L. Hall & H. S. Sharp (Eds.), *Wolf and man*. New York: Plenum Press, 1978.

Harrop, A. E. Some observations on canine semen. *Veterinary Records*, 1955, 67, 494–498.

Harvey, E. B., & Rosenberg, L. E. An apocrine gland complex in the pika. *Journal of Mammalogy*, 1960, 41(2), 213–220.

Hatt, R. T. A large beaver-felled tree. *Journal of Mammalogy*, 25, 313, 1944.

Haugen, A. O. Breeding records of captive white-tailed deer in Alabama. *Journal of Mammalogy*, 1959, 40(1), 108–113.

Hediger, H. Mammalian territories. *Bij dragen tot de Dierkunde*, 1949, 28, 172–184.

Hediger, H. *The psychology and behaviour of animals in zoos and circuses*. New York: Dover, 1955.

Henry, J. D. The use of urine marking in the scavenging behavior of the red fox *(Vulpes vulpes)*. *Behaviour*, 1976, 34, 217–227.

Heptner, W. G., & Nasimovitsch, A. A. *Der Elch*. Wittenberg: Ziemsen Verlag, 1968.

Heptner, W. G., Nasimovitsch, A. A., & Bannikov, A. G. *Mammals of the Soviet Union*. Jena: VEB Gustav Fischer-Verlag, 1961.

Hertel, H. Hydrodynamics of swimming and wave-riding dolphins. In H. T. Andersen (Ed.), *The biology of marine mammals*. New York: Academic Press, 1969.

Herter, K. Das Verhalten der Insektivoren. *Handbuch Zoologie*, 1957, 8(10), 1–50.

Hesterman, R. R., & Mykytowycz, R. Some observations on the odours of anal gland secretions from the rabbit. *CSIRO Wildlife Research*, 1968, 13, 71–81.

Hill, W. C. O. *Evolutionary biology of the primates*. New York: Academic Press, 1972.

Hill, W. C. O. *Primates: Comparative anatomy and taxonomy*. New York: Wiley, 1966.

Hinde, R. A. *Animal behavior: A synthesis of ethology and comparative psychology* (2nd ed.). New York: McGraw-Hill, 1970.

Hingston, R. W. G. *The meaning of animal colour and adornment*. London: Methuen, 1933.

Hirth, D. H. *Social behavior of white-tailed deer in relation to habitat*. Unpublished doctoral dissertation, University of Michigan, 1973.

Hirth, D. H., & McCullough, D. R. Evolution of alarm signals in ungulates with special reference to white-tailed deer. *American Naturalist*, 1977, 111, 31–42.

Hockett, C. F. Logical considerations in the study of animal communication. In W. E. Lanyon & W. N. Tavolga (Eds.), *Animal sounds and communication*, Washington, D.C.: American Institute of Biological Science, 1960.

Hockett, C., & Ascher, R. The human revolution. *Current Anthropology*, 1964, 5(3), 135–168.

Hoese, H. D. Dolphin feeding out of water in a salt marsh. *Journal of Mammalogy*, 1971, 52(1), 222–223.

Hubbs, C. L. Dolphin protecting dead young. *Journal of Mammalogy*, 1953, *34*, 498.

Huffman, M. *Communication in vervet monkeys.* Unpublished manuscript, 1976. (Available from R. Peters, Department of Psychology, Fort Lewis College, Durango, Colo. 81301.)

Jerison, H. J. Fossil evidence of the evolution of the human brain. *Annual Review of Anthropology*, 1975, *4*, 27–58.

Johnson, C. S. Sound detection thresholds in marine mammals. In W. N. Tavolga (Ed.), *Marine bio-acoustics.* New York: Pergamon Press, 1967.

Johnson, D. E. Biology of the elk calf *(Cervus canadensis). Journal of Wildlife Management*, 1951, *15*(4), 396–410.

Johnson, D. R. Diet and reproduction of Colorado pikas. *Journal of Mammalogy*, 1967, *48*, 311–315.

Joslin, P. W. B. *Summer activities of two timber wolf packs in Algonquin Park.* Unpublished master's thesis, University of Toronto, 1966.

Joslin, P. W. B. Movements and home sites of timber wolves in Algonquin Park. *American Zoologist*, 1967, *1*, 279–288.

Kalmus, H. Animal behaviour and theories of games and language. *Animal Behaviour*, 1969, *17*(4), 607–617.

Kaplan, S. Cognitive maps in perception and thought. In R. G. Golledge & G. T. Moore (Eds.), *Image and environment.* Chicago: Aldine, 1973.

Kawamichi, T. Annual cycle of behavior and social patterns of the Japanese pika. *Journal of the Faculty of Science of Hokkaido University*, 1971, *18*(1), 173–185.

Kawamichi, T. Hay territories and dominance rank of pikas. *Journal of Mammalogy*, 1976, *57*(1), 133–148.

Kellogg, W. N. *Porpoises and sonar.* Chicago: University of Chicago Press, 1961.

Kellogg, W. N., & Rice, C. E. Visual discrimination and problem solving in a bottlenose dolphin. In K. S. Norris (Ed.), *Whales, dolphins, and porpoises.* Berkeley: University of California Press, 1966.

Kile, T. L., & Marchinton, R. L. White-tailed deer rubs and scrapes: Spatial, temporal and physical characteristics and social role. *American Midland Naturalist*, 1977, *97*(2), 257–266.

Kilham, L. Territorial behavior in pikas. *Journal of Mammalogy*, 1958, *39*(2), 307.

Kirkpatrick, C. M. Rabbit. *World book encyclopedia* (Vol. 16). Chicago: World Book–Childcraft International, 1971.

Kleiman, D. Scent marking in the Canidae. *Symposia of the Zoological Society of London*, 1966, *18*, 167–177.

Knappe, H. Zur Funktion des Jacobonschen Organs. *Zoologischer Garten*, 1964, *28*, 188–194.

Knight, R. R. The Sun River elk herd. *Wildlife Monographs*, 1970, *23*, 1–66.

Kortlandt, A. Discussion contribution on "On the essential morphological basis for human culture" by A. L. Bryon. *Current Anthropology*, 1965, *6*, 320–326.

Krames, L. Responses of female rats to the individual body odors of male rats. *Psychonomic Science*, 1970, *20*, 274-275.

Krames, L., Carr, W. J., & Bergmann, B. A pheromone associated with social dominances among male rats. *Psychonomic Science*, 1969, *16*(1), 11–12.

Krear, H. R. An ecological and ethological study of the pika in the Front Range of Colorado (Doctoral dissertation, University of Colorado, 1965). *Dissertation Abstracts International*, 1966, *26*(10), 2781. (University Microfilms No. 66–2808)

Kruuk, H. *The spotted hyena*. Chicago: University of Chicago Press, 1972.

Kucera, T. Social behavior and breeding system of the desert mule deer. *Journal of Mammalogy*, 1978, *59*(3), 463-476.

Lancaster, J. B. Play-mothering: The relations between juvenile females and young infants among free-ranging vervet monkeys. *Folia Primatologica*, 1971, *15*, 116–132.

Lang, H. Caudal and pectoral glands of African elephant shrews. *Journal of Mammalogy*, 1923, *4*, 261–263.

Lang, T. G., & Smith, H. A. P. Communication between dolphins in separate tanks by way of an electronic acoustic link. *Science*, 1965, *150*, 1839–1843.

Lechleitner, R. R. *Wild mammals of Colorado*. Boulder, Colo.: Pruett, 1969.

Leighton, A. H. Notes on the relations of beavers to one another and to the muskrat. *Journal of Mammalogy*, 1933, *14*, 27–35.

Le Magnen, J. Étude des phenomenes olfacto-sexuels chez le rat blanc. *Comptes Rendus de la Société de Biologie de Paris*, 1951, *145*, 850–861.

Lenneberg, E. *Biological foundations of language*. New York: Wiley, 1967.

Lent, P. C. Mother–infant relationships in ungulates. In V. Geist & F. Walther (Eds.), *The behaviour of ungulates and its relation to management*. Morges, Switzerland: IUCN Publication No. 24, 1974.

Leroi-Gourhan, A. *The art of prehistoric man in Western Europe*. London: Thames and Hudson, 1968.

Leyhausen, P. Das Verhalten der Katzen. *Handbuch Zoologie*, 1956, 8, 17–34. (a)

Leyhausen, P. Verhaltenstudien an Katzen. *Zeitschrift für Tierpsychologie*, 1956, *2*, 2–120. (b)

Leyhausen, P. The communal organization of solitary mammals. *Symposia of the Zoological Society of London*, 1965, *14*, 249–263. (a)

Leyhausen, P. Über die Funktion der relativen Stimmunghierarchie. *Zeitschrift für Tierpsychologie*, 1965, *22*, 249–263. (b).

Leyhausen, P., & Wolff, R. Das Revier einer Hauskatze. *Zeitschrift für Tierpsychologie*, 1959, *16*, 666–670.

Lilly, J. Vocal behavior of the bottlenose dolphin. *Proceedings of the American Philosophical Society*, 1962, *106*, 520–529.

Lilly, J. Distress call of the bottlenose dolphin: Stimuli and evoked behavioral responses. *Science*, 1963, *139*, 116–118.

Lilly J. *The mind of the dolphin*. New York: Doubleday, 1967.

Lilly J., & Miller, A. Sounds emitted by the bottlenose dolphin. *Science*, 1961, *133*, 1689–1693. (a)

Lilly, J., & Miller, A. Vocal exchanges between dolphins. *Science*, 1961, *134*, 1873–1876. (b)

Lilly, J., & Miller, A. Operant conditioning of the bottlenose dolphin with electrical stimulation of the brain. *Journal of Comparative and Physiological Psychology*, 1962, *55*, 73–79.

Lincke, M. *Das Wildkaninchen*. Berlin: Parey, 1943.

Lind, J. The infant cry. *Proceedings of the Royal Society of Medicine*, 1971, *64*, 468.

Linsdale, J. M., & Tomich, P. Q. *A herd of mule deer*. Berkeley: University of California Press, 1953.

Lockwood, R. Dominance in wolves: Useful construct or bad habit? In E. Klinghammer (Ed.), *The behavior and ecology of wolves*. New York: Garland, 1979.

Locy, W. A. *The story of biology*. New York: Garden City Publishing, 1925.

Long, W. S. Response of a pika to a weasel. *Journal of Mammalogy*, 1938, *19*, 250.

Lorenz, K. The comparative method in studying innate behaviour patterns. *Symposia of the Society for Experimental Biology*, 1950, *4*, 221–268.

Lorenz, K. *King Solomon's ring: New light on animal ways*. London: Methuen, 1952.

Lorenz, K. *On aggression*. London: Methuen, 1966.

Lutton, L. M. Notes on territorial behavior and response to predators of the pika. *Journal of Mammalogy*, 1975, *56*, 231–234.

MacFarlane, A. *The psychology of childbirth*. Cambridge, Mass.: Harvard University Press, 1977.

Markham, O. D. Pikas. *Colorado Outdoors*, 1975, *7*(4), 31–34.

Markham, O. D., & Whicker, F. W. Notes on the behavior of the pika in captivity. *American Midland Naturalist*, 1973, *89*, 192–199.

Marler, P. Characteristics of some animal calls. *Nature*, 1955, *176*, 6–7.

Marr, J. N., & Gardner, L. E., Jr. Early olfactory experience and later social behavior in the rat: Preference, sexual responsiveness, and care of the young. *Journal of Genetic Psychology*, 1965, *107*, 167–174.

Martin, R. D. Tree shrews: Unique reproductive mechanism of systematic importance. *Science*, 1966, *152*, 1402–1404.

Martin, R. D. Reproduction and ontogeny in tree shrews (*Tupaia belangeri*) with reference to their general behavior and taxonomic relationships. *Zeitschrift für Tierpsychologie*, 1968, *25*(4), 409–495; *25*(5), 505–532.

Martinka, C. J. Population ecology of elk in Jackson Hole, Wyoming. *Journal of Wildlife Management*, 1969, *33*(3), 465–481.

Martins, T., & Valle, J. R. Hormonal regulation of the micturition pattern of the dog. *Journal of Comparative and Physiological Psychology*, 1948, *41*, 301–311.

Masters, W. H., & Johnson, V. E. *Human sexual response*. Boston: Little, Brown, 1966.

McBride, A. F., & Hebb, D. Behavior of the captive bottlenose dolphin, *Tursiops truncatus. Journal of Comparative and Physiological Psychology,* 1948, *41,* 111–123.

McBride, A. F., & Kritzler, H. Observations on pregnancy, parturition, and post-natal behavior in the bottlenose dolphin. *Journal of Mammalogy,* 1951, *32*(3), 251–266.

McCullough, D. R. The tule elk: Its history, behavior, and ecology. *University of California Publications in Zoology,* 1969, 88, 1–209.

McGuire, M. *Progress report on eight months of primatological studies on St. Kitts.* Behavioral Sciences Foundation Scientific Report No. 3, April 15, 1971.

McGuire, M. T. The St. Kitts vervet. *Contributions to Primatology,* 1974, *1,* 1–199.

McKenna, M. C. Paleontology and the origin of the primates. *Folia Primatologica,* 1966, *4,* 1–25.

Mech, L. D. *The wolves of Isle Royale* (Fauna Series No. 7). Washington, D.C.: U.S. Park Service, 1966.

Mech, L. D. *The wolf: The ecology and behavior of an endangered species.* Garden City, N.Y.: Natural History Press, 1970.

Mech, L. D. Spacing and possible mechanisms of population regulation in wolves. *American Zoologist,* 1972, *12*(4), 9. (abstract)

Mech, L. D. *Wolf numbers in the Superior National Forest of Minnesota* (USDA Forest Service Research Paper No. NC–97). St. Paul: North Central Forest Experiment Station, 1973.

Mech, L. D. Current techniques in the study of elusive wilderness carnivores. Proceedings of the 11th International Congress of Game Biologists. Stockholm: National Swedish Environment Protection Board, 315–322, 1974.

Mech, L. D. Personal communication, March 22, 1977. (a)

Mech, L. D. Productivity, mortality, and population trend in wolves from northeastern Minnesota. *Journal of Mammalogy,* 1977, *58*(4), 559–574. (b)

Mech, L. D. Wolf pack buffer zones as prey reservoirs. *Science,* 1977, *198,* 320–321. (c)

Mech, L. D., & Frenzel, L. D. Continuing timber wolf studies. *Naturalist,* 1969, *20*(1), 30–35.

Mech, L. D., & Frenzel, L. D. Ecological studies of the timber wolf in northeastern Minnesota. USDA Forest Service Research Paper NC–52. St. Paul, Minn.: North Central Forest Experiment Station, 1971.

Mech, L. D., & Knick, S. T. Sleeping distance in wolf pairs in relation to the breeding season. *Behavioral Biology,* 1978, *23,* 521–525.

Mech, L. D., Rothman, R. J., Colin, J., & Seal, U. S. Olfactory effects in reproduction in mice and wolves: Similar systems? 1978. Unpublished manuscript available from L. D. Mech, U. S. Bureau of Sport Fisheries and Wildlife, Folwell Ave., St. Paul, Minn.

Meester, J. Shrews in captivity. *African Wild Life,* 1960, *14,* 57–63.

Mehrabian, A. *Silent messages.* Belmont, Calif.: Wadsworth, 1971.

Mello, N. K., & Peterson, N. J. Behavioral evidence for color discrimination in cats. *Journal of Neurophysiology,* 1964, *27,* 323–333.

Melrose, D. R., Reed, H. C., & Patterson, R. L. S. Androgen steroids associated with boar odor as an aid to the detection of estrus in pig artificial insemination. *British Veterinary Journal,* 1971, *127,* 497–501.

Michael, R. P. Observations upon the sexual behaviour of the domestic cat *(Felis catus)* under laboratory conditions. *Behaviour,* 1961, *18,* 1–24.

Michael, R. P., & Bonsall, R. W. Chemical signals and primate behavior. In D. Müller-Schwarze & M. Mozell (Eds.), *Chemical signals in vertebrates.* New York: Plenum Press, 1977.

Michael, R. P., Bonsall, R. W., & Warner, P. Human vaginal secretions: Volatile fatty acid content. *Science,* 1974, *186,* 1217–1219.

Michael, R. P., Keverrene, E. B., & Bonsall, R. W. Pheromones: Isolation of male sex attractants from a female primate. *Science,* 1971, *172,* 964–966.

Miller, G., Galanter, G., & Pribram, K. *Plans and the structure of behavior.* New York: Holt, Rinehart & Winston, 1960.

Moelk, M. Vocalization in the house-cat: A phonetic and functional study. *American Journal of Psychology,* 1944, *57,* 184–205.

Moore, W. G., & Marchinton, R. L. Marking behavior and its social function in white-tailed deer. In V. Geist & F. Walther (Eds.), *The behaviour of ungulates and its relation to management.* Morges, Switzerland: IUCN, 1974, No. *24,* 447–456.

Morris, D. *The mammals.* New York: Harper & Row, 1965.

Morris, D. *Intimate behaviour.* New York: Random House, 1971.

Morris, D. *Manwatching.* New York: Abrams, 1977.

Morrison, J. A. Characteristics of estrus in captive elk. *Behaviour,* 1960, *16,* 84–92.

Müller-Schwarze, D. Pheromones in black-tailed deer *(Odocoileus hemionus columbianus).* *Animal Behaviour,* 1971, *19,* 141–152.

Müller-Schwarze, D. Social significance of forehead rubbing in black-tailed deer. *Animal Behaviour,* 1972, *20,* 788–797.

Müller-Schwarze, D. Olfactory recognition of species, groups, individuals, and physiological states among mammals. In M. Birch (Ed.), *Pheromones.* Amsterdam: North Holland, 1974.

Müller-Schwarze, D. Complex mammalian behavior and pheromone bioassay in the field. In D. Müller-Schwarze & M. Mozell (Eds.), *Chemical signals in vertebrates.* New York: Plenum Press, 1977.

Müller-Schwarze, D., Silverstein, R. M., Müller-Schwarze, C., Singer, A. G., & Volkman, N. Y. Responses to a mammalian pheromone and its geometric isomer. *Journal of Chemical Ecology,* 1976, *2,* 389–398.

Murie, O. J. Elk calls. *Journal of Mammalogy,* 1932, *13*(4), 331–336.

Murie, O. J. *The wolves of Mt. McKinley* (Fauna Series No. 5). Washington, D.C.: U.S. National Park Service, 1944.

Murie, O. J. *The elk of North America.* Harrisburg, Penna.: Stackpole, 1951.

Myers, K., Hale, C. S., Mykytowycz, R., & Hughes, L. The effects of varying density and space on sociality and health in animals. In A. H. Esser (Ed.), *Behavior and environment*. New York: Plenum Press, 1971.

Myers, K., & Mykytowycz, R. Social behaviour in the wild rabbit. *Nature*, 1958, *181*, 1515–1566.

Myers, K., & Poole, W. E. A study of the biology of the wild rabbit, *Oryctolagus cuniculus* (L.), in confined populations. II. The effects of season and population increase on behavior. *Commonwealth Scientific and Industrial Research Organization Wildlife Research*, 1961, *3*, 7–25.

Mykytowycz, R. Social behaviour of an experimental colony of wild rabbits, *Oryctolagus cuniculus* (L.), I, II, III. *Commonwealth Scientific and Industrial Research Organization: Wildlife Research*, 1958, *3*, 7–25; 1959, *4*, 1–13; 1960, *5*, 1–20.

Mykytowycz, R. Territorial marking by rabbits. *Scientific American*, 1968, *218*(5), 116–126.

Mykytowycz, R. The role of skin glands in mammalian communication. In J. W. Johnston, D. G. Moulton, & A. Turk (Eds.), *Advances in chemoreception*. New York: Appleton-Century-Crofts, 1970.

Mykytowycz, R. Aggressive and protective behaviour of adult rabbits towards juveniles. *Behaviour*, 1972, *43*, 97–120.

Mykytowycz, R. Reproduction of mammals in relation to environmental odours. *Journal of Reproduction and Fertility*, 1973, *19*, 433–446.

Mykytowycz, R. Odour in the spacing behavior of mammals. In M. Birch (Ed.), *Pheromones*. Amsterdam: North Holland, 1974.

Mykytowycz, R., & Dudzinski, M. L. A study of the weight of odoriferous and other glands in the wild rabbit. *Commonwealth Scientific and Industrial Research Organization Wildlife Research*, 1966, *11*, 31–47.

Mykytowycz, R., & Dudzinski, M. L. Aggressive and protective behavior of adult rabbits, *Oryctolagus cuniculus* (L.), towards juveniles. *Behaviour*, 1972, *43*, 97–120.

Mykytowycz, R., & Ward, M. Some reactions of nestlings of the wild rabbit when exposed to natural conspecific odors. *Forma et Function*, 1971, *4*, 137–148.

Neff, W., & Hind, J. Auditory thresholds of the cat. *Journal of the Acoustic Society of America*, 1955, *8*, 63–71.

Neisser, U. *Cognitive psychology*. New York: Appleton-Century-Crofts, 1966.

Nelson, E. W. *Wild animals of North America*. Washington, D.C.: National Geographic Society, 1930.

Nichol, A. A. Experimental feeding of deer. *University of Arizona College Agricultural Station Technical Bulletin*, 1938, *75*, 1–39.

Nikol'skii, A. A. Basic modifications of the mating call in males of the Turkestan red deer. *Zoologicheskii Zhurnal*, 1975, *54*, 1897–1900.

Noirot, E. Ultrasounds in young rodents. II. Changes with age in albino rats. *Animal Behaviour*, 1968, *16*, 129–134.

Norris, K. S. (Ed.). *Whales, dolphins, and porpoises*. Berkeley: University of California Press, 1966.

Norris, K. S. Aggressive behavior in Cetacea. In C. D. Clemente & D. B. Lindsley (Eds.), *Aggression and defense: Neural mechanisms and social patterns.* Berkeley: University of California Press, 1967.

Norris, K. S., & Prescott, J. H. Observations of Pacific cetaceans of California and Mexican waters. *University of California Publications in Zoology,* 1961, 63(4), 291–402.

Norris, K. S., Prescott, J. H., Asa-Dorian, P. V., & Perkins, P. An experimental demonstration of echolocation behavior in the porpoise. *Biology Bulletin,* 1961, *120,* 163–176.

Olson, S. F. Organization and range of the pack. *Ecology,* 1938, *19,* 168–170.

Orr, R. T. *The little-known pika.* New York: Macmillan, 1977.

Orsulak, P. J., & Gawienowski, A. M. Olfactory preferences for the rat preputial gland. *Biology of Reproduction,* 1972, *6,* 219.

Ozoga, J. J. Aggressive behavior of white-tailed deer at winter cuttings. *Journal of Wildlife Management,* 1972, 36(3), 861–868.

Paget, R. *Human speech.* New York: Harcourt Brace Jovanovich, 1930.

Palen, G. F., & Goddard, G. V. Catnip and oestrous behaviour in the cat. *Animal Behaviour,* 1966, *14,* 372–377.

Patterson, F. Conversations with a gorilla. *National Geographic,* 1978, 154(4), 438–466.

Pearson, O. P. Scent glands of the short-tailed shrew. *Anatomical Record,* 1946, *94,* 615–625.

Pepper, R. L., & Beach, F. A. Preliminary investigations of tactile reinforcement in the dolphin. *Cetology,* 1972, *7,* 1–8.

Perez, J. M., Dawson, W. W., & Landon, D. Retinal anatomy of the bottlenose dolphin *(Tursiops truncatus). Cetology,* 1972, *11,* 1–11.

Peters, R. P. Cognitive maps in wolves and men. In W. F. Preisser (Ed.), *Environmental design research.* Stroudsburg, Pa.: Dowden, Hutchinson, and Ross, 1973.

Peters, R. P. *Wolf-sign: Scents and space in a wide-ranging predator.* Unpublished doctoral dissertation, University of Michigan, 1974.

Peters, R. P. Communication, cognitive mapping and strategy in wolves and hominids. In R. L. Hall & H. S. Sharp (Eds.), *Wolf and man.* New York: Academic Press, 1978.

Peters, R. P. Mental maps in wolf territoriality. In E. Klinghammer (Ed.), *The behavior and ecology of wolves.* New York: Garland, 1979.

Peters, R. P., & Mech, L. D. Scent-marking in wolves: A field study. *American Scientist,* 1975, 63, 628–637.

Pilleri, G., & Knuckey, J. Behavior patterns of some Delphinidae observed in the western Mediterranean. *Zeitschrift für Tierpsychologie,* 1969, 26(1), 48–72.

Pimlott, D. H. Review of F. Mowat's *Never cry wolf. Journal of Wildlife Management,* 1966, 30, 236–237.

Pliny the Elder. *The natural history of Pliny* (P. Holland, trans.). New York: McGraw-Hill, 1962.

Poduschka, W. Insectivore communication. In T. A. Sebeok (Ed.), *How animals communicate.* Bloomington: Indiana University Press, 1977.

Poduschka, W., & Firbas, W. Das selbstbespeicheln des Igels. *Zeitschrift für Säugetierkunde*, 1968, *33*, 160–172.

Poirier, F. E. The St. Kitts green monkey *(C. aethiops jabaeus)*: Ecology, population dynamics and selected behavioral traits. *Folia Primatologica*, 1972, *17*, 20–55.

Premack, D. Language in chimpanzee? *Science*, 1971, *172*, 802–822.

Price, R. Urine-marking and the response to fresh versus aged urine in wild versus domestic Norway rats. *Journal of Chemical Ecology*, 1977, *3*(1), 27–31.

Pruitt, W. O. Rutting behavior of the whitetail deer. *Journal of Mammalogy*, 1954, *55*, 129–130.

Puente, A. E., & Dewsbury, D. A. Courtship and copulatory behavior of the bottlenose dolphin *(Tursiops truncatus)*. *Cetology*, 1976, *21*, 1–9.

Quay, W. B. Comparative survey of the sebaceous and sudoriferous glands of the oral lips and angle in rodents. *Journal of Mammalogy*, 1965, *43*, 303–310.

Quay, W. B., & Müller-Schwarze, D. Functional histology of integumentary glandular regions in black-tailed deer. *Journal of Mammalogy*, 1970, *51*, 675–694.

Rabb, G., Woolpy, J. H., & Ginsburg, B. Social relationships in a group of captive wolves. *American Zoologist*, 1967, *7*, 305–311.

Ralls, K. Mammalian scent marking. *Science*, 1971, *171*, 443–449.

Randall, J. H. *Aristotle*. New York: Columbia University Press, 1960.

Rankin, J. Notes on the ecology, capture, and behaviour in captivity of the elephant shrew. *Zoology of Africa*, 1965, *1*, 73–80.

Rausch, R. L. Some aspects of the population ecology of wolves, Alaska. *American Zoologist*, 1967, *7*, 253–265.

Read, C. *The origin of man*. New York: Cambridge University Press, 1923.

Reed, C. A. The copulatory behaviour of small mammals. *Journal of Comparative and Physiological Psychology*, 1946, *39*, 185–206.

Reiff, M. Über territoriumsmarkierung bei Hausratten und Hausmäusern. *Verhalten Schweizer Naturforschen Geschichte Luzern*, 1952, *1*, 150–151.

Reyniers, J. A., & Ervin, R. F. Breeding rats. *LOBUND Reports*, 1946, *1*, 1–84.

Richards, D. B., & Stevens, D. A. Evidence for marking with urine in rats. *Behavioral Biology*, 1974, *12*, 517–523.

Romer, A. *Man and the vertebrates*. Baltimore: Penguin, 1954.

Rosenblatt, J. S. Views on the onset and maintenance of maternal behavior in the rat. In L. R. Aronson, E. Tobach, D. S. Lehrman, & J. S. Rosenblatt (Eds.), *Development and evolution of behavior*. San Francisco: Freeman, 1970.

Rosenblatt, J. S. Learning in newborn kittens. *Scientific American*, 1972, *227*(6), 18–25.

Rosenblatt, J. S., & Schneirla, T. C. Behaviour of the cat. In E. S. Hafez (Ed.), *The behaviour of domestic animals*. London: Ballière, Tindall & Cox, 1962.

Rothman, R. J., & Mech, L. D. Scent-marking in lone wolves and newly formed pairs. *Animal Behaviour*, 1979, *27*, 750–760.

Rowell, T. E. Organization of caged groups of *Cercopithecus aethiops*. *Animal Behaviour*, 1971, *19*(4), 625–645.

Rowley, L., & Mollison, C. Copulation in the wild rabbit. *Behaviour*, 1955, *8*, 81–84.

Rue, L. L. *The world of the beaver*. Philadelphia:Lippincott, 1964.

Rumbaugh, D. M., & Gill, T. V. Language and the acquisition of language-like skills by a chimpanzee *(Pan)*. *Annals of the New York Academy of Science*, 1976, *270*, 90–123.

Russell, M. J. Human olfactory communication. *Nature*, 1976, *200*, 520–522.

Rutter, R. J., & Pimlott, D. H. *The world of the wolf*. Philadelphia: Lippincott, 1968.

Saayman, G. S., Tayler, C. K., & Bower, D. Diurnal activity cycles in captive and free-ranging Indian Ocean dolphins. *Behaviour*, 1973, *44*, 212–233.

Sales, G. D. Functional aspects of ultrasound production by infant albino rats. *Animal Behaviour*, 1972, *20*, 175–185. (a)

Sales, G. D. Ultrasound and aggressive behaviour in rats and other small mammals. *Animal Behaviour*, 1972, *20*, 88–100. (b)

Sandburg, C. Rat riddles (Poem). *The complete poems of Carl Sandburg*. New York: Harcourt Brace Jovanovich, 1950.

Saprykin, V. A., Kovtuenko, S. V., Korolev, V. P., Dmitrieva, E. S., Ol'sjamslo, V. I., & Becker, I. V. Invariability of auditory perception with respect to time signal transformations in the dolphin, *T. truncatus*. *Journal of Evolutionary Biochemistry and Physiology*, 1977, *12*(3), 246–249.

Schaffer, J. *Die Hautdrusenorgane der Saügetiere*. Berlin: Urban and Schwartzenburg, 1940.

Schalken, A. P. Pheromones in the domestic rabbit. *Journal of Chemical Senses and Flavor*, 1976, *2*(2), 139–155.

Schaller, G. *The mountain gorilla*. Chicago: University of Chicago Press, 1963.

Schaller, G. *The deer and the tiger*. Chicago: University of Chicago Press, 1967.

Schaller, G. *The Serengeti lion*. Chicago: University of Chicago Press, 1972.

Schapiro, S., & Sales, M. Behavioral response of infant rats to maternal odor. *Physiology and Behavior*, 1970, *5*, 815–817.

Schenkel, R. Ausdrücks-Studien an Wölfen. Gefangenschafts-Beobachten. *Behaviour*, 1947, *1*(2), 319–329.

Schenkel, R. Submission: Its features and function in the wolf. *American Zoologist*, 1967, *7*(2), 319–329.

Schneirla, T. C., Rosenblatt, J. S., & Tobach, E. Maternal behavior in the cat. In H. L. Reingold (Ed.), *Maternal behavior in mammals*. New York: Wiley, 1963.

Schönberner, D. Observations of the reproductive biology of the wolf. *Zeitschrift für Saügetierkunde*, 1965, *30*(3), 171–178.

Schwaier, A. The breeding stock of *Tupaias* at the Batelk-Institut. *Laboratory Animal Handbook*, 1975, 6, 141–149.

Scott, J. P. The evolution of social behavior in dogs and wolves. *American Zoologist*, 1967, 7, 373–381.

Scott, J. P. *Animal behavior.* Chicago: University of Chicago Press, 1972.

Scott, J. P., & Fuller, J. L. *Genetics and social behavior of the dog.* Chicago: University of Chicago Press, 1965.

Scott, P. P., & Lloyd-Jacob, M. A. Some interesting features in the reproductive cycle of the cat. *Studies in Fertility*, 1955, 7, 123–129.

Sebeok, T. Communication among social bees; porpoises and sonar; man and dolphin. *Language*, 1963, 39(3), 448–466.

Sebeok, T. *Animal communication.* Bloomington: Indiana University Press, 1968.

Sebeok, T. *Perspectives in zoosemiotics.* The Hague: Mouton, 1972.

Sebeok, T., & Umiker-Sebeok, D. J. *Speaking of apes.* New York: Plenum Press, 1980.

Seton, E. T. *Life histories of northern mammals.* New York: Scribner's, 1909.

Seton, E. T. *Lives of game animals* (Vol. 3). New York: Literary Guild, 1929.

Severeid, J. H. The gestation period of the pika. *Journal of Mammalogy*, 1950, 31, 356–357.

Seward, J. P. Aggressive behavior in the rat. *Journal of Comparative and Physiological Psychology*, 1945, 38, 213–224.

Shelford, V. E. *The ecology of North America.* Urbana-Champaign: University of Illinois Press, 1963.

Shillito, J. F. Field observations on the growth, reproduction, and activity of a woodland population of the common shrew. *Proceedings of the Zoological Society of London*, 1963, 140, 99–114.

Shoemaker, H. W. *Extinct Pennsylvania mammals.* Altoona, Pa., 1917. (Cited in Young, 1944.)

Siebenaler, J. B. & Caldwell, D. K. Cooperation among adult dolphins. *Journal of Mammology*, 1956, 37(1), 126–128.

Skinner, B. F. *Verbal behavior.* New York: Appleton-Century-Crofts, 1957.

Slijper, E. J. *Whales and dolphins* (J. Drury, trans.). Ann Arbor: University of Michigan Press, 1977.

Smith, W. J. Messages of vertebrate communication. *Science*, 1969, 165, 145–150.

Smithers, R. H. Cat of the pharaohs. *Animal Kingdom*, 1968, 69, 163–167.

Sokolov, V. E., & Kuznetsov, V. B. Chemoreception in the Black Sea dolphin (*Tursiops truncatus*). *Dokady Akademia Nauk SSR*, 1971, 201, 998–1000.

Somers, P. Dialects in southern Rocky Mountain pikas, *Ochotona princeps. Animal Behaviour*, 1973, 21(1), 124–138.

Somers, P. Personal communication, January 3, 1979.

Sorenson, M. W. Behavior of tree shrews. In L. A. Rosenblum (Ed.), *Primate behavior.* New York: Academic Press, 1970.

Southern, H. N. The ecology and population dynamics of the wild rabbit. *Annals of Applied Biology*, 1940, 27, 509–526.

Southern, H. N. Sexual and aggressive behaviour in the wild rabbit. *Behaviour*, 1948, *1*, 173–194.

Steiniger, F. Beitrage zur Soziologie und sonstiger Biologie der Wanderatte. *Zeitschrift für Tierpsychologie*, 1950, *7*, 356–379.

Stenlund, M. H. A field study of the timber wolf (*Canis lupus*) in the Superior National Forest, Minnesota. *Minnesota Department of Conservation Technical Bulletin*, 1955, *4*, 1–55.

Stevens, D. A., & Koster, E. P. Open-field responses of rats to odors from stressed and non-stressed conspecifics. *Behavioral Biology*, 1972, *7*, 519–525.

Stokoe, W. C. *Semiotics and human sign languages.* The Hague: Mouton, 1972.

Strauss, J. S., & Ebling, F. J. Control and function of skin glands in mammals. In G. H. Benson & J. G. Phillips (Eds.), *Hormones and the environment.* Cambridge: Cambridge University Press, 1970.

Struhsaker, T. Auditory communication among vervet monkeys. In S. A. Altman (Ed.), *Social communication among primates.* Chicago: University of Chicago Press, 1967. (a)

Struhsaker, T. Behavior of elk (*Cervus canadensis*) during rut. *Zeitschrift für Tierpsychologie*, 1967, *24*, 80–114. (b)

Struhsaker, T. Behavior of vervet monkeys and other cercopithecines. *Science*, 1967, *156*, 1197–1203. (c)

Struhsaker, T. Behavior of vervet monkeys (*Cercopithecus aethiops*). *University of California Publications in Zoology*, 1967, *82*, 1–74. (d)

Struhsaker, T. Ecology of vervet monkeys (*C. aethiops*) in the Masai-Amboseli Game Reserve, Kenya. *Ecology*, 1967, *48*, 891–903. (e)

Tavolga, M. C. Behavior of the bottlenose dolphin. In K. S. Norris (Ed.), *Whales, dolphins, and porpoises.* Berkeley: University of California Press, 1966.

Tavolga, M. C., & Essapian, F. S. The behavior of the bottlenose dolphin: Mating, pregnancy, parturition, and mother–infant behavior. *Zoologica*, 1957, *42*(1), 11–36.

Tavolga, W. N. (Ed.). *Marine bio-acoustics.* New York: Pergamon Press, 1964.

Tavolga, W. N. Levels of interaction in animal communication. In L. Aronson, E. Tobach, D. S. Lehrman, & J. S. Rosenblatt (Eds.), *Development and evolution of behavior.* San Francisco: Freeman, 1970.

Tayler, C. K., & Saayman, G. S. Imitative behavior by Indian Ocean bottlenose dolphins in captivity. *Behaviour*, 1973, *44*(3), 286–298.

Tembrock, G. Acoustic behavior of mammals. In R. Busnell (Ed.), *Acoustic behavior of animals.* London: Elsevier, 1963.

Terrace, H. S., Petitto, L. A., Sanders, R. J., & Bever, T. G. Can an ape create a sentence? *Science*, 1979, *206*, 891–902.

Tevis, L. Summer behavior of a family of beavers in New York State. *Journal of Mammalogy*, 1950, *31*, 40–65.

Theberge, J. B., & Falls, J. B. Howling as a means of communication in timber wolves. *American Zoologist*, 1967, *7*, 331–338.

Theissen, D. Footholds for survival. *American Scientist*, 1973, *61*(3), 346–351.

Theissen, D., & Rice, M. Mammalian scent-gland marking and social behavior. *Psychological Bulletin*, 1976, 83(4), 505–539.

Thomas, J. W., Robinson, R. M., & Marburger, R. G. Social behavior in a white-tailed deer herd containing hypergonadal males. *Journal of Mammalogy*, 1965, 46, 314–327.

Thompson, D. W. (Trans.). Aristotle's *Historia animalia*. Oxford: Clarendon Press, 1910.

Thompson, H. V., & Worden, A. N. (Eds.). *The rabbit*. London: Willmer, 1956.

Thorndike, E. L. A proof of the law of effect. *Science*, 1933, 77, 173–175.

Thorpe, W. H. *Bird-song*. Cambridge: Cambridge University Press, 1961.

Tinbergen, N. "Derived" activities: Their causation, biological significance, origin, and emancipation during evolution. *Quarterly Review of Biology*, 1952, 27(1), 1–32.

Tomkins, S. S. *Affect, imagery, consciousness* (Vol. 2). New York: Springer, 1963.

Valenta, J. G., & Rigby, M. K. Discrimination of the odor of stressed rats. *Science*, 1968, 161, 599–601.

Verberne, G., & DeBoer, J. Chemo-communication among domestic cats. *Zeitschrift für Tierpsychologie*, 1976, 42(1), 86–109.

Von Holst, D. Social stress in the tree shrew. In R. D. Martin, G. A. Doyle, & A. C. Walker (Eds.), *Prosimian biology*. Pittsburgh: University of Pittsburgh Press, 1974.

Von Uexkühl, J., & Sarris, E. G. Das Duftfleck des Hundes. *Zeitschrift für Hundeforschung*, 1931, 1, 55–68.

Voss, G. Zwillingsgeburt beim grossohr Hirsch, *O. hemionus. Zeitschrift für Saügetierkunde*, 1965, 30, 20–33.

Vowles, D. M. Neuroethology, evolution, and grammar. In L. R. Aronson, E. Tobach, D. S. Lehrman, & J. S. Rosenblatt (Eds.), *Development and evolution of behavior*. San Francisco: Freeman, 1970.

Walther, F. R. Zum Liegeverhalten des Weissschawanzgnus (*Connochaetes gnou*). *Zeitschrifte für Sügetierkunde*, 1966, 31, 1–16.

Walther, F. R. Artiodactyla. In T. Sebeok (Ed.), *How animals communicate*. Bloomington: University of Indiana Press, 1977.

Wells, A. *Copulatory behavior in laboratory rats*. Unpublished manuscript, 1977. (Available from R. Peters, Department of Psychology, Fort Lewis College, Durango, Colo. 81301.)

Wemmer, C., & Scow, K. Communication in the Felidae with emphasis on scent marking and contact patterns. In T. Sebeok (Ed.), *How animals communicate*. Bloomington: Indiana University Press, 1977.

White, S. Age differences in reaction to stimulus variation. In O. J. Harvey (Ed.), *Experience structure and adaptability*. New York: Springer, 1966.

Wickler, W. Socio-sexual signals and their intra-specific imitation among primates. In D. Morris (Ed.), *Primate ethology*. Chicago: Aldine, 1967.

Williams H. W., Sorenson, M. W., & Thompson, P. Antiphonal calling of the tree shrew, *Tupaia palawanensis*. *Folia Primatologica*, 1969, 11(3), 200–205.

Wilson, E. O. Sociobiology. Cambridge: Belknap, 1975.

Wilson, J. R., Adler, N., & Le Boeuf, B. The effects of intromission frequency on successful pregnancy in the female rat. *Proceedings of the National Academy of Science,* 1965, *53,* 1392–1395.

Wilsson, L. *My beaver colony.* New York: Doubleday, 1968.

Wittgenstein, L. *Philosophical Investigations* (G. E. M. Anscombe, Ed. and trans.). New York: Macmillan, 1953.

Wolff, P. H. The natural history of crying and other vocalizations in early infancy. In B. M. Foss (Ed.), *Determinants of infant behavior* (Vol. 4). London: Methuen, 1969.

Wood, F. G. *Marine mammals and man.* Washington, D.C.: Luce, 1973.

Woolpy, J. H. The social organization of wolves. *Natural History,* 1968, *77*(5), 46–55.

Wundt, W. *Völker Psychologie.* Leipzig: Engleman, 1900.

Wursig, B., & Wursig, M. Photographic determination of group size, composition and stability of coastal porpoises *(T. truncatus). Science,* 1977, *198,* 755–756.

Wynne-Edwards, V. C. *Animal dispersion in relation to social behavior,* New York: Hafner, 1962.

Young, S. P. *The wolves of North America* (Vol. 1). New York: Dover, 1944.

Zajonc, R. Attraction, affiliation, and attachment. In J. F. Eisenberg & W. S. Dillon (Eds.), *Man and beast: Comparative social behavior.* Washington, D.C.: Smithsonian Institution Press, 1971.

Zarrow, M. X., Denenberg, V. H., & Anderson, C. O. Rabbit: Frequency of suckling in the pup. *Science,* 1965, *150,* 1835–1836.

Zeuner, F. E. *A history of domesticated animals.* New York: Harper & Row, 1963.

Zimen, E. *Wölfe und Königspüdel.* Munich: Piper, 1971.

Zuckerman, S. *The social life of monkeys and apes.* New York: Harcourt Brace Jovanovich, 1932.

NAME INDEX

Alberts, J. R., 57, 309
Alcock, J., 17, 309
Aleksiuk, M., 68, 73, 74, 76, 309
Allee, W. C., 20, 21, 309
Allin, J. T., 54, 309
Altmann, J., 164, 309
Altmann, M., 110, 112–114, 122, 130, 309
Altmann, S. A., 164, 309
Andrew, R. J., 192, 203, 309
Aristotle, 14–16, 29, 30, 309

Barash, D., vi, 95–97, 99, 100, 309
Barfield, R., 59, 64, 309
Barker, R. G., 21, 309
Barnett, S. A., 52, 53, 55–57, 59, 60–62, 64, 309
Bartlett, D., 66, 69, 75, 309
Bartlett, J., 66, 69, 75, 309
Bastian, J., 190, 193, 198, 206, 310, 314
Bateson, G., 8, 13, 201, 212, 273, 298, 310, 324
Beer, J. R., 67, 68, 310
Behse, J. H., 54, 310
Bekoff, M., 237, 310
Benedict, F. G., 55, 310
Bentley, M., 278, 310
Berg, I. A., 236, 310
Berkowitz, L., 241, 310
Bermant, G., 64, 310
Berne, E., 12, 310
Bertin, L., 31, 310
Birdwhistell, R., 268, 271, 310
Blount, W. P., 93, 310
Bohlken, H., 215, 310
Bolten, T. R., 116, 311
Bolwig, N., 175, 311
Bossert, W. H., 226, 311
Bourlière, F., 77, 81, 311
Bowyer, T., 108, 110–114, 116, 118, 121, 122, 124–128, 131, 132–134, 139, 153, 156, 158, 311
Bradt, G. W., 68, 311
Brain, C. K., 163, 166–167, 169, 170–171, 173, 311
Braine, M., 274, 311

Brannigan, C. R., 268, 311
Brind, B., 219, 311
Broadbrooks, H. F., 94, 98, 100, 311
Browman, L. G., 158, 311
Brown, A. M., 64, 311
Brown, B. A., 137, 311
Brown, D. H., 192, 311
Brown, R., 277–278, 311
Browne, T., 16, 311
Brownlee, R. G., 139, 145, 311
Bubenik, A. B., 113, 311
Burghardt, G. M., 299, 312
Burt, U. H., 20, 31, 66–68, 312

Cahalane, V. H., 109, 113, 120, 312
Caldwell, D. K., 187, 190, 197, 201–206, 209–211, 312
Caldwell, M. C., 187, 190, 197, 201–206, 209, 210–211, 312
Calhoun, J. B., 53, 55, 63, 312
Campbell, B. G., 24, 312
Carpenter, C. R., 20, 312
Chalmers, N. R., 167, 312
Chance, M. R., 57
Chapman, R. C., 216, 312
Cheatum, E. L., 156, 312
Chesler, P., 225, 312
Chomsky, N., 275, 312
Christian, J. J., 55, 312
Claesson, A., 139, 313
Colin, J., 244, 321
Collias, N. E., 286, 290, 300, 301 (table), 313
Comfort, A., 272, 313
Coon, C., 279, 313
Cousteau, J., 190, 194, 313
Cowan, I. M., 139, 152–153, 156, 313
Crabtree, D. E., 280, 313
Crisler, L., 224, 231, 313
Crowcroft, P., 31, 32, 34, 313

Dagg, A. I., 132, 157, 313
Darling, F., 20, 111, 114–119, 120, 124, 127–131, 133, 313
Darwin, C., 17–20, 128, 270, 271, 294, 313

331

Davis, D. E., 110, 313
DeMore, P. P., 164–165, 313
Deutsch, J. A., 83, 313
DeVore, I., 42, 314
De Vos, A. P., 124, 146, 148, 150, 152, 154, 156, 313
Dewsbury, D. A., 210, 325
Donovan, C. A., 229, 245, 313
Doty, R. L., 245, 272, 313
Downing, R. L., 141, 314
Dreher, J. J., 194–197, 203, 205, 209, 211, 314
Dryden, G. L., 34, 38, 314
Dudzinski, M. L., 83–85, 90, 314

Eadie, W. R., 38, 314
Eibl-Eibesfeldt, I., 21, 269, 272, 314
Eimerl, S., 42, 314
Eiseley, L., 185, 314
Eisenberg, J. F., 35, 36, 227, 314
Eliot, T. S., 213, 314
Errington, P. L., 251, 314
Essapian, F. S., 197, 199, 200, 202, 206, 210, 314
Estes, R. D., 157, 314
Etkin, W., 279, 314
Evans, W. E., 190, 194, 195, 198, 203, 205–206, 314–315
Ewer, R. F., 21, 30, 53, 55, 62, 87, 112, 126, 142, 237, 250, 252–257, 259, 263, 283, 287, 288, 294, 296, 299, 300, 315

Fedigan, L., 168–169, 315
Fentress, J., 228, 231, 315
Field, R., 231, 315
Finstad, G., 116, 315
Fisher, A. E., 64, 315
Fisher, R. A., 296, 315
Fouts, R., vi, 7, 275, 315
Fox, M., 21, 220–222, 224–225, 228–229, 235–238, 247, 315
Frenzel, D., 218, 321
Frings, H., 6, 8, 315
Frings, M., 6, 8, 315
Fuller, J. L., 219, 220, 227, 230, 247, 315, 327

Gardner, B., 193, 275, 315
Gardner, R., 193, 275, 315
Gartlan, J. S., 164–165, 181, 315
Gawienowski, A. M., 56, 58, 62, 63, 315
Geist, V., 12, 21, 117, 124, 126, 150, 152–154, 156–157, 313, 315
Ginsburg, B., 217, 325
Goldman, E. A., 216, 315
Golley, F. B., 159, 315

Goodall, J., 279, 316
Goodrich, B. S., 85, 316
Goodwin, G. G., 81, 315
Goodwin, M., 245, 316
Gould, E., 32, 39, 316
Graf, W., 111, 117, 131, 156, 316
Grant, E. C., 57, 60, 62, 316
Grant, U. S., 243, 316
Grundlach, H., 117, 316
Grzimek, B., 90, 316
Guhl, A. M., 237, 316
Guthrie, R. D., 154, 316

Haber, G., 247
Hafez, E. S. E., 93, 316
Haga, R., 96, 316
Hall, R. L., 214, 316
Hamilton, W. D., 110, 136, 316
Hamilton, W. J., 273, 297, 316
Harder, W., 54, 316
Harper, J. A., 114, 316
Harrington, F., 227, 229, 231, 233, 238, 242, 317
Harrop, A. E., 247, 317
Harvey, E. B., 98, 317
Hatt, R. T., 66, 317
Haugen, A. D., 139, 151, 156, 159, 317
Hebb, D., 191–192, 198, 199, 201, 203, 206–209, 321
Hediger, H., 20, 90, 317
Henry, J. D., 8, 317
Heptner, W. G., 123, 317
Hertel, H., 187, 317
Herter, R., 27–30, 32–34, 38, 39, 62, 317
Hesterman, R. R., 88, 89, 317
Hill, W. C. D., 42, 44, 163, 317
Hinde, R. A., 55, 317
Hirth, D. H., 137, 143, 154, 155, 317
Hockett, C. F., 275, 278–279, 317
Hoese, H. D., 189, 317
Hubbs, C. L., 205, 318
Huffman, M., vi, 171, 174, 181, 318

Jaynes, J., 54, 310
Jerison, H. J., 8, 280, 318
Johnson, C. S., 187, 318
Johnson, D. E., 112, 113, 129, 318
Johnson, D. R., 96, 318
Joslin, P. W. B., 220, 223, 229, 238, 242, 318

Kalmus, H., 280, 318
Kawamichi, T., 95, 96, 99, 100, 101, 318
Kellogg, W. N., 189, 192, 194, 318
Kile, T. L., 148, 318
Kilham, L., 96, 98, 99, 100, 318
Kirkpatrick, C. M., 81, 82, 318

Kleiman, D., 226–228, 318
Knappe, H., 132, 318
Knick, S., 244, 247, 321
Knight, R. R., 111, 318
Kortlandt, A., 214, 298, 318
Krames, L., 57–58, 62, 319
Krear, H., 96, 100, 319
Kruuk, H., 22, 319
Kucera, T., 149, 150–152, 154–159, 319

Lancaster, J. B., 164, 168, 180, 319
Lang, H., 35, 319
Lang, T. G., 194–195, 197, 202, 211, 319
Lankester, E. R., 14, 15
Le Boeuf, B., 64, 330
Lechleitner, R. R., 116, 319
Lehrman, D., 21
Leighton, A. H., 68–69, 70–72, 319
Le Magnen, J., 63, 319
Lenneberg, E., 274, 275, 319
Lent, P. C., 140, 141, 319
Leroi-Gourhan, A., 14, 319
Leyhausen, P., 21, 249–251, 253, 255, 256–261, 263–265, 319
Lilly, J., 193–194, 197, 201–205, 319, 320
Lind, J., 269, 320
Linsdale, J. M., 137–143, 146–159, 320
Lockwood, R., 236, 237, 320
Locy, W. A., 14–15, 320
Long, W. S., 97, 320
Lorenz, K., 2, 13, 21, 34, 232, 278–279, 294, 320
Lutton, L. M., 96, 320

MacFarlane, A., 269, 320
Markham, O. D., 94, 97, 100, 320
Marler, P., 229, 320
Marr, J. N., 54, 320
Martin, R. D., 42, 43, 83, 320
Martinka, C. J., 108, 109, 320
Martins, T., 236, 320
Masters, W. H., 273, 320
McBride, A. F., 190–192, 198, 199, 201, 203, 206, 207–209, 321
McCullough, D. R., 108–111, 113–115, 117–128, 130–133, 143, 154, 317, 321
McGuire, M., 162, 165, 174, 321
McKenna, M. C., 24, 321
Mech, L. D., 22, 215, 217–220, 223–225, 227, 229–232, 235–238, 241–242, 244–247, 279, 317, 321, 324, 326
Meester, J., 33, 321
Mehrabian, A., 268, 269, 322
Mello, N. K., 250, 322
Michael, R. P., 181, 246, 263–264, 272, 322
Miller, A., 193, 194, 197, 202, 203, 320

Miller, G., 279, 322
Moelk, M. 251–252, 253, 254, 256, 258, 261, 263, 264, 322
Moore, M., 161
Moore, W. G., 138, 149, 150–151, 154, 157–158, 322
Morris, D., 25, 34, 106, 268, 269, 271, 322
Morrison, J. A., 133–134, 322
Müller-Schwarze, D., 9, 56, 137–140, 145–152, 157, 287, 322, 325
Murie, A., 112–114, 117, 127, 128, 131, 133, 217–221, 224, 230, 322
Myers, K., 82, 90, 92, 323
Mykytowycz, R., 22, 80, 82–87, 89, 314, 316, 317, 320

Neff, W., 250, 323
Neisser, U., 275, 323
Nelson, E. W., 108, 124, 323
Nichol, A. A., 141, 323
Nikol'skii, A. A., 128, 323
Noirot, E., 54, 323
Norris, K. S., 192, 194, 205, 207, 208, 311–312

Olson, S. F., 217, 324
Orr, R. T., 80, 94, 324
Orsulak, P. J., 63, 324
Ozoga, J. J., 148, 324

Paget, R., 276–277, 324
Palen, G. F., 246, 264, 324
Pepper, R. L., 204, 324
Perez, J. M., 187, 324
Peters, R. P., 226–228, 236–238, 242, 244–246, 272, 280, 290, 324
Pilleri, G., 190–191, 203, 205, 324
Pimlott, D. H., 224, 225, 243, 324, 326
Pliny, 16, 324
Poduschka, W., 25, 27, 29, 30, 324
Poirier, F. E., 165, 179, 180, 325
Premack, D., 275, 325
Price, H., 58, 325
Pruitt, W. O., 150, 325
Puente, A. E., 210, 325
Pura, H., 138

Quay, W. B., 28, 147, 325

Rabb, G., 217, 326, 241, 244–247, 325
Ralls, K., 290, 325
Randall, J. H., 16, 325
Rankin, J., 32, 325
Rausch, R. L., 217, 325
Read, C., 214, 325
Reed, C. A., 77, 325
Reiff, M., 56, 57, 325

Reyniers, J. A., 54, 325
Richards, D. B., 58, 325
Romer, A., 24, 325
Rosenblatt, J. S., 54, 252–254, 283, 287, 325
Rothman, R., 218, 237, 244–246, 326
Rowell, T., 164, 181, 326
Rowley, L., 92, 326
Rue, L., 66–67, 141, 326
Rumbaugh, D., 275, 326
Russell, M. J., 269, 326
Rutter, R. J., 224, 243, 326

Saayman, G. S., 189, 190, 192, 201, 203–204, 207–210, 326
Sales, M., 54, 60, 61, 64, 326
Sandburg, C., 51, 326
Saprykin, V. A., 197, 326
Schaffer, J., 129, 238, 263, 326
Schalken, A. P., 86, 326
Schaller, G., 22, 257, 326
Schapiro, S., 54, 326
Schenkel, R., 20, 217, 220, 227, 229, 231–238, 241, 244, 247, 326
Schneirla, T. C., 21, 252–254, 325, 326
Schönberner, D., 219, 326
Schwaier, A., 42, 47, 326
Scott, J. P., 110, 219–220, 227, 230–231, 234, 283, 299, 326
Scott, P. P., 252, 327
Seal, U. S., 244, 321
Sebeok, T., 6, 8, 9, 22, 276, 327
Seton, E. T., 72, 74–75, 141, 158, 226, 327
Severeid, J. H., 96, 101, 327
Seward, J. P., 60, 327
Shakespeare, W., 1, 267
Sharp, H. S., 214, 316
Shelford, V. E., 108, 327
Shillito, J. F., 32, 327
Shoemaker, H. W., 224, 327
Siebenaler, J. B., 205, 327
Skinner, B. F., 274, 327
Slijper, E. J., 190, 210, 327
Smith, W. J., 299, 300, 327
Smithers, R. H., 259, 327
Sokolov, V. E., 188, 327
Somers, P., 97, 100–101, 327
Sorenson, M. W., 42, 44–47, 327
Southern, H. N., 81–83, 85–86, 89–91, 327–328
Steiniger, F., 63, 328
Stenlund, M. H., 217, 328
Stevens, D. A., 58, 328
Stokoe, V. C., 276, 328
Struhsaker, T., 109, 111, 113, 116–117, 123–124, 126–128, 130, 132, 133–134, 162–183, 278, 328

Tavolga, M. C., 191–192, 199, 200, 201, 206, 207, 210, 328
Tavolga, W. N., 6, 8, 200, 328
Tayler, C. K., 189, 192, 328
Tembrock, G., 223, 229, 238, 328
Terrace, H. S., 276, 328
Tevis, L., 67–72, 75–76, 328
Theberge, J. B., 227, 328
Theissen, D., 22, 99, 328–329
Thomas, J. W., 152–154, 329
Thompson, D. W., 14, 329
Thorndike, E., 277, 329
Thorpe, W. H., 100, 329
Tinbergen, N., 13, 329
Tomilin, A., 193
Tomkins, S. S., 271, 329

Valenta, J. G., 55, 329
Verberne, G., 263, 329
Von Holst, D., 45, 329
Von Uexküll, J., 227, 329
Voss, G., 138, 329
Vowles, D. M., 280, 329

Walther, F. R., 106, 125, 126, 129, 150, 154, 159, 294, 329
Wells, A., 64, 329
Wemmer, C., 253, 256–257, 259, 260, 261, 315, 329
White, S., 192, 329
Wickler, W., 175, 181, 183, 329
Williams, H. W., 44, 329
Wilson, E. O., 2, 22, 75, 165, 167, 203, 226, 256, 282, 283, 299, 300 (table), 301, 311, 330
Wilson, J. R., 64, 330
Wilsson, L., 67–69, 71–76, 330
Wittgenstein, L., 281, 330
Wolff, P. H., 269, 286, 330
Wolff, R., 251, 257, 319
Wood, F. G., 187, 193, 209, 330
Woolpy, J. H., 217, 218, 236, 241, 330
Wundt, W., 276, 330
Wursig, B., 189, 190, 330
Wursig, M., 189, 190, 330
Wynne-Edwards, V. C., 73, 260, 330

Young, S. P., 217, 223–224, 245, 330

Zajonc, R., 12, 55, 137, 230–231, 247, 289, 330
Zarrow, M. X., 83, 330
Zeuner, F. E., 249, 250, 330
Zimen, E., 215, 330
Zuckerman, S., 20, 330

SUBJECT INDEX

Action-specific energy, 305
Adaptation, 305
Affiliative messages, general, 9, 287, 289, 298, 299
  in cats, 258, 289
  in beavers, 71, 289
  in deer, 144, 289
  in dolphins, 203–204, 289
  in humans, 270
  in rabbits, 86
  in rats, 55–56, 289
  in tree shrews, 44, 45, 289
  in vervets, 169–171, 289
  in wapiti, 116, 289
  in wolves, 230–232, 235, 289
Agonism, 305
Agonistic messages, 292–295 (see also the individual species)
Alarm, 12, 289, 297, 298
  in beavers, 71
  in deer, 146–148
  in dolphins, 205–206
  in humans, 270
  in rabbits, 87
  in rats, 55
  in tree shrews, 45
  in vervets, 171–173
  in wapiti, 117–118
  in wolves, 229
Allelomimesis, 299, 300
Allogrooming:
  in beavers, 70–71, 289
  in cats, 256–258
  in deer, 144, 289
  defined, 9, 305
  in dolphins, 203–204
  as epimeletic behavior, 299
  in rabbits, 289
  in rats, 60, 289
  in tree shrews, 44, 47, 289
  in vervets, 167, 169–170, 289
  in wapiti, 116
  in wolves, 289

Alloparenting, 113, 167, 199, 220, 253, 287, 305
Altricial, mammal, 305
American Sign Language, 276
Anal glands, 84, 86, 88, 91, 245, 256, 257
Analogue coding, 298
Analogues, 305
Antithesis, 19, 154, 206, 235, 272, 294
Antlers, 106, 108, 119, 121–124, 125, 131, 135, 150–151, 153–157
Antorbital gland, 146
Apocrine glands, 98, 129, 132, 238, 245, 272
Appetitive behavior, 305
Artiodactyl, 106, 108, 208, 305
Assembly message, 12, 224, 271, 292, 297

Baboons, 164, 282
Bachelor group, 109, 128
Barking, 118
Beavers (Castor canadensis), 66–67
  agonistic messages, 72–76
  alarm, 71
  castoreum, 69, 73
  ecology, 67
  integrative messages, 69–72, 288
  neonatal messages, 68–69, 283–284, 286
  scent-marking, 69, 73, 76
  scrotum, 16
  sexual messages, 76–77
  social organization, 67–68
Behavior, 305
Bighorn sheep (Ovis canadensis), 154, 282
Biome, 305
Birth, 112, 138, 165, 198, 219, 252
Black-tailed deer (Odocoileus hemionus columbianus), see Deer
Bottlenosed dolphin, see Dolphin
Boxing:
  in beavers, 69
  general, 294
  in hedgehogs, 29
  in humans, 271

Boxing (continued)
in rabbits, 90
in rats, 55, 62
in tree shrews, 47
in vervets, 168
in wapiti, 120
Bugling, 122, 127–129

Canid, 214, 305
Carnivore, 186, 214
Castor, see Beaver
Castoreum, 16, 69, 73–75, 292
Cats (Felis catus):
agonistic messages, 259–263
ecology, 250
general, 249–250
integrative messages, 255–259, 288, 289
neonatal messages, 252–255, 284, 287
purring, 254, 258, 264
scent-marking, 256–257, 263
sexual messages, 263–265
social system, 251
Cercopithecid, 305
Cervid, 106, 297, 395
Cervus canadensis, see Wapiti
Cetacean, 186, 305
Chimpanzee, 275–276, 279, 282
Chinning, 84
Clicks, 193–194, 202, 205–206
Cloaca, 66, 73, 305
Cognitive map, 280
Communication:
defined, 6–8
nonverbal, in man, 8
phasic, 7
signal, 7
symbolic, 7
Conspecifics, 305
Consummatory behavior, 305
Consummatory face, 229, 270
Contact messages:
in cats, 256
in deer, 142–143
in dolphins, 197, 202
general, 9, 288–289, 297, 301
in hedgehogs, 27
in humans, 269
in pikas, 97
in shrews, 35
in tree shrews, 44
in vervets, 173
in wapiti, 115–116
in wolves, 225
Coprophagy, 54, 305
Copulatory signal, 247–273, 297
Courtship:
in beavers, 70, 76–77, 297
in cats, 264–265, 296, 297

Courtship (continued)
in deer, 156–159, 296
defined, 12
in dolphins, 208–210, 296
general, 297
in hedgehogs, 30, 296
in humans, 273
in pikas, absent, 296
in rabbits, 91–92, 296
in rats, 63
in shrews, 38–40
in tree shrews, 47–48
in vervets, 182–183
in wapiti, 132–133
in wolves, 246–247, 296, 297
Critical period, 306
Cro-Magnons, 13–14
Cuban solenodon, 25, 35
Cud, 106, 108

Deer (Odocoileus hemionus and virginianus):
agonistic messages, 149–156
ecology, 135–136
general, 135, 269
integrative messages, 141–149, 288
neonatal messages, 138–141, 283, 286, 287
scent-marking, 148–149, 150–151, 157
sexual messages, 156–159
social system, 136–138
Dialects, 97
Diastema, 52, 80, 306
Digital coding, 298, 306
Dimorphism, 306 (see also Sexual dimorphism)
Direct action of nervous system, 19
Displacement, 90, 91, 92, 237, 241, 260, 266, 306
as a feature of language, 275
Distress message:
in beavers, 292
in cats, 256, 264, 292
in deer, 147, 292
defined, 12, 292
in dolphins, 195, 197, 204–205, 209, 292
general, 286
in humans, 269, 271
in pikas, 97, 292
in rabbits, 87
in shrews, 34
in vervets, 166–167, 171
in wapiti, 113, 119
in wolves, 229, 292
Dogs, 192, 193, 214, 215, 224, 230 (see also Wolves)
Dolphin (Tursiops spp.):
agonistic messages, 206–209
echolocation, 189

Dolphin (*continued*)
  ecology, 189–190
  general, 186–188
  integrative messages, 200–206, 288, 289
  intelligence, 191–193
  language, 193
  neonatal messages, 197–200, 283, 286
  sexual messages, 208–210
  social system, 190–191
Dominance:
  in beavers, 75–76
  in cats, 251, 259, 294
  common, 297
  in deer, 149–152
  defined, 12, 294
  in dolphins, 191, 206–207
  general, 29, 298, 301
  in humans, 272
  in rabbits, 89
  in rats, 62
  in shrews, 36
  in tree shrews, 46
  in vervets, 174–175
  in wapiti, 124–125
  in wolves, 233, 235–236, 294
Dorsal surface, 306
Drive, 306
Duality of patterning, 275

Echolocation, 8, 189, 193, 202, 306
Elephants, 15, 17, 192, 291
Elephant shrews (Family Macroscelididae), 4
    (*table*), 35
Elk, *see* Wapiti
Emancipation, 13
Enfleurage, 140
Enuration:
  defined, 62, 306
  in rabbits, 85, 90, 92
  in rats, 62
Epimeletic behavior, 223, 306
Estrus, 63, 80, 91, 127, 156, 181, 209, 244,
    246–247, 263, 306
Et-epimeletic behavior, 306
Ethogram, 306
Ethology, 306

Fallow deer, 293
Familiarization message:
  in beavers, 73
  in deer, 148–149, 290
  defined, 12
  in hedgehogs, 28
  in humans, 270
  and μ-function, 298
  in pikas, 97, 99, 290
  in rabbits, 85, 290
  in rats, 56, 290

Familiarization message (*continued*)
  and semilattice, 301
  in shrews, 35
  in tree shrews, 45
  in wolves, 228, 290
Feces, *see* Scent-marking
Felid, 214, 306
Feral, 306
Fighting:
  in beavers, 76, 292
  in cats, 262–263
  common, 297
  in deer, 155–156
  defined, 11, 292
  in dolphins, 207–208
  in hedgehogs, 29
  in pikas, 100, 292
  in rabbits, 90
  in rats, 62
  in shrews, 36
  in tree shrews, 47
  in wapiti, 122, 123–124, 292
  in wolves, 241–242, 292
Flehmen, 14, 28, 132, 157, 258, 263, 296,
    306
Functional explanation, 18

Grammar, 274–275
Green monkey, *see* Vervet monkey

Harem, 109, 110, 114, 126, 127, 128, 132,
    149, 297
Hedgehogs (*Erinaceas* spp.):
  agonistic messages, 28–29
  "boxing," 28–29
  copulation by, 15, 30
  distress, 27
  general, 23, 26
  integrative messages, 27–28, 288
  mounting, 27
  neonatal messages, 26–27, 283, 286
  play, 27
  "roundabout," 29
  scent-marking, 28, 29
  self-salivation, 28
  senses, 27
  sexual messages, 29–30
  social system, 26, 297
Herd, 110, 136
Homoiothermy, 4, 306
Homology, 183, 306
Homosexuality, 183, 200–201, 207
Howling, 216, 224–225, 227, 242, 243
Humans (*Homo sapiens*), *see also* Language
  agonistic messages, 271–272
  complexity of nonverbal communication,
      268
  integrative messages, 269–271

Humans (*continued*)
  neonatal messages, 269
  sexual messages, 272–273

Ibex, 293
Identity message:
  in cats, 257–258
  in deer, 145
  defined, 12, 291
  in dolphins, 202, 290
  in humans, 269
  in pikas, 290
  in rabbits, 290
  in rats, 290
  and semilattice, 301
  in vervets, 173, 290
  in wolves, 227–228, 290
Imprinting:
  in deer, 138
  defined, 42, 306
  general, 287
  in humans, 269
  in rabbits, 84
  sexual consequences, 48
  in tree shrews, 42, 48
Inciting, 278–279
Individual distance, 90
Induced ovulation:
  in cats, 265
  defined, 306
  in hedgehogs, 30
  in lagomorphs, 80
  in rabbits, 93
Information, 282
Inguinal odor, 84
Inguinal region, 306
Insectivores (Order Insectivora), 24–25, 306
Integrative message, 287–292 (*see also*
    Message system *and the individual species*)
Integument, 306
Intelligence, 192, 298
Interdigital glands, 143, 144

Killer whale, *Orcinus orca,* 189
Kin selection, 18

Lagomorph, 80, 306
Lamarckianism, 20
Language:
  characteristics, 7, 273–276
  in chimpanzees, 275–276
  in dolphins, 191, 193
  evolution, 273, 276–280
  general, 269–270
Laughing, 270
Leadership, 116
Leporid, 82, 306

Linear-dominance heirarchy, 306
Lions (*Panthera Leo*), 15, 16
Lip retraction, 119
Lordosis, 63, 92, 133, 159, 307
Low-stretch, 307

Mammals:
  adaptations, 3–6
  characteristics, 3–6
  included in book (*table*), 4–5
  number of species, 3
  similarities in communication, 2, 285–297
Man, *see* Humans
Masturbation, 183
Maternal absenteeism, 83
Menstruation, 181
Message system:
  agonistic, 12
  defined, 9–12, 303
  integrative, 9
  neonatal, 9
  sexual, 12
Message types, 9, 10–11 (*table*), 12, 284,
  303
Metacommunication, 8
Metatarsal gland, 129, 144, 147
Mice, 16, 282
Milk tread, 54, 253, 259
Mimicry, 203
Mnemonic, 307
Moles, 25
Monkey, 18, 161 (*see also* Vervet monkey)
Morpheme, 274, 278, 307
Morphology, 307
Mounting, 13, 62, 152 (*see also* Dominance,
  Sexual messages)
Mu (μ)-function, 298
Mule deer, *see* Deer
Murids, 52, 307

Naming, 276–278
Natural selection, 17–18
Neonatal communication, 283–287, 301,
  307 (*see also the individual mammals*)
Nursery group, 109, 116

Ochotona, see Pika
Olfaction, 307
Omnivore, 307
Onomatopoeia, 277
Open-ness, 275, 278
Oryctolagus, see Rabbit
Osmollaxis, 56
Otara, 80
Ownership, 207, 235

Paralinguistic signal, 269, 270

Parental investment, 192, 307
Pelage, 307
Perianal, 307
Perineum, 307
Phasic interactions, 6, 283
Phatic function, 269–270
Pheromones, 28, 54, 64, 118, 307 (see also Scent-marking)
Phonation, 187, 193–197, 201, 204, 307
Phonemic pattern, 274, 307
Phonetic pattern, 274
Phylogeny, 307
Pikas (Ochotona princeps):
    agonistic messages, 99–100
    ecology, 94–95
    general, 94–95, 299
    integrative messages, 97–99, 289
    neonatal messages, 96, 286
    scent-marking, 97–99, 101
    sexual messages, 100–101
    social system, 95–96, 297
    territorial advertisement, 99–100
Pithecanthropines, 279
Placenta, 68, 112, 166, 198, 219, 252
Play:
    in beavers, 69
    in cats, 255–256
    in deer, 141–142
    in dolphins, 200–201
    general, 9, 10 (table), 287–288, 297
    in hedgehogs, 27
    in humans, 269
    in rabbits, 86
    in rats, 55
    in shrews, 34
    in tree shrews, 43–44
    in vervets, 168–169
    in wapiti, 114–115
    in wolves, 221
Play face, 168, 221
Polygyny, 83, 106
Porpoise, see Dolphin
Prairie dogs, 290
Precocial mammal, 307
Prehensile, 307
Primate, 162
Punning, 299, 301
Purring, see Cat

Quack, 197, 198, 201

Rabbit (Oryctolagus cuniculus):
    agonistic messages, 87–91
    alarm, 87
    ecology, 81–82
    general, 80–81
    integrative messages, 85–87, 288, 289

Rabbit (continued)
    neonatal messages, 83–85, 286, 287
    play, 86–87
    scent-marking, 84–85, 86, 87–89, 92
    sexual messages, 91–93
    social organization, 82–83
    territorial advertisement, 87–89
Rats (Rattus rattus and R. norvegicus):
    agonistic messages, 57–62
    ecology, 52
    general, 52, 269
    imprinting, 54
    integrative messages, 55–57, 288
    mounting by, 13
    neonatal messages, 54–55, 286, 387
    play, 55
    scent-marking, 56, 63
    sexual messages, 63–65
    social organization, 53
Red deer, see Wapiti
Redirected aggression, 180
Redundancy, 299
Refection, 31, 307
Reversal-shift discrimination, 192, 307
Ritualization, 118, 119, 307
Rodents, 52
Rub-urination, 139, 151–152, 158
Rump patch:
    in deer, 141, 142
    described, 17
    evolution, 17–18
    in wapiti, 107, 115, 118
Rut, 109, 117, 127–131, 149, 151, 156–159, 307

Satisfaction message, 12, 229, 258, 270, 286, 292, 297
Scent-marking:
    in beavers, 69, 73, 76
    in cats, 256–257, 263
    in deer, 148–151, 157, 290
    defined, 307
    in dogs, 15, 16
    general, 8, 270, 289, 294, 295, 299, 301
    in hedgehogs, 28, 29
    in pikas, 97–99, 101, 290
    in rabbits, 84, 85, 87–89, 92, 290
    in rats, 56, 63, 290
    in shrews, 35, 38
    in tree shrews, 43, 45, 46, 287
    in wapiti, 117, 129, 131
    in wolves, 226–227, 228, 242–243, 245–246, 290
Schwanzstrauben (SST), 45
Sebaceous glands, 129, 139, 148, 238, 245, 257, 263, 264, 272
Semilattice, 301–303

Semiotics, 9, 307
Serviceable associated habits, 18, 119
Sexual advertisement:
  in beavers, 76
  in cats, 263–264, 295
  in deer, 156–158, 295
  defined, 12
  in dogs, 15
  in dolphins, 208–209, 295
  in hedgehogs, 29, 295
  in humans, 272–273
  in pikas, 272–273
  in rabbits, 91, 295
  in rats, 63, 295
  and semilattice, 301
  in shrews, 38
  in tree shrews, 47
  in vervets, 181
  in wapiti, 127–132, 295
  in wolves, 244–246, 295
Sexual dimorphism, 106, 162, 183, 307
Sexual messages, 295–297 (see also the
    individual species, Courtship, and Sexual
    advertisement)
Sexual suppression, 12, 244, 297
Sexual synchronization, 12, 91, 244, 297
Shrews (Family Soricidae):
  agonistic messages, 36–38
  caravans, 33–34
  ecology 31–32
  general, 31
  integrative messages, 34–35, 288
  neonatal messages, 32–34, 283, 286, 287
  sexual messages, 38–40
  social system, 32
Smiling, 270
Snuffling:
  in tree shrews, 48
  in wolves, 231–232
Solicitation message:
  in beavers, 69–70
  in cats, 256, 292
  in dolphins, 201
  general, 12, 292
  in humans, 271
  in tree shrews, 47
  in vervets, 169–171, 179
  in wolves, 221–223, 235, 292
Solomon, 2
Specific presence, 301–303
Spronking, see Stotting
Staring, 125, 154, 176, 206, 235, 236, 272
Stotting, 14, 118, 147, 307
Submission:
  as affiliation, 301
  in beavers, 76, 294
  in cats, 259

Submission (continued)
  in deer, 154–155
  defined, 11 (table), 12, 294
  in humans, 272
  in rabbits, 90
  in rats, 61, 294
  in tree shrews, 44, 294
  in vervets, 175–176, 294
  in wapiti, 125
  in wolves, 232–235, 294
Supplantation, 46, 124, 174

Tail gland, 129, 143
Tail switch, 142–143
Tarsal gland, 139–141, 144, 145, 147, 158
Taxon, 307
Teat order, 253
Tending bond, 137, 297
Tenrecs, 25, 282
Territory advertisement:
  in beavers, 73–74
  in cats, 263
  defined, 11, 294
  in humans, 271
  in pikas, 99–100
  in rabbits, 87–89
  in rats, 58
  and semilattice, 301
  in shrews, 38
  in vervets, 177–180
  in wolves, 242–244
Thrash-urinate, 130
Threats:
  in beavers, 72
  in cats, 255, 260–261
  common, 297
  in deer, 152–154
  defensive, 12, 294
  defined, 12
  in dolphins, 207
  in humans, 270–272
  offensive, 12, 294
  in rabbits, 89–90
  in rats, 61
  in shrews, 36, 38
  in tree shrews, 45–47
  in vervets, 176–177
  in wapiti, 119–122
  in wolves, 221, 237–240
Tonic interactions, 6, 283
Transformational grammar, 275
Tree-like hierarchies, 301–303
Tree shrews (Family Tupaiidae):
  agonistic messages, 45–47
  ecology, 24, 42
  fighting, 47
  general, 24, 41–42

Tree shrews (*continued*)
  integrative messages, 43–44, 288
  male parental care, 48
  neonatal messages, 42–43, 287
  scent-marking, 43, 45, 46
  sexual messages, 47–48
  social system, 42
  submission, 47
Troop, 164
Tupaiid, 308 (*see also* Tree shrew)
Tursiops, see Dolphin
Twist, 308

Ungulate, 186–187, 207, 216 (*see also* Deer,
    Wapiti)
Urination, *see* Scent-marking

Vacuum activity, 308
Vegetative interactions, 6
Ventral surface, 308
Vervet monkey (*Cercopithecus aethiops*):
  agonistic messages, 174–180
  ecology, 163–164
  general, 162, 269, 298
  integrative messages, 168–173, 288, 289
  neonatal messages, 165–168, 286
  sexual messages, 180–183
  social system, 164–165, 297
Viviparous, 308
Vocalization, 308
Vomeronasal organ, 28, 132

Wallowing, 121, 130–131, 158
Wapiti (*Cervus elaphus*):
  agonistic messages, 119–126
  ecology, 108
  general, 106–108, 269
  integrative messages, 114–119, 288, 289
  neonatal messages, 112–114, 286, 287
  scent-marking, 117, 129, 131
  sexual messages, 126–134
  social system, 108–111
Whistles, 194–197
Whitetail deer, *see* Deer
Wolves (*Canis lupus*):
  agonistic messages, 232–244, 298
  dominance, 13
  ecology, 216–217
  general, 108, 124, 214–216, 269, 272,
    298
  howling, 216
  integrative messages, 220–232, 288–289
  neonatal messages, 218–220, 286–287
  scent-marking, 226–229, 242–243
  sexual messages, 244–247
  social system, 217–218, 297
  submission, 19, 21
  threats, 19, 20

Zoosemantics, 9, 282, 301
Zoosemiotics, 2, 308